SEROTONIN—
NEW VISTAS

Biochemistry and Behavioral and Clinical Studies

Advances in Biochemical Psychopharmacology
Volume 11

Advances in Biochemical Psychopharmacology

Series Editors:

Erminio Costa, M.D.
Chief, Laboratory of Preclinical Pharmacology
National Institute of Mental Health
Washington, D. C., U.S.A.

Paul Greengard, Ph.D.
Professor of Pharmacology
Yale University School of Medicine
New Haven, Connecticut, U.S.A.

Serotonin — New Vistas

Biochemistry and Behavioral and Clinical Studies

Advances in Biochemical Psychopharmacology Volume 11

Editors

E. Costa, M.D.
Chief, Laboratory of Preclinical Pharmacology
National Institute of Mental Health
Washington, D.C., U.S.A.

G. L. Gessa, M.D.
Professor of Pharmacology
Cagliari University School of Medicine
Cagliari, Sardinia, Italy

Merton Sandler, M.D.,
The Bernhard Baron Memorial
Research Laboratories
Queen Charlotte's Maternity Hospital
London, England

Raven Press ■ New York

International Standard Book Number 0-911216-69-3
Library of Congress Catalog Card Number 73-91166

PHOTOGRAPHS: 1, 2, B.B. Brodie; 3, A. Carlsson; 4, B.B. Brodie and G.L. Gessa; 5, R. Paoletti; 6, E. Costa. (Photography by Giorgio Biolchini.)

Preface

The vast growth in biological knowledge in modern times has not been a continuous process but has varied in pace depending on the impetus given to it by certain individuals of exceptional talent and ability. One might classify them into two categories — "heroes" and "great men of biology." The former make history by some single great leap forward for which the time is ripe; the latter excel as promoters of biological progress which they foster by their creative foresight and ability to fire the imagination of young scientists. By this yardstick, B. B. Brodie is one of the great men of modern pharmacology.

Branching off from physiology in the early part of this century, pharmacology for some time kept close connections with its parent discipline. B. B. Brodie who, by formal training, was an organic chemist, had a major role in promoting a methodological cross-fertilization between pharmacology and biochemistry. As a result of this merger, many of the fundamental principles of drug metabolism were established. It would not be an exaggeration to say that Dr. Brodie largely created the new discipline of biochemical pharmacology; his pupils, more than 200 over the years, have carried the message to many parts of the world. This new breed of pharmacologists have in turn influenced their clinical colleagues to an awareness that drug metabolism is a variable in the response to drugs which cannot be ignored. In consequence, the mechanisms of a number of unwanted drug actions and interactions have been elucidated to the greater benefit of mankind.

In the late 1950's, Dr. Brodie moved into the field of neuropharmacology and here too he provided a new impetus. With his colleagues he helped to establish the fundamental principle that drugs may act on neurons by modifying neurotransmitter dynamics, a guiding assumption in present-day neuropsychiatric research. The seminal observation in this area was that reserpine changes the steady-state conditions of an important brain constituent, 5-hydroxytryptamine (for review, see B. B. Brodie, and P. A. Shore, *Ann. N.Y. Acad. Sci.*, 66:631, 1957). It was Brodie who suggested that 5-hydroxytryptamine functions as a neurotransmitter, regulating brain processes involved in the control of rest and activity. Now that the amine has become more firmly established in this role and its association with rest and sleep underlined, it was only fitting to dedicate this meeting (and the resulting two-volume publication) to Dr. Brodie. During its course, his friends and disciples derived great pleasure from the award of an honorary degree by the University of Cagliari, Sardinia, to this great pharmacologist.

E. Costa
G. L. Gessa
M. Sandler

Contents

BIOCHEMISTRY

BEHAVIORAL AND CLINICAL STUDIES

CONTENTS OF VOLUME 10

Histochemistry

Pharmacology

Advances in Biochemical Psychopharmacology, Vol. 11
Raven Press, New York © 1974

Tryptophan-5-Hydroxylase: Function and Control

E. Martin Gál

Neurochemical Research Laboratory, Department of Psychiatry, College of Medicine, University of Iowa, Iowa City, Iowa 52240

It is ten years almost to the month since we presented the first evidence of the existence of cerebral tryptophan-5-hydroxylase *in vivo* (Gál, Poczik, and Marshall, 1963; Gál and Marshall, 1964; Gál, Morgan, Chatterjee, and Marshall, 1964). Further reports described hydroxylation of L-tryptophan by brain tissue *in vitro* (Grahame-Smith, 1964; Gál, 1965; Green and Sawyer, 1966; Gál, Armstrong, and Ginsberg, 1966; Lovenberg, Jéquier, and Sjoerdsma, 1967). At that time the enzyme that catalyzed conversion of L-tryptophan to L-hydroxytryptophan had already been demonstrated by carefully documented research (Clark, Weissbach, and Udenfriend, 1954; Udenfriend, Titus, Weissbach, and Peterson, 1956). The elucidation of the reaction mechanism, the so-called "NIH shift," we also owe to Udenfriend and his colleagues (Renson, Daly, Weissbach, Witkop, and Udenfriend, 1966).

In spite of concerted efforts from several laboratories, including our own, only a tenfold purification of cerebral tryptophan-5-hydroxylase has thus far attained (Friedman, Kappelman, and Kaufman, 1972). Lack of a pure homogenous enzyme preparation of high specific activity greatly hampers understanding the nature of this regulatory enzyme and thus contributes much to controversial interpretations of its regulation *in vivo*.

Tryptophan-5-hydroxylase is one of the monooxygenases, catalyzing the irreversible reaction

$$\text{L-Tryp} + BH_4 + O_2 \rightarrow \text{5-HTP} + BH_2 + H_2O$$

This enzyme catalyzes the rate-limiting reaction in the synthesis of 5-hydroxyindoles and is dependent on many factors for its optimal activity. Of course, the definition of "optimal activity" which is attainable *in vitro* is somewhat subjectively dependent on the choice of assay. An illustrative example that comes to mind is the various values quoted in the literature for tryptophan-5-hydroxylase activity for a given cerebral area.

Until recently, several assays of tryptophan-5-hydroxylase activity were employed (Lovenberg et al., 1967; Lovenberg, Bensinger, Jackson, and Daly, 1971; Ichiyama, Nukamura, Nishizuka, and Hayaishi, 1970), which gave a fair estimate of the enzyme activity but were fraught either with technical difficulties such as preparation of uniformly resolving columns, availability of pure 5-^{3}H-tryptophan, or were dependent on coupled reaction with 5-HTP decarboxylase, relying on complete decarboxylation of 5-HTP hopefully without involving that of L-tryptophan. The disadvantage of this latter method in particular is that it cannot be applied to crude tissue fractions nor to any studies where the substrate concentration must be over 2×10^{-5} M.

The rapid, nonisotopic methods (Friedman et al., 1972; Gál and Patterson, 1973) obviate these difficulties and permit direct measurement of either 5-HTP or total 5-hydroxyindoles.

At present, the requirements for optimal assay of cerebral tryptophan-5-hydroxylase can be classed into two categories: obligatory and optional (Table 1).

Let us look at these requirements in more detail, as well as at the method of preparation of the enzyme, since this is relevant to understanding the controversy concerning the intracellular state of the enzyme. Cerebral tryptophan-5-hydroxylase is easily extractable as soluble enzyme following break-up of the tissue with hypotonic buffer

TABLE 1. *Requirements for optimal assay of cerebral tryptophan-5-hydroxylase activity*

Obligatory	pH 7.4–7.6, buffer other than phosphate
	O_2 (preferably at 1 atm barometric pressure)
	Tetrahydropterin-cofactor (preferably BH_4 or 6-MPH_4 at K_M level)
	L-tryptophan (preferably at K_M-level consistent with the cofactor employed)
	Enzyme protein: 0.1–3.0 mg (depending on stage of purity)
	Dithiothreitol or mercaptoethanol in buffer
Optional	Ferrous ion*
	Catalase
	$NADPH_2$
	Stimulating factor
Inhibitory	Presence of sucrose, ficol; sonication in one minute (loss 70% activity)
Enhancement	Pretrypsinization of tissue

*IMPORTANT for the enzyme assay from most cells and pineal (Lovenberg et al., 1967).

solutions (Robinson, Lovenberg, and Sjoerdsma, 1968); on the other hand, in hypertonic or isotonic sucrose, ficol, or mannitol solution, most of the enzyme activity of the homogenized tissue appears to reside mainly in the mitochondrial fraction (Gál et al., 1966) or so-called particulate fraction (Knapp and Mandell, 1972). Our observation that this particulate-bound enzyme, unlike the soluble one, is unresponsive to exogenous cofactor (Gál et al., 1966) was recently confirmed (Knapp and Mandell, 1972). We have recently noted that subsequent treatments of the "particulate" from 0.25 to 0.33 M sucrose with hypotonic buffer solutions failed to solubilize the enzyme. At present the nature of the change in the particulate structure brought about by isotonic or hypertonic sucrose is unclear. For instance, tryptophan-5-hydroxylase activity from rat septum is wholly particulate bound with isotonic sucrose, but entirely soluble upon treatment with 0.05 M tris-acetate at pH 7.6.

It is therefore premature to assume that we are dealing with different forms of the enzyme (isoenzymes) or with an enzyme which, when structure bound, is complexed with saturating amounts of the cofactor. The proposition that certain cerebral areas contain no soluble but only particulate-bound enzyme (Knapp and Mandell, 1972) is by and large arbitrary. In my view, an answer to whether cerebral tryptophan-5-hydroxylase is soluble, particulate bound, or both in the cell must be deferred until brain tissue can be explored with peroxidase-labeled antibody to pure tryptophan-5-hydroxylase by electron microscopy.

On reviewing the controls of tryptophan-5-hydroxylase an absolute requirement for oxygen is obvious, and there is also good agreement that an increase in pO_2 (Fisher and Kaufman, 1972) *in vivo* and hypoxia *in vivo* (Davis and Carlsson, 1973) is inhibitory to the hydroxylases.

However, a point of certain contention is whether there is enough substrate, L-tryptophan, available to subserve the need for the synthesis of 5-hydroxytryptophan in the brain.

Most of the available data refer to levels obtained from brain of rats; these, therefore, will serve as the basis of the present discussion.

It is known that L-tryptophan and 5-hydroxytryptamine (5-HT) levels in the brain and plasma are subject to diurnal rhythms (Wurtman and Fernstrom, 1972). We may therefore assume that an average cerebral content of this amino acid is about 5 μg/g (or about 25 nmole/g), although 45 nmole/g values have also been reported (McKean, Boggs, and Peterson, 1968). This represents a 4.5 to 2.5 $\times$ 10^{-5} M cerebral concentration. Most workers in this area have confirmed and accepted a K_m of 3 $\times$ 10^{-4} M L-tryptophan for crude or partially purified tryptophan-5-hydroxylase obtained with 2-amino-4-hydroxy-6,7-dimethyltetrahydropteridine (DMPH$_4$) as cofactor (Jéquier, Robinson,

Lovenberg, and Sjoerdsma, 1969; Ichiyama et al., 1970). This led to an immediate and somewhat incautious conclusion that such K_m was rather high compared to the cerebral level of free tryptophan, thus leaving the enzyme in a state of unsaturation. Naturally, these facts were rather perplexing and have led to many interesting observations (Tagliamonte, Tagliamonte, Perez-Cruet, and Gessa, 1971; Tagliamonte, Tagliamonte, Di Chiara, Gessa, and Gessa, 1972), albeit to debatable interpretation. For instance, these authors have stated that "indirect evidence indicates that the rate-limiting step in the synthesis of brain 5-HT is the concentration of tryptophan in brain and not, as previously considered (Green and Sawyer, 1966), tryptophan-hydroxylase" (Tagliamonte, Biggio, Vargiu, and Gessa, 1973).

However, such a conclusion ought to be consistent with data obtained from using naturally occurring cofactors rather than a model structure such as $DMPH_4$. Therefore, a comparison of the above statement as to its validity within the context of the results of a recent study (Friedman et al., 1972) will be most instructive. Friedman et al. confirm and expand earlier observations obtained with phenylalanine-4-hydroxylase (Kaufman, 1970) which reveal significant differences in the kinetic parameters dependent on the pterin cofactor used. A summary of the apparent K_m values (Table 2) does not support the idea that tryptophan-5-hydroxylase *in vivo* operates under substrate limitation. "Soluble" or "particulate" enzyme functions within a K_m of 2 to 5×10^{-5} M L-tryptophan with the probable natural cofactor BH_4, consequently an average of 2.5×10^{-5} M cerebral tryptophan, is within the concentration range corresponding to its K_m value. In addition, greater than 2×10^{-4} M tryptophan with BH_4 was shown to be inhibitory to the enzyme *in vitro* (Friedman et al., 1972).

Inhibition of the enzyme by its substrate *in vivo* can be assessed by a survey of studies (Table 3) on the effect of tryptophan loading. It has been repeatedly demonstrated that an increase in cerebral tryptophan levels results in increased synthesis of cerebral 5-hydroxytryptamine. However, several relevant facts must be considered. First, the total cerebral 5-hydroxyindole concentration in control animals represents about 25% conversion of the available free tryptophan (25 nmoles/g) in the brain. This means that under steady-state conditions approximately 0.7 μg of tryptophan is channeled into the synthesis of cerebral 5-HT. This amount of tryptophan comes from a total concentration within the 2 to 5×10^{-5} M range of the apparent K_m with BH_4 as cofactor. The actual K_m value *in vivo* may be even lower than 2×10^{-5} M. Second, tryptophan loading, with or without pretreatment with a monoamineoxidase inhibitor, will produce a 10- to 15-fold increase in cerebral free tryptophan concentration without, however, proportionately increasing the concentration of 5-HT in the brain. This seems

TABLE 2. Apparent K_M values for different substrates of
tryptophan-5-hydroxylase from rat brain

Enzyme	Substrate	Cofactor		
		BH_4	6-MPH_4	$DMPH_4$
"Solubilized"	L-TRP	3.8; 5.0[a] 2.5*	8.9; 7.8[a]	30; 29[a]
	L-AMTP	–	11.7	38.4
"Particulate"	L-TRP	2[c]; 2.2[d]	–	–
"Solubilized"	Pterin (with L-TRP)	3.5; 3.1[a] 0.5[b]	9.7; 6.7[a]	13; 13[a] 3[b]
	(with L-AMTP)	–	11.2	–
	Oxygen	2.5%[a]	–	20%[a]

[a]Friedman et al. (1972) rabbit brain.
[b]Jéquier et al. (1969).
[c]Ichiyama et al. (1970) guinea pig brain.
[d]Peters et al. (1968).
Other values from the author's laboratory. All the values (except for oxygen)
are at $\times 10^{-5}$ M concentration.
*With stimulating factor (Gál and Roggeveen, 1973).

to be indicative of a substrate-produced enzyme inhibition *in vivo*.
Furthermore, even at 8.3×10^{-5} M saturating concentration of the
enzyme by free tryptophan, only a 50% increment in 5-HT concentration takes place (Table 3).

It was therefore suggested that tryptophan-5-hydroxylase might be
regulated through "feedback" inhibition by 5-HT (Macon, Sokoloff,
and Glowinski, 1971). A negative correlation between generally labeled
5-HT [^{3}H-G]-5-HT accumulation from [^{3}H-G] tryptophan and endogenous concentration of 5-HT after monoamineoxidase inhibition was
adduced as evidence for feedback inhibition. This, however, is
consistent with the stimulatory effect of elevated plasma corticosteroids on hydroxylation of tryptophan in the brain, which will occur
during the first 60 to 90 min following injection of a MAO inhibitor
(Millard and Gál, 1971). There are several factors which militate against
the concept of a feedback inhibition produced by 5-HT. Even at levels
of 10^{-4} M to 10^{-3} M (in the presence of $DMPH_4$ or BH_4), 5-HT does
not inhibit tryptophan-5-hydroxylase *in vitro*. In our experience, using
uniformly labeled ^{3}H-tryptophan led to tritium exchange with hydro-
gen in the protein. Over 20% label could be recovered from amino acids
other than tryptophan following acid hydrolysis of samples of brain

TABLE 3. *Correlation of available free tryptophan to levels of 5-hydroxyindoles in rat brain*

Condition	TRP (mg/kg)	TRP	TA	5-HTP (nmoles/g)	5-HIPA (wet weight)	5-HT	5-HIAA	Reference
Control	—	15–35[a]	0.12	0.18–0.32	0.22[b]	2.3–3.7**	2.1–3.1	Millard et al., 1971[b] Wurtman and Fern-strom, 1972[a]
TRP-load	125	330	—	—	—	4.5(2.8)	—	Wurtman and Fern-strom, 1972
TRP-load + TCP*	30	83	—	—	—	3.5(2.3)	—	Graham-Smith, 1971
TRP-load + TCP*	200	539	—	—	—	6.8	—	Graham-Smith, 1971
TRP-load	400	666	—	—	—	3.5(1.8)	3.4(1.2)	Eccleston, Ashcroft, and Crawford, 1965

*Animals were given tranylcypromine (TCP) 20 mg/kg, 30 min later L-TRP, and sacrificed 1 hr later.

**5-HT values represent data obtained by various methods of analysis. Numbers in brackets refer to 5-HT content from control rats.

[a],[b]See Table 2.

TABLE 4. *Rate of synthesis of 5-HT and NE in rat brain*

Treatment	5-HT (nmole/g)	$k_{5\text{-HT}}$ (hr^{-1})	Rate of synthesis (nmole/g/hr)	Reference
Control (7)	2.8 ± 0.1	0.70 ± 0.07	1.9 ± 0.2	Lin et al. (1969)
Pargyline (7)	6.5 ± 0.2	0.30 ± 0.03	1.9 ± 0.2	
Control (12)	2.2 ± 0.1	0.60 ± 0.02	1.4 ± 0.14	
Pargyline (12)	4.9 ± 0.3	0.24 ± 0.03	1.2 ± 0.10	Millard et al. (1972)
	NE (nmole/g)	k_{NE}	nmole/g/hr	
Control (5)	3.0 ± 0.2	0.25 ± 0.02	0.73 ± 0.05	
Pargyline (5)	4.2 ± 0.2	0.079 ± 0.007	0.33 ± 0.02	Lin et al. (1969)

Substrates ([^{14}C] L-tryptophan and [^{14}C] L-tyrosine) were given by intravenous infusion.

protein from animals sacrificed 3 hr after intracerebral [^{3}H-G] tryptophan. Another objection to an interpretation of feedback inhibition derives from lack of available data (Macon et al., 1971) on changes in specific activity of tryptophan against time. Taken in all, it appears that tryptophan-5-hydroxylase, a rate-limiting regulatory enzyme, is not inhibited by 5-HT either *in vitro* (Jéquier et al., 1969) or *in vivo* (Millard and Gál, 1971). Further studies could not confirm any negative feedback effect on tryptophan hydroxylase by increased concentration of 5-HT (Table 4) (Lin, Neff, Ngai, and Costa, 1969; Millard, Costa, and Gál, 1972).

Another problem concerns localization and axoplasmic transport of tryptophan-5-hydroxylase. It has been reported that tryptophan-5-hydroxylase is localized in the synaptosomes (Grahame-Smith, 1967). Subcellular fractionation revealed that 40% of the enzyme activity was associated with the crude mitochondrial fraction (p_2), and about two-thirds of this activity was recoverable upon density gradient analysis from the synaptosomal fraction (B) and one-third from the myelin fraction (A). However, it is noteworthy that the 17,000 × g supernatant (S_2) contained almost as much activity, mostly as soluble enzyme, as the synaptosomal fraction (B), and that upon rupture of the synaptosomes, the tryptophan-5-hydroxylase activity became responsive to addition of DMPH$_4$. This suggests the presence of both soluble and particulate-bound enzyme in the synaptosomal fraction. This is not unexpected, because the enzyme must reach the nerve endings from the perikaryal region by axoplasmic flow since there is no convincing evidence that there is synaptosomal synthesis of hydroxylases. Indeed, relevant studies indicate that all the soluble proteins of the nerve endings are transported from the cell body (Barondes, 1969). Recently, the lack of inhibition of septal tryptophan-5-hydroxylase in parachlorophenylalanine (*p*-CP)-treated rats was given as evidence for a slow axoplasmic flow of this enzyme from the cell bodies of the raphe to septum, an area of "high serotonergic synaptosomal terminals" (Knapp and Mandell, 1972). To explain this in terms of the "particulate" nature of the septal enzyme, one must either postulate axoplasmic flow of mitochondria or the flow of the soluble enzyme which, when it ultimately reaches its destination in the nerve endings, will have to be incorporated into the existing particulate structures in a manner leading to profound changes in kinetic and physicochemical characteristics of the enzyme. If, indeed, the septum partially or totally depends on supply of the enzyme from the raphe, then 10 to 12 days following transection of the medial forebrain bundle (MFB) there should be an effect on the tryptophan-5-hydroxylase content of the septum. Since the approximate half-life of the enzyme is 2 to 3 days (Meek and Neff, 1972) or 5 days (Knapp and Mandell, 1972), 10 days were deemed to be a sufficient period to assess any changes in the septum due to MFB lesions. Available evidence does not indicate that MFB lesions even after the tenth postoperative day will alter the pattern of tryptophan-5-hydroxylase activity in the septum (Table 5). Intraperitoneal injec-

TABLE 5. *Effect of medial forebrain bundle (MFB) lesions and p-chlorophenylalanine (300 mg/kg) on tryptophan-5-hydroxylase*

Brain region	Treatment	Sham-control	MFB-lesion
		Enzyme activity (nmole/mg/hr)	
Septum	Saline	2.9 ± 0.24 (12)	2.5 ± 0.28 (12)
	p-CP	2.0 ± 0.27 (17)	1.9 ± 0.03 (17)
Colliculi	Saline	7.9 ± 0.51 (12)	3.8 ± 0.62 (12)
	p-CP	1.4 ± 0.42 (17)	0.82 ± 0.08 (17)
Hippocampus	Saline	0.56 ± 0.12 (6)	0.45 ± 0.06 (6)
	p-CP	—	—
Pons	Saline	4.4 ± 0.35 (12)	5.3 ± 0.39 (12)
	p-CP	0.76 ± 0.17 (17)	0.51 ± 0.04 (17)
Telencephalon	Saline	1.2 ± 0.13 (6)	0.48 ± 0.06 (6)
	p-CP	—	—

Number of animals in brackets. *p*-CP was administered intraperitoneally on the sixth postoperative day.

tion of *p*-CP (300 mg/kg) on the sixth postoperative day did not lead to significant irreversible inactivation of the enzyme, even though the presence of ^{14}C-*p*-CP was still 218 dpm/area (Harvey and Gál, 1974). These experiments support our view that (1) the septum has a very active synthesis of its own tryptophan-5-hydroxylase; (2) the contribution to the level of septal hydroxylase by axoplasmic flow is minimal; and (3) the septal response to *p*-CP in terms of irreversible inactivation (Jéquier et al., 1969; Gál, Roggeveen, and Millard, 1970) of the enzyme is only 25 to 30%. Perhaps, as suggested by others (Knapp and Mandell, 1972), some of the septal and hippocampal enzyme could be an isoenzyme of tryptophan-5-hydroxylase.

This presentation is not intended to dampen enthusiasm or belittle anyone's effort, but rather as an expression of my faith in our common goal: a search for facts.

REFERENCES

Barondes, S. H. (1969): Two sites of synthesis of macromolecules in neurons. In: *Cellular Dynamics of the Neuron*, edited by S. H. Barondes. Academic Press, New York.

Clark, C. T., Weissbach, H., and Udenfriend, S. (1954): 5-Hydroxytryptophan decarboxylase: Preparation and properties. *Journal of Biological Chemistry*, 210:139-148.

Davis, J. N., and Carlsson, A. (1973): Effect of hypoxia on tyrosine and tryptophan hydroxylation in an anesthetized rat brain. *Journal of Neurochemistry*, 20:913-915.

Eccleston, D., Ashcroft, G. W., and Crawford, T. B. B. (1965): 5-Hydroxyindole metabolism in rat brain. A study of intermediate metabolism using the technique of tryptophan loading-II. *Journal of Neurochemistry*, 12:493-503.

Fisher, D. B., and Kaufman, S. (1972): The inhibition of phenylalanine and tyrosine hydroxylases by high oxygen levels. *Journal of Neurochemistry*, 19:1359-1365.

Friedman, P. A., Kappelman, A. H., and Kaufman, S. (1972): Partial purification and characterization of tryptophan hydroxylase from rabbit hindbrain. *Journal of Biological Chemistry*, 247:4165-4173.

Gál, E. M., (1965): *In vitro* hydroxylation of tryptophan by brain tissue. *Federation Proceedings*, 24:580.

Gál, E. M., Armstrong, J. C., and Ginsberg, B. (1966): The nature of *in vitro* hydroxylation of L-tryptophan by brain tissue. *Journal of Neurochemistry*, 13:643-654.

Gál, E. M., and Marshall, F. D. (1964): The hydroxylation of tryptophan by pigeon brain *in vivo*. In: *Progress in Brain Research*, edited by H. E. Himwich and W. A. Himwich. Elsevier, Amsterdam.

Gál, E. M., Morgan, M., Chatterjee, S. K., and Marshall, F. D. (1964): Hydroxylation of tryptophan by brain tissue *in vivo* and related aspects of 5-hydroxytryptamine metabolism. *Biochemical Pharmacology*, 13:1639-1653.

Gál, E. M., and Patterson, K. (1973): Rapid nonisotopic assay of tryptophan-5-hydroxylase activity in tissues. *Analytical Biochemistry*, 52:625-629.

Gál, E. M., Poczik, M., and Marshall, F. D. (1963): Hydroxylation of tryptophan to 5-hydroxytryptophan by brain tissue *in vivo*. *Biochemical and Biophysical Research Communications*, 12:39-43.

Gál, E. M., and Roggeveen, A. E. (1973): Cerebral hydroxylases: Stimulation by a new factor. *Science*, 179:809-811.

Gál, E. M., Roggeveen, A. E., and Millard, S. A. (1970): DL-(2-^{14}C-)*p*-Chloro-phenylalanine as an inhibitor of tryptophan-5-hydroxylase. *Journal of Neurochemistry*, 17:1224-1235.

Grahame-Smith, D. G. (1964): Tryptophan hydroxylation in brain. *Biochemical and Biophysical Research Communications*, 16:586-592.

Grahame-Smith, D. G. (1967): The biosynthesis of 5-hydroxytryptamine in brain. *Biochemical Journal*, 105:351-360.

Grahame-Smith, D. G. (1971): Studies *in vivo* on the relationship between brain tryptophan, brain 5-HT synthesis and hyperactivity in rats treated with a monoamineoxidase inhibitor and L-tryptophan. *Journal of Neurochemistry*, 18:1053-1066.

Green, H., and Sawyer, J. L. (1966): Demonstration, characterization and assay procedure of tryptophan hydroxylase in the rat brain. *Analytical Biochemistry*, 15:53-64.

Harvey, J. A., and Gál, E. M. (1974): Effect of medial forebrain bundle lesions and *p*-CP on tryptophan-5-hydroxylase and 5-hydroxyindoles in rat brain. *Science (in press)*.

Ichiyama, A., Nukamura, S., Nishizuka, Y., and Hayaishi, O. (1970): Enzyme studies on the biosynthesis of serotonin in mammalian brain. *Journal of Biological Chemistry*, 245:1699-1709.

Jéquier, E., Robinson, D. S., Lovenberg, W., and Sjoerdsma, A. (1969): Further studies on tryptophan hydroxylase in rat brain stem and beef pineal. *Biochemical Pharmacology*, 18:1071-1081.

Kaufman, S. (1970): A protein that stimulates rat liver phenylalanine hydroxylase. *Journal of Biological Chemistry*, 245:4751-4759.

Knapp, S., and Mandell, A. J. (1972): Parachlorophenylalanine—Its three phase sequence of interactions with two forms of brain tryptophan hydroxylase. *Life Sciences*, 16:761-771.

Lin, R. C., Neff, N. H., Ngai, S. H., and Costa, E. (1969): Turnover rates of serotonin and norepinephrine in brain of normal and pargyline treated rats. *Life Sciences*, 8:1077-1084.

Lovenberg, W., Bensinger, R. E., Jackson, R. L., and Daly, J. W. (1971): Rapid analysis of tryptophan hydroxylase in rat tissue using 5-H^3-tryptophan. *Analytical Biochemistry*, 43:269-274.

Lovenberg, W., Jéquier, E., and Sjoerdsma, A. (1967): Tryptophan hydroxylation: Measurement in pineal gland, brain stem, and carcinoid tumor. *Science*, 155:217-219.

Macon, J., Sokoloff, L., and Glowinski, J. (1971): Feedback control of rat brain 5-hydroxytryptamine synthesis. *Journal of Neurochemistry*, 18:323-331.

McKean, C. M., Boggs, D. E., and Peterson, N. A. (1968): The influence of high phenylalanine and tyrosine on the concentration of essential amino acids in brain. *Journal of Neurochemistry*, 15:235-241.

Meek, J. L., and Neff, N. H. (1972): Tryptophan-5-hydroxylase: Approximation of half-life and rate of axonal flow. *Journal of Neurochemistry*, 19:1519-1525.

Millard, S. A., Costa, E., and Gál, E. M. (1972): On the control of brain serotonin turnover rate by end product inhibition. *Brain Research*, 40:545-551.

Millard, S. A., and Gál, E. M. (1971): The contribution of 5-hydroxyindolepyruvic acid to cerebral 5-hydroxyindole metabolism. *International Journal of Neuroscience*, 1:211-218.

Peters, D. A. V., McGeer, P. L., and McGeer, E. G. (1968): The distribution of tryptophan hydroxylase in cat brain. *Journal of Neurochemistry*, 15:1431-1435.

Renson, J., Daly, J., Weissbach, H., Witkop, B., and Udenfriend, S. (1966): Enzymatic conversion of 5-tritiotryptophan to 4-tritiotryptophan. *Biochemical and Biophysical Research Communications*, 25:504-513.

Robinson, D., Lovenberg, W., and Sjoerdsma, A. (1968): Subcellular distribution and properties of rat brain tryptophan hydroxylase. *Archives of Biochemistry and Biophysics*, 123:419-420.

Tagliamonte, A., Biggio, G., Vargiu, L. and Gessa, G. L. (1973): Increase of brain tryptophan and stimulation of serotonin synthesis by salicylate. *Journal of Neurochemistry*, 20:909-912.

Tagliamonte, A., Tagliamonte, P., Di Chiara, G., Gessa, R., and Gessa, G. L. (1972): Increase of brain tryptophan by electroconvulsive shock in rats. *Journal of Neurochemistry*, 19:1509-1512.

Tagliamonte, A., Tagliamonte, P., Perez-Cruet, J., and Gessa, G. L. (1971): Increase of brain tryptophan caused by drugs which stimulate serotonin synthesis. *Nature*, 229:125-126.

Udenfriend, S., Titus, E., Weissbach, H., and Peterson, R. E. (1956): Biogenesis and metabolism of 5-hydroxyindole compounds. *Journal of Biological Chemistry*, 219:335-344.

Wurtman, R. J., and Fernstrom, J. D. (1972): L-tryptophan, L-tyrosine, and the control of brain monoamine biosynthesis. In: *Perspectives in Neuropharmacology*, edited by S. H. Snyder. Oxford University Press, London.

Advances in Biochemical Psychopharmacology, Vol. 11
Raven Press, New York © 1974

Purification of Pig Brainstem Tryptophan Hydroxylase and Some of Its Properties

Moussa B. H. Youdim, Michel Hamon, and Sylvie Bourgoin

MRC Unit and Department of Clinical Pharmacology, University of Oxford, Radcliffe Infirmary, Oxford, England
Groupe, N.B., Laboratoire de Biologie Moleculaire, Collège de France, Paris, France

The first enzymatic step in the formation of the neurotransmitter 5-hydroxytryptamine (serotonin) is the hydroxylation of tryptophan to 5-hydroxytryptophan. The enzyme responsible for this reaction is tryptophan hydroxylase (tryptophan 5-mono-oxygenase, EC 1.1.99 1.4). It is only recently that this enzyme has been identified in the mammalian central nervous system (Grahame-Smith, 1964; Gál, 1965), the highest activity being observed in the brainstem and pineal body (Lovenberg, Jéquier, and Sjoerdsma, 1968).

Subcellular fractionation of pig brainstem according to the method of Gray and Whittaker (1962) indicates that a large part of enzyme activity (65%) is associated with the particulate fraction, with a little more than 25% in the soluble fraction. Attempts to solubilize more of the particulate enzyme fraction with the use of an ultrasonic disintegrator or treatment with 0.1% Triton® X-100 (a nonionic detergent) failed, with accompanying loss of hydroxylase activity. However, homogenization of brainstem in 0.05 M tris-acetate buffer, pH 7.4, containing 1 mM dithiothreitol and 10 mM L-tryptophan, and followed by ultrasonic oscillation resulted in increased total activity in the soluble fraction (50%).

Heat inactivation of the subcellular fractions, including that of the homogenate, at $45°C$ showed first-order denaturation and an identical rate of loss of activity for all fractions. Sonic oscillation of the particulate and soluble fraction in the presence of 1 mM 2-amino-4-hydroxy-6,7-dimethyltetrahydropteridine ($DMPH_4$) resulted in the protection of enzyme activity only in the soluble fraction.

Attempts to purify and characterize brain tryptophan hydroxylase have failed mainly because of its low activity, instability, and the lack of a rapid enzyme assay. Using the assay methods of Green and Sawyer

(1966), a method has been developed (Youdim, Hamon, and Bourgoin, 1973) for the preparation of a moderately pure enzyme. Pig brainstem, including the caudate nucleus, was homogenized in 0.05 M tris-acetate buffer (pH 7.4) containing 1 mM dithiothreitol, using an Ultratrax homogenizer. L-Tryptophan was added to the homogenate at a final concentration of 10 mM and the mixture was sonicated for 5 to 10 min at 4°C. The preparation was centrifuged at 17,000 × g for 90 min and the supernatant used for further purification of the enzyme. Ammonium sulphate fractionation, Sephadex G-25 Column chromatography, calcium phosphate (hydroxy-lapatite) fraction, and Sephadex G-100 or G-200 column chromatography resulted in a 40-fold purification, with specific activity of 0.5 nmole/min/mg protein (Table 1). This is a considerably higher specific activity than that reported recently by Friedman, Kappelman, and Kaufman (1972). However, polyacrylamide gel electrophoresis indicates that the enzyme is still too impure to carry out final chemical characterization. The loss in enzyme activity was about 50% within 24 hr, and all the steps in the purification were carried out within this period. The molecular weight of the enzyme as measured by gel filtration was 55,000 to 60,000, which is similar to that reported for phenylalanine hydroxylase (Guroff and Ito, 1965; Kaufman, 1971).

Preliminary enzyme kinetic studies indicate a change in K_m when the enzyme is solubilized. This change is from 0.02 mM for the enzyme bound to particles to 0.4 mM for the purified enzyme. The reason for this change is not understood; however, it may be associated with the cofactor requirement of the enzyme which is present in the particulate fraction. The molecular structure of the natural cofactor is not known,

TABLE 1. *Purification of tryptophan hydroxylase*

Step	Protein (mg)	Units	Specific activity[a] (unit/mg protein)	Yield (%)
Extract (homogenate)	7200	93.6	0.013	100
Ammonium sulfate (25−60%)	1420	21.3	0.015	23
Sephadex G 25	350	34.6	0.099	37
Calcium phosphate	125	22.9	0.183	24.5
Sephadex G 100	17	6.8	0.400	7
Sephadex G 200	15	7.5	0.499	8

[a]Specific activity is defined as nanomoles of 5-hydroxytryptamine formed per minute per milligram of protein.

and DMPH$_4$ has been used in all enzyme assays as a substitute for it. Some authors (Ichiyama, Nakamura, Nishizuka, and Hayaishi, 1970) have taken these results to indicate the presence of two enzymic forms. In our studies, neither the rate of heat inactivation of the soluble and particulate fractions nor the gel filtration of the purified enzyme indicated the presence of more than one molecular form of tryptophan hydroxylase. In their recent studies, Friedman et al. (1972) found a change in K_m depending on whether DMPH$_4$ or tetrahydrobiopterin (BH$_4$) was used as a cofactor. Unless the enzyme is further purified to the extent that it can be used for either electrophoretic separation or amino acid analysis, the presence of different tryptophan hydroxylases must not be taken seriously.

Earlier studies on the partially purified enzyme indicated that tryptophan hydroxylase may be an iron-containing protein. This opinion derived from studies on the stimulation of enzyme activity by ferrous iron (Lovenberg et al., 1968) and the inhibition of tryptophan hydroxylase activity by iron chelators (Lovenberg et al., 1968; Ichiyama et al., 1970). Purified pig brainstem tryptophan hydroxylase activity was stimulated by neither 1 mM ferrous iron nor chelating agents such as 8-hydroxyquinoline. O-Phenanthroline or EDTA at 1 mM inhibited only moderately. Not enough evidence is available at present to implicate iron as a cofactor (Table 2). During the purification procedure a sevenfold increase in enzyme activity was observed after chromatography on Sephadex G-25 (Table 1). This

TABLE 2. *Activity of tryptophan hydroxylase in the presence of various chemicals*

Chemical	Concentration (mM)	Percent activity remaining
Fe^{2+}	0.5	88
1-10 Phenanthroline	1.0	65
8-Hydroxyquinoline	1.0	59
EDTA	1.0	52.5
5-Hydroxytryptamine	0.1	85
	0.5	163
Tyramine	0.5	94
Tryptamine	0.5	88
Dopamine	0.1	51

Each value is the mean of at least three separate estimations. DMPH$_4$ at 1 mmole final concentration was used as the cofactor.

phenomenon may be due to the removal of excess ammonium sulphate. It has also been suggested that this increase in enzyme activity may be due to the removal of a dialyzable natural inhibitor (Lovenberg, Bensinger, Jackson, and Daly, 1971). We have some further evidence to support the presence of such an inhibitor. Heat-inactivated (100°C for 5 min) 30% (w/v) pig brainstem homogenate or 17,000 $\times$ g supernatant, when added to the native particulate or purified enzyme, produces a 50% inhibition of enzyme activity.

Monoamine oxidase inhibitors are regularly used in the tryptophan hydroxylase assay when 5-hydroxytryptamine is the end product to be estimated. Since monoamine oxidase inhibitors are known to exert nonspecific inhibiting effects on a number of enzyme systems, their action on the activity of tryptophan hydroxylase was examined. It appears that the hydrazine inhibitors and chlorgyline interfere with the production of 5-hydroxytryptamine. It is not known whether the decrease in 5-hydroxytryptamine formation is due to the inhibition of tryptophan hydroxylase or 5-hydroxytryptophan decarboxylase.

ACKNOWLEDGMENTS

This work was supported by a traveling fellowship from the Wellcome Trust and INSERM to M. B. H. Youdim. We thank Dr. J. Glowinski, in whose laboratory this work was carried out.

REFERENCES

Friedman, P. A., Kappelman, A. H., and Kaufman, S. (1972): Partial purification and characterization of tryptophan hydroxylase from rabbit hindbrain. *Journal of Biological Chemistry*, 247:4165-4173.

Gál, E. M. (1965): *In vitro* Hydroxylation of trytophan by brain tissue. *Federation Proceedings*, 241:580.

Grahame-Smith, D. G. (1964): Tryptophan hydroxylation in brain. *Biochemical and Biophysical Research Communications*, 16:586-592.

Gray, E. G., and Whittaker, V. P. (1962): The isolation of nerve endings from brain: An electron-microscopic study of cell fragments derived by homogenization and centrifugation. *Journal of Anatomy (London)*, 96:79-88.

Green, H., and Sawyer, J. L. (1966): Demonstration, characterization and assay procedure of tryptophan hydroxylase in rat brain. *Analytical Biochemistry*, 15:53-64.

Guroff, G., and Ito, T. (1965): Purification of phenylalanine hydroxylase. *Journal of Biological Chemistry*, 240:1175-1183.

Ichiyama, A., Nakamura, S., Nishizuka, Y., and Hayaishi, O. (1970): Enzymatic studies on the biosynthesis of serotonin in mammalian brain. *Journal of Biological Chemistry*, 245:1699-1709.

Kaufman, S. (1971): The phenylalanine hydroxylating system from mammalian liver. *Advances in Enzymology*, 35:245-319.

Lovenberg, W., Bensinger, R. E., Jackson, R. L., and Daly, J. (1971): Rapid analysis of tryptophan hydroxylase in rat tissue using 5-^{3}H-tryptophan. *Analytical Biochemistry*, 43:269-274.

Lovenberg, W., Jéquier, E., and Sjoerdsma, A. (1968): Tryptophan hydroxylation in mammalian systems. *Advances in Pharmacology,* 6A:21-36.

Youdim, M. B. H., Hamon, M., and Bourgoin, S. (1973): Properties of purified pig brain stem tryptophan hydroxylase. *(Submitted for publication).*

Advances in Biochemical Psychopharmacology, Vol. 11
Raven Press, New York © 1974

On the Regulation of Tryptophan Hydroxylase in Brain

B. Zivkovic, A. Guidotti, and E. Costa

Laboratory of Preclinical Pharmacology, National Institute of Mental Health, Saint Elizabeths Hospital, Washington, D.C. 20032

I. INTRODUCTION

Although regulation of serotonin (5-HT) biosynthesis in brain tissue of living animals has been extensively studied, our knowledge of this process is still incomplete. Brain and plasma concentrations of tryptophan, the biological precursor of 5-HT, are currently considered to play a role in the control of serotonin biosynthesis (Tagliamonte, Tagliamonte, Perez-Cruet, Stern, and Gessa, 1971; Fernstrom and Wurtman, 1971; Knott, Joseph, and Curzon, 1973). The emphasis on this control mechanism has derived from the finding that brain concentrations of tryptophan are smaller than the apparent K_m of tryptophan hydroxylase for its natural substrate. However, recent kinetic studies indicated that tryptophan hydroxylase, in the presence of its presumptive natural cofactor (tetrahydrobiopterin), has a K_m for tryptophan which is close to the concentration of tryptophan in brain (Friedman, Kappelman, and Kaufman, 1972). Obviously, these discrepancies cannot be resolved until the actual concentration of tryptophan at the site of serotonin synthesis is measured.

Since the possibility that tryptophan concentration regulates brain 5-HT synthesis is still an open question, we thought it would be of interest to study whether tryptophan hydroxylase activity of brain homogenates changes in response to situations that would impose a long-lasting strain on the function of serotonergic nerves. This chapter presents data concerning increases and decreases of tryptophan hydroxylase activity in brain homogenates of rats receiving chronically drugs which change some aspects of serotonergic neuronal function.

II. INCREASE OF TRYPTOPHAN HYDROXYLASE ELICITED BY RESERPINE

Reserpine causes profound changes in serotonin metabolism. It impairs 5-HT storage (Brodie, Tomich, Kuntzman, and Shore, 1957), facilitates 5-HT availability to monoamine oxidase (MAO), and,

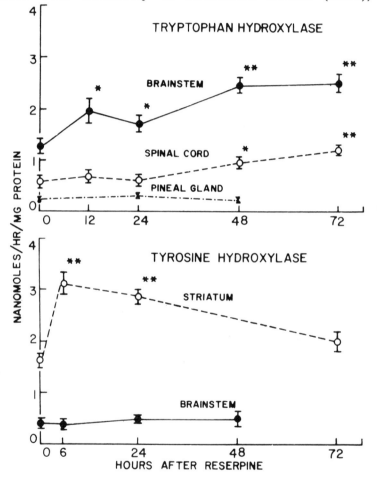

FIG. 1. Activities of tryptophan hydroxylase in brainstem, spinal cord, and pineal gland and tyrosine hydroxylase in brainstem and striatum at different time intervals after treatment with reserpine (5 mg/kg, i.p.). Activity of tryptophan hydroxylase was measured in 30,000 × *g* supernatant of brainstem and spinal cord homogenate by the method of Gál and Patterson (1973). Activity of pineal tryptophan hydroxylase was assayed in crude homogenate in the presence of 6-methyltetrahydropteridine by the modified method of Ichiyama, Nakamura, Nishizuka, and Hayaishi (1970). Activity of pineal enzyme is expressed per gland. Tyrosine hydroxylase was measured in 9,000 × *g* supernatant by the method of Waymire et al. (1971). Proteins were determined by Folin-phenol reaction (Lowry, Rosebrough, Farr, and Randall, 1951). Each point represents the mean of five to six experimental values. Vertical bars indicate standard errors. *$p < 0.05$. **$p < 0.01$.

consequently, accelerates brain 5-HT turnover rate (Tozer, Neff, and Brodie, 1966). We reasoned that if compensatory mechanisms were available to regulate biosynthesis of tryptophan hydroxylase, they might be activated during the persistent action of reserpine on 5-HT storage processes. We found that reserpine injections produce long-term increases of tryptophan hydroxylase in rat CNS (Zivkovic, Guidotti, and Costa, 1973). Figure 1 (upper part) shows that a single i.p. injection of reserpine (5 mg/kg) increases the tryptophan hydroxylase activity of brainstem homogenates by 60% at 12 hr and by 100% at 48 and 72 hr. Reserpine also increases tryptophan hydroxylase activity in homogenates of spinal cord; however, the appearance of this increase is delayed with respect to that of the brainstem enzyme (Fig. 1). In contrast, the activity of pineal tryptophan hydroxylase remains unchanged at various time intervals after reserpine administration (Fig. 1). Neither the apparent K_m for tryptophan (0.072 mM) nor that for 6-methyltetrahydropteridine (0.060 mM) of the crude homogenates from brainstem of rats receiving reserpine was different from the affinity constants of crude homogenates from brainstem of normal rats. However, the maximum velocity of the enzyme is considerably higher in brainstem homogenates of reserpine-treated (V_{max} = 4.5 nmoles/ hr/mg of protein) than in the same homogenates of normal animals (V_{max} = 3.5 nmoles/hr/mg of protein). Dialysis for 6 hr does not abolish the increase of tryptophan hydroxylase activity elicited by reserpine. The addition of reserpine (from 10^{-9} to 10^{-5} M) to homogenates of rat brainstem fails to change the tryptophan hydroxylase activity. From these results we have inferred that reserpine may increase the synthesis of enzyme protein molecules rather than increase the activity of the enzyme.

To ascertain whether the increase of tryptophan hydroxylase activity after reserpine is due to an increased number of enzyme protein molecules, we studied whether the effects of reserpine were inhibited by a protein synthesis inhibitor. Rats received 450 μg cycloheximide intraventricularly; 18 hr later ^{14}C-L-leucine was injected intravenously and its incorporation into brainstem protein estimated. It appeared that brainstem protein synthesis was reduced by 50 to 70% in animals receiving cycloheximide intraventricularly. When cycloheximide was injected after reserpine, this protein synthesis inhibitor greatly reduced the increase of tryptophan hydroxylase activity elicited by reserpine (Fig. 2).

The increase of tryptophan hydroxylase activity elicited by reserpine may reflect a general action of this drug on brain protein synthesis. Such a possibility may not be significant, because the tyrosine hydroxylase activity of brainstem is not changed by reserpine (Fig. 1, lower part). The data reported in Fig. 1 also show that the tyrosine

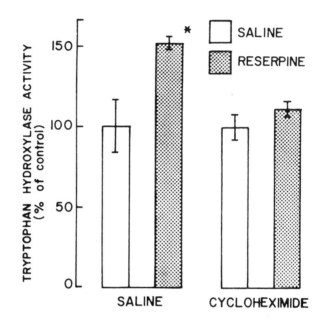

FIG. 2. Effect of cycloheximide on reserpine-elicited increase of brainstem tryptophan hydroxylase. Cycloheximide was injected intraventricularly (450 μg) 12 hr after treatment with reserpine (5 mg/kg, i.p.). The enzyme activity was assayed 18 hr after application of cycloheximide by the method of Gál and Patterson (1973). Enzyme activity of reserpine-treated animals is expressed as percent of respective controls (saline 1.2 ± 0.24 nmoles/hr/mg of protein and cycloheximide 1.4 ± 0.13 nmoles/hr/mg of protein). Each bar represents the mean ± SE of six to seven animals. *$p < 0.05$.

hydroxylase activity of striatum is increased by a single dose of reserpine (5 mg/kg, i.p.). However, the time course of this increase differs from that of the tryptophan hydroxylase activity. The increase of striatal tyrosine hydroxylase activity reaches a maximum at about 6 hr and decreases thereafter. The increment of enzyme activity is completely obliterated 72 hr after reserpine, when the increase of tryptophan hydroxylase is still maximal in both brainstem and spinal cord (Fig. 1). Keeping in mind that the part of the brain termed brainstem contains the cell bodies of dopaminergic and serotonergic neurons and that striatum and spinal cord contain the respective nerve endings, we can reason that since the increase of tyrosine hydroxylase activity in striatum is not preceded by an increase of the enzyme activity in brainstem, presumably the increment of tyrosine hydroxylase activity elicited by reserpine is not due to an increase in protein synthesis. The latter should appear first in cell bodies which possess the mechanisms to control initiation of protein synthesis.

It is well established that the activity of tyrosine hydroxylase may be regulated by end-product inhibition (Neff and Costa, 1966; Nagatsu, Levitt, and Udenfriend, 1964), whereas that of tryptophan hydroxylase is not (McGeer and Peters, 1969; Jequier, Robinson, Lovenberg and Sjoerdsma, 1969; Lin, Neff, Ngai, and Costa, 1969). Since the rapid increase of striatal tyrosine hydroxylase activity elicited by reserpine is closely related to the time course of the dopamine depletion caused by the alkaloid, one might infer that the rapid increase of tyrosine hydroxylase activity elicited by reserpine may be due to a decrease in the efficiency of end-product repression. In support of this interpretation we may adduce the following indirect evidence: (1) when striatal homogenates of normal and reserpine-treated rats are dialyzed, the tyrosine hydroxylase activity of tissue homogenates of normal rats increases while that of rats receiving reserpine does not; these results indicate that a dialyzable inhibitor is present in striatum of normal rats but not in that of reserpine-treated animals; (2) treatment with cycloheximide does not affect the increase of tyrosine hydroxylase activity elicited by reserpine; (3) when the depletion of dopamine caused by reserpine is inhibited by deprenyl, the increase of tyrosine hydroxylase activity induced by this alkaloid is also prevented. Thus, when the dopamine stores are depleted by reserpine, the control of tyrosine hydroxylase by product inhibition is decreased and dopamine is synthesized at a fast rate in the attempt to compensate the depleting effect of reserpine. In the serotonergic neurons, where end-product inhibition is not operative, the compensation for the depletion of serotonin stores caused by reserpine depends exclusively on either an increased synthesis of enzyme molecules or an increased availability of the serotonin precursor, tryptophan. Unfortunately, the extent by which the latter mechanism is involved cannot be assessed at this time.

The data in Fig. 1 show that the increase of tryptophan hydroxylase activity in nerve terminals (spinal cord) is delayed with respect to the increase of enzyme activity seen in cell bodies (brainstem) (Fig. 1, upper part). This different time course suggests that the enzyme is synthesized in cell bodies and then transported down the axons by an active axonal transport into the nerve terminals. Calculation shows that the delay documented in Fig. 1 is in agreement with the estimated rate of axonal transport of tryptophan hydroxylase (5 to 7 mm/day) (Meek and Neff, 1972).

III. POSSIBLE MECHANISMS OF THE INCREASE OF TRYPTOPHAN HYDROXYLASE ACTIVITY AFTER RESERPINE

The triggering event that is associated with the increase of tryptophan hydroxylase activity elicited by reserpine might involve: (1) a control of enzyme synthesis by intraneuronal concentrations of serotonin; (2) an activation of interneuronal loops when synaptic 5-HT receptors are unoccupied for a long time; and (3) a humoral regulation (corticosteroids). These three possibilities will be analyzed and discussed sequentially.

If an increase of tryptophan hydroxylase activity after reserpine were to depend upon the depletion of intraneuronal serotonin stores, then norfenfluramine, which causes marked depletion of brain serotonin without affecting brain catecholamine stores (Costa and Revuelta, 1972; Morgan, Lofstrandh, and Costa, 1971), should also increase the activity of tryptophan hydroxylase. However, repeated intraperitoneal injections of norfenfluramine (10 mg/kg/4 days) failed to change tryptophan hydroxylase activity of brainstem homogenates.

If the increase of tryptophan hydroxylase activity after reserpine were to be mediated by an interneuronal feedback mechanism triggered by the lack of occupancy of serotonergic postsynaptic receptors, it could be presumed that catecholaminergic neurons may be involved in this neuronal loop that controls tryptophan hydroxylase biosynthesis. Rats receiving intracisternally doses of 6-hydroxydopamine that lowered brain catecholamine content and tyrosine hydroxylase activity in striatum exhibited the increase of tryptophan hydroxylase activity when challenged with reserpine (Table 1). The data of Table 1 also show that the increase of tyrosine hydroxylase activity elicited by reserpine is prevented in the rats that had received 6-hydroxydopamine.

TABLE 1. *Effect of 6-hydroxydopamine on the increase of tryptophan and tyrosine hydroxylase activities elicited by reserpine*

Treatment	Tryptophan hydroxylase (nmoles/hr/mg of protein)	Tyrosine hydroxylase (nmoles/hr/mg of protein)
Saline	1.4 ± 0.11	1.8 ± 0.10
Reserpine	1.8 ± 0.05[a]	2.7 ± 0.23[a]
6-Hydroxydopamine	1.3 ± 0.08	1.3 ± 0.07[b]
6-Hydroxydopamine + reserpine	1.8 ± 0.09[b]	1.7 ± 0.14

6-Hydroxydopamine or saline was given intracisternally according to the schedule of Bloom, Algeri, Groppetti, Revuelta, and Costa (1969). Reserpine was injected i.p. (5 mg/kg) on the fifth day after the last injection of 6-hydroxydopamine. Tryptophan hydroxylase activity of brainstem homogenates (Gál and Patterson, 1973) and tyrosine hydroxylase activity of striatal homogenates (Waymire, Bjur, and Weiner, 1971) were assayed 48 hr after reserpine treatment. Each value represents the mean ± SE of five animals.

[a] $p < 0.02$ as compared with rats treated with saline.
[b] $p < 0.01$ as compared with rats treated with saline.

Recently Yang and Neff (1974) have shown that deprenyl is a selective inhibitor *in vitro* and *in vivo* of a molecular form of monoamine oxidase termed type B. This enzyme metabolizes dopamine but not serotonin or the other catecholamines (Yang and Neff, 1974). Deprenyl was injected before reserpine and prevented the depletion of brain dopamine elicited by this alkaloid. This treatment failed to inhibit the increase of tryptophan hydroxylase activity elicited by reserpine (Fig. 3). This suggests that depletion of brain dopamine is not involved in the action of reserpine on tryptophan hydroxylase.

Reserpine causes an intraneuronal release of serotonin. The amine released from storage sites is immediately metabolized by MAO. Consequently, in the brain of rats treated with reserpine the post-synaptic receptor contiguous to a serotonergic nerve ending may not be sufficiently stimulated by serotonin, its physiological agonist. This reduced receptor occupancy could somehow signal the serotonergic neurons to increase the synthesis of tryptophan hydroxylase as a compensatory device. We reasoned that if this mechanism were operational, continuous stimulation of postsynaptic serotonergic receptors in reserpine-treated rats may in turn decrease the amount of

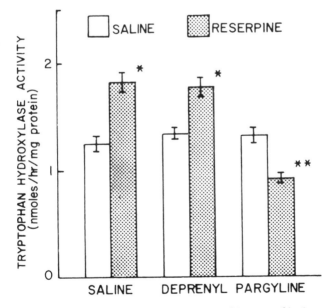

FIG. 3. Effect of deprenyl and pargyline on reserpine-elicited increase of brainstem tryptophan hydroxylase. Deprenyl (5mg/kg, i.v.) or pargyline (15 mg/kg, i.v.) was injected 5 min before reserpine (5 mg/kg, i.p.). The enzyme was assayed 48 hr after treatment with drugs (Gál and Patterson, 1973). Each value represents the mean of four to five animals. *$p < 0.02$. **$p < 0.01$.

tryptophan hydroxylase synthesized. To produce a prolonged stimulation of serotonergic postsynaptic receptors, we injected reserpine (5 mg/kg, i.p.) and the MAO inhibitor, pargyline (15 mg/kg, i.v.) together. When 48 hr had elapsed, instead of the increase of tryptophan hydroxylase activity which is usually elicited by reserpine, a significant decrease of enzyme activity was observed (Fig. 2). We also proved that the decrease after pargyline and reserpine is not due to the presence of an inhibitory substance, since after mixing this homogenate with that from normal rats, no inhibition was observed. Moreover, the apparent K_m for tryptophan of the crude homogenate is not affected by the simultaneous injection of pargyline and reserpine.

D-Lysergic acid diethylamide (LSD) decreases the turnover rate of brain serotonin (Lin, Ngai, and Costa, 1969) and lowers the firing rate of serotonergic cells in raphe nuceli (Aghajanian, 1972). Presumably, this action is due to a stimulation of brain postsynaptic receptors of serotonergic neurons by D-lysergic acid diethylamide. The 2-bromo-D-lysergic acid diethylamide (BOL) fails to decrease brain serotonin turnover rate (Lin, Ngai, and Costa, 1969) and does not change the firing rate of serotonergic cells in raphe nucleus (Aghajanian, 1972). We have found that rats given LSD (4 mg/kg/2 days) have a reduction of tryptophan hydroxylase activity without any change of the tyrosine hydroxylase activity of brainstem (Table 2); in contrast, a similar treatment with BOL fails to change tryptophan hydroxylase activity. This finding supports the hypothesis that regulation of tryptophan hydroxylase may be mediated by the level of stimulation of post-

TABLE 2. *Effect of D-lysergic acid diethylamide (LSD) on tryptophan and tyrosine hydroxylase activities of rat brainstem*

Treatment	Tryptophan hydroxylase (nmoles/hr/mg of protein)	Tyrosine hydroxylase (nmoles/hr/mg of protein)
Saline	1.3 ± 0.05	0.72 ± 0.06
BOL	1.3 ± 0.12	0.67 ± 0.02
LSD	1.0 ± 0.06[a]	0.63 ± 0.03

D-Lysergic acid diethylamide, 2-bromo-D-lysergic acid diethylamide (BOL) in 150mM NaCl (4 mg/kg; 5 ml/kg) or equivalent volume of NaCl solution were injected i.p. 48 hr and 24 hr before killing the animals. Each value represents the mean ±SE of five to eight animals.

[a] $p < 0.01$ when compared with saline-treated animals and $p < 0.02$ when compared with BOL-treated animals.

synaptic serotonergic receptors.

Corticosteroids have been shown to take part in the regulation of some enzymes in tyrosine (Govier, Lovenberg, and Sjoerdsma, 1969) and catecholamine metabolism (Wurtman and Axelrod, 1966). It has also been shown that adrenalectomy decreases the tryptophan hydroxylase activity of brainstem homogenates (Azmitia and McEwen, 1969) and the serotonin turnover rate in brainstem (Azmitia, Algeri, and Costa, 1970). Moreover, Millard, Costa, and Gál (1972) have shown that corticosterone is involved in the action of pargyline on serotonin turnover rate. Thus it appears that corticosteroids might be also involved in the control of the biosynthesis of tryptophan hydroxylase. In fact, if adrenalectomized rats are given corticosterone, the reduction of tryptophan hydroxylase activity can be partially overcome (Azmitia and McEwen, 1969). Therefore, we investigated whether the change of tryptophan hydroxylase elicited by reserpine given alone or associated with MAO inhibitors is paralleled by a change of corticosterone concentrations in blood plasma. In agreement with previous reports (Martel, Westermann, and Maickel, 1962), we showed (Table 3) that reserpine produces a striking increase of plasma corticosterone and that this increase can be completely blocked with pargyline. However, this

TABLE 3. *Effect of various drugs on corticosterone levels in rat plasma*

Treatment	Plasma corticosterone (μg/100 ml of plasma)	t-Test
Unhandled	28 ± 3.0	—
Saline	28 ± 1.9	N.S.[a]
Reserpine (5 mg/kg, i.p.)	85 ± 8.3	$p < 0.01$[b]
Pargyline (15 mg/kg, i.v.) + reserpine (5 mg/kg, i.p.)	23 ± 5.8	N.S.[b]
Deprenyl (5 mg/kg, i.v.) + reserpine (5 mg/kg, i.p.)	90 ± 4.6	$p < 0.01$[b]
Norfenfluramine (10 mg/kg, i.p.)	35 ± 5.0	N.S.[b]
BOL (4 mg/kg, i.p.)	25 ± 4.9	N.S.[b]
LSD (4 mg/kg, i.p.)	25 ± 2.8	N.S.[b]

Corticosterone concentration in plasma was analyzed 2 hr after treatment with drugs by the method of Zenker and Bernstein (1958). Animals were killed between 4 and 6 p.m. Each value represents the mean ± SE of four animals.

[a]As compared with unhandled controls.

[b]As compared with saline-treated animals.

increase is not affected by deprenyl (Table 3), which does not change the increment of tryptophan hydroxylase activity elicited by reserpine. Finally, norfenfluramine depletes brain serotonin but increases neither tryptophan hydroxylase activity nor plasma corticosterone levels. From these data we might infer that the increase of tryptophan hydroxylase activity elicited by reserpine correlates with the increase of corticosterone levels in plasma. However, the decrease of tryptophan hydroxylase activity after combined treatment with pargyline and reserpine or after LSD cannot be explained by an action on adrenal cortex, since plasma concentrations of corticosterone were normal. This finding is not entirely understood. Provisionally, we propose that two mechanisms can control synthesis rates of tryptophan hydroxylase—one mediated by corticosteroids initially described by Azmitia and McEwen (1969) and confirmed by the present experiments, the other mediated through an interneuronal loop involving the increase and, perhaps, the decrease in the functional activity of serotonergic postsynaptic receptors.

IV. CONCLUSIONS

Reserpine produces a selective, long-lasting increase of tryptophan hydroxylase activity in cell bodies and nerve terminals of serotonergic neurons of rat brain. In contrast, the activation of tyrosine hydroxylase caused by reserpine is present only in dopaminergic terminals and appears to be related to the depletion of an inhibitory dialyzable factor (dopamine?). The relationship between the increase of tryptophan hydroxylase activity in brain and the elevation of plasma corticosterone concentrations after treatment with reserpine suggests that pituitary-adrenal activation may play an important role in the regulation of tryptophan hydroxylase.

Tryptophan hydroxylase activity is decreased in rats receiving either a combined treatment with pargyline and reserpine or chronic treatment with D-lysergic acid diethylamide. Since both treatments produce hyperstimulation of postsynaptic receptors of serotonergic neurons, we propose as a working hypothesis that tryptophan hydroxylase might also be regulated through a feedback interneuronal loop which is triggered by the persistent occupancy of serotonergic receptors by an agonist.

REFERENCES

Aghajanian, G. K. (1972): Influence of drugs on the firing of serotonin-containing neurons in brain. *Federation Proceedings*, 31:91−96.
Azmitia, E. C., Jr., Algeri, S., and Costa, E. (1970): *In vivo* conversion of ^{3}H-L-tryptophan into ^{3}H-serotonin in brain areas of adrenalectomized rats.

Science, 169:201–203.

Azmitia, E. C., Jr., and McEwen, B. S. (1969): Corticosterone regulation of tryptophan hydroxylase in midbrain of the rat. *Science*, 166:1274–1276.

Bloom, F. E., Algeri, S., Groppetti, A., Revuelta, A., and Costa, E. (1969): Lesions of central norepinephrine terminals with 6-OH-dopamine: Biochemistry and fine structure. *Science*, 166:1284–1286.

Brodie, B. B., Tomich, E. G., Kuntzman, R., and Shore, P. A. (1957): The mechanism of action of reserpine: Effect of reserpine on capacity of tissue to bind serotonin. *Journal of Pharmacology and Experimental Therapeutics*, 119:461–467.

Costa, E., and Revuelta, A. (1972): Norfenfluramine and serotonin turnover rate in the rat brain. *Biochemical Pharmacology*, 21:2385–2393.

Fernstrom, J. D., and Wurtman, R. J. (1971): Brain serotonin content: Physiological dependence on plasma tryptophan levels. *Science*, 173:149–151.

Friedman, P. A., Kappelman, A. S., and Kaufman, S. (1972): Partial purification and characterization of tryptophan hydroxylase from rabbit hindbrain. *Journal of Biological Chemistry*, 247:4165–4173.

Gál, E. M., and Patterson, K. (1973): Rapid non-isotopic assay of tryptophan-5-hydroxylase activity in tissues. *Analytical Biochemistry*, 52:625–629.

Govier, W. C., Lovenberg, W., and Sjoerdsma, A. (1969): Studies on the role of catecholamines as regulators of tyrosine aminotransferase. *Biochemical Pharmacology*, 18:2661–2666.

Ichiyama, A., Nakamura, S., Nishizuka, Y., and Hayaishi, O. (1970): Enzymic studies on the biosynthesis of serotonin in mammalian brain. *Journal of Biological Chemistry*, 245:1699–1709.

Jequier, E., Robinson, D. S., Lovenberg, W., and Sjoerdsma, A. (1969): Further studies on tryptophan hydroxylase in rat brain stem and beef pineal. *Biochemical Pharmacology*, 18:1071–1081.

Knott, P. J., Joseph, M. H., and Curzon, G. (1973): Effects of food deprivation and immobilization on tryptophan and other amino acids in rat brain. *Journal of Neurochemistry*, 20:249–251.

Lin, R. C., Neff, N. H., Ngai, S. H., and Costa, E. (1969): Turnover rates of serotonin and norepinephrine in brain of normal and pargyline-treated rats. *Life Sciences*, 8:1077–1084.

Lin, R. C., Ngai, S. H., and Costa, E. (1969): Lysergic acid diethylamide: Role in conversion of plasma tryptophan to brain serotonin (5-hydroxytryptamine). *Science*, 166:237–239.

Lowry, O. H., Rosebrough, N. J., Farr, A. L., and Randall, R. J. (1951): Protein measurement with the Folin phenol reagent. *Journal of Biological Chemistry*, 193:265–275.

McGeer, E. G., and Peters, D. A. V. (1969): *In vitro* screen of inhibitors of rat brain serotonin synthesis. *Canadian Journal of Biochemistry*, 47:501–506.

Martel, R. R., Westermann, E. O., and Maickel, R. P. (1962): Dissociation of reserpine-induced sedation and ACTH hypersecretion. *Life Sciences*, No. 4:151–155.

Meek, J. L., and Neff, N. H. (1972): Tryptophan-5-hydroxylase: Approximation of half life and rate of axonal transport. *Journal of Neurochemistry*, 19:1519–1525.

Millard, S. A., Costa, E., and Gál, E. M. (1972): On the control of brain serotonin turnover rate by end product inhibition. *Brain Research*, 40:545–551.

Morgan, D., Lofstrandh, S., and Costa E. (1972): Amphetamine analogues and brain amines. *Life Sciences*, 11:83–96.

Nagatsu, T., Levitt, M., and Udenfriend, S. (1964): Tyrosine hydroxylase: The initial step in norepinephrine biosynthesis. *Journal of Biological Chemistry*, 239:2910–2917.

Neff, N. H., and Costa, E. (1966): The influence of monoamine oxidase inhibition on catecholamine synthesis. *Life Sciences*, 5:951–959.

Tagliamonte, A., Tagliamonte, P., Perez-Cruet, J., Stern, S., and Gessa, G. L. (1971): Effect of psychotropic drugs on tryptophan concentration in the rat brain. *Journal of Pharmacology and Experimental Therapeutics*, 177:475–480.

Tozer, T. N., Neff, N. H., and Brodie, B. B. (1966): Application of steady-state kinetics to the synthesis rate and turnover time of serotonin in the brain of normal and reserpine treated rats. *Journal of Pharmacology and Experimental Therapeutics*, 153:177–182.

Waymire, J. C., Bjur, R., and Weiner, N. (1971): Assay of tyrosine hydroxylase by coupled decarboxylation of dopa formed from 1-^{14}C-L-tyrosine. *Analytical Biochemistry*, 43:588–600.

Wurtman, R. J., and Axelrod, J. (1966): Control of enzymatic synthesis of adrenaline in the adrenal medulla by adrenal cortical steroids. *Journal of Biological Chemistry*, 241:2301–2305.

Yang, H.-Y.T., and Neff, N. H. (1974): β-Phenylethylamine: A specific substrate for type B monoamine oxidase of brain. *Journal of Pharmacology and Experimental Therapeutics* (*in press*).

Zenker, N., and Bernstein, D. E. (1958): The estimation of small amounts of corticosterone in rat plasma. *Journal of Biological Chemistry*, 231:695–701.

Zivkovic, B., Guidotti, A., and Costa, E. (1973): Increase of tryptophan hydroxylase activity elicited by reserpine. *Brain Research*, 57:522–526.

Advances in Biochemical Psychopharmacology, Vol. 11
Raven Press, New York © 1974

Current Status of 5-Hydroxytryptophan in Brain

**M. H. Aprison, K. H. Tachiki, J. E. Smith, J. D. Lane, and
W. J. McBride**

*Section of Neurobiology, The Institute of Psychiatric Research and Departments of
Biochemistry and Psychiatry, Indiana University Medical Center, Indianapolis,
Indiana 46202*

I. INTRODUCTION

Since 5-hydroxytryptophan (5-HTP) is the immediate precursor of
5-hydroxytryptamine (serotonin or 5-HT), it is important to know its
content in the CNS and its subdivisions. The content of 5-HTP in brain
is very small, and attempts to measure the level of this amino acid have
met with difficulty. Unfortunately, there are fewer reports of such
experiments (Wiegand and Scherfling, 1962; Eccleston, Ashcroft, and
Crawford, 1965; Lindqvist, 1971; Fischer and Aprison, 1972; Carlsson,
David, Kehr, Lindqvist, and Atack, 1972) than there are on the
measurement of 5-HT concentrations in brain. The present chapter
attempts to review briefly the current status of the "5-HTP story" in
brain in the hope that more work will be done on this important
compound. We cite some of the problems involved in the measurement
of 5-HTP in brain and offer possible solutions.

II. EXPERIMENTAL

A. Tissue Preparation

Wistar adult male rats were used in this study. The animals had free
access to food and water and were housed in group cages (12 animals
per cage). On the day of the experiment the rats were killed by
decapitation and the brains rapidly (less than 2 min) removed and
immersed in liquid nitrogen (−197°C). The frozen brains were
pulverized under liquid nitrogen in a stainless steel mortar placed on
dry ice (Takahashi and Aprison, 1964).

B. Analytical Procedure

The pulverized tissue was extracted with 20 parts of cold 1 N formic acid/acetone (15/85, v/v) according to the method of Toru and Aprison (1966) and then treated as described by Fischer, Kariya, and Aprison (1970) to the point where the brain extract is dried under dry N_2 at $37°C$. In the latter procedure, the tissue lipids were removed prior to the drying step by extracting three volumes of the formic acid/acetone supernatant with ten volumes of a mixture of 8:1 (v/v) heptane/ chloroform.

After the dried brain extract was dissolved in 450 μl of H_2O (adjusted to pH 4), a 400-μl portion was added to 350 μl of 0.2 M phosphate buffer, pH 6.1. The pH was adjusted (if necessary) to 6.1. A 700-μl portion was allowed to pass on to a Bio Rex 70 column (1.5 X 0.6 cm) before adding 1.0 ml of 0.02 M phosphate buffer, pH 6.1. The eluate was passed through a second Bio Rex 70 column, thus assuring the complete removal of 5-HT and similar compounds. The second column was washed with an additional 0.5 ml of 0.02 M phosphate buffer. The total eluate from the second column was adjusted to pH 1-2 with 5 N HCl and then extracted twice with butyl acetate (3 ml) in the presence of 1.4 g of NaCl to remove 5-hydroxyindoleacetic acid (5-HIAA). The 5-HTP in the remaining aqueous phase was extracted into butanol (7.5 ml) by shaking with a mechanical shaker. An aliquot of the butanol layer (7.0 ml) was removed and back-extracted with 0.3 ml of 0.1 N HCl and 15 ml of heptane. The butanol-heptane layer was removed and the aqueous fraction was transferred to a small test tube. These samples were used in the chemical assays for 5-HTP. However, in the chromatographic experiments, the samples were dried under N_2 at $37°C$ to volumes less than 100 μl and then absolute ethanol was added to precipitate the excess salts that were present. The ethanol phase was removed and taken to dryness. The residue was dissolved in 50 μl of 85% aqueous acetone (v/v) and then spotted on the chromatographic paper (Whatman No. 1). An additional 50 μl of 85% acetone was used to quantitate the transfer from the test tube to the paper.

Chemical Assays. The fluorescence measurements in 3 N HCl were made as described by Fischer and Aprison (1972), except that a Corning 3-73 filter was used in place of the 3-69 filter between the cuvet and the second monochrometer in the Farrand spectrophoto-fluorometer. A Corning 7-54 filter was employed between the light source and the first monochrometer.

In other experiments, the method of Maickel and Miller (1966) was scaled down to measure the fluorescent products formed by the reaction of indole derivatives and certain other compounds with o-phthalaldehyde (OPT). To 50 μl of sample contained in small test tubes we added 100 μl of 10 N HCl and 5 μl of OPT (0.05%) in absolute methanol. The tubes were capped with parafilm (a pin hole was made with a needle) and then placed in water at 85°C for 60 min; after the tubes cooled, the fluorescence was read at uncorrected wavelengths of 360 nm (activation) and 480 nm (emission). In addition, complete fluorescence spectra were determined for 5-HTP standards, tissue samples, internal standards plus tissue, reagent blanks, and tissue blanks.

Chromatographic Method. A unidirectional chromatographic procedure employing a solvent containing butanol:water:glacial acetic acid (120/50/30) was used to separate the components in the 5-HTP fraction. This method is essentially that of Jepson (1955). The chromatograms (Whatman No. 1 paper precut into strips but with the ends still intact) were developed with a descending technique for 39 hr. The spotted samples were bracketed with known standards of 5-HTP on adjacent strips. After the paper was air dried, the spots due to the standards were located with ultraviolet light or by development of the characteristic color produced after the reaction with Ehrlich's reagent. The R_f for each component of interest was determined. The spot due to 5-HTP was separated from all other compounds which were tested (see RESULTS). Based on these data, the 5-HTP spot or area on the strip chromatographed with the tissue sample was cut out, eluted with 0.1 N HCl, and used in the chemical assays.

III. RESULTS

Since the content of 5-HTP in rat brain as reported by Fischer and Aprison (1972) was considerably higher than that reported by several other groups, we examined the reasons for this discrepancy. As shown in Fig. 1, the use of Corning filter 3-69 in reading the fluorescence in 3 N HCl produced a false peak; this is due to the sharp cutoff of this filter, which is too close to the region of maximum fluorescence (535 nm, uncorrected). If a 3-73 filter is used, the peak does not occur, and the fluorescence continues to increase at wavelengths lower than 535 nm. These data indicate that some other compounds present in the 5-HTP fraction may contribute to the 535-nm fluorescence. However, if 5-HTP is added directly to the sample of brain tissue, or if a brain is obtained from a rat which was injected with 50 mg/kg of D,L-5-HTP, the 5-HTP fraction as isolated by the procedure described by Fischer and Aprison (1972) shows the characteristic increase of fluorescence at 535 nm (uncorrected). These data are shown in Fig. 2. We can confirm the data of Fischer and Aprison (1972) that (^{3}H)-5-HTP or (^{14}C)-5-HTP is found in the 5-HTP fraction as isolated by their procedure, and that there is no significant contamination of this

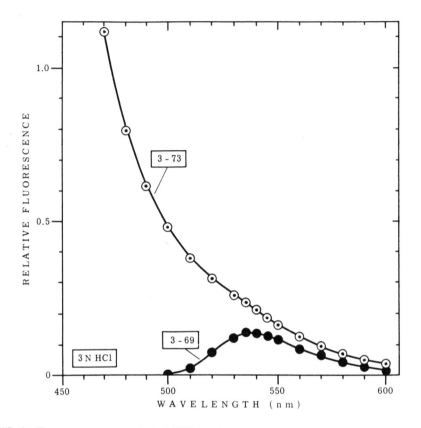

FIG. 1. Fluorescence spectra of the 5-HTP fraction isolated from rat brain using the procedure of Fischer and Aprison (1972); the spectra were determined using either one of two different filters between the sample and second monochromator. The activation wavelength was 300 nm; a Corning 7-54 filter was employed between the light source and the first monochromator.

fraction by either 5-HT or 5-HIAA. Thus, under certain experimental conditions, it appears that this method can be used to measure radioactively labeled 5-HTP as well as tissue levels of 5-HTP if present in high enough amounts. However, under certain experimental conditions an investigator must establish that the radioactivity in the 5-HTP fraction is due solely to 5-HTP. For example, if labeled tryptophan is used, it will also be found in the fraction containing 5-HTP. Furthermore, the method as previously described does not appear to measure accurately normal endogenous levels of 5-HTP in samples of brain from rats.

In an attempt to reduce the fluorescence contribution at lower wavelengths from other compounds present in the 5-HTP fraction and possibly obtain a better purification of the 5-HTP fraction, this sample

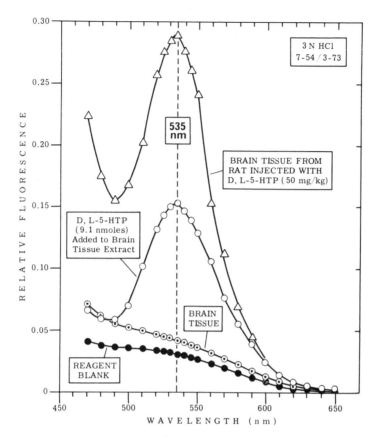

FIG. 2. Fluorescence spectra of the 5-HTP fraction isolated by the procedure of Fischer and Aprison (1972): (a) from brain of a rat injected with 50 mg/kg of D,L-5-HTP, (b) with 9.1 nmoles of D,L-5-HTP added to a brain tissue extract of a normal rat, (c) from brain tissue extract of a normal rat, and (d) the reagent blank. Tissue weights of the brain samples used in (b) and (c) were the same whereas in the (a), the weight was greater (2X). All readings were made in 3 N HCl with filters 7-54 (excitation wavelength set at 300 nm) and 3-73 (fluorescence).

was extracted into butanol and then back-extracted into acid after the addition of heptane (see EXPERIMENTAL). With such a procedure, only compounds with p*K* values similar to that of 5-HTP should be present in the 0.1 N HCl layer. Such a sample produced the fluorescence spectrum shown in Fig. 3. Note that the difference spectrum in the upper right-hand corner of this figure shows a characteristic 5-HTP peak at 535 nm (uncorrected). These data also indicate that (1) there still is one or more other compounds present and (2) the sensitivity of this method does not appear to be adequate to measure this amino acid quantitatively.

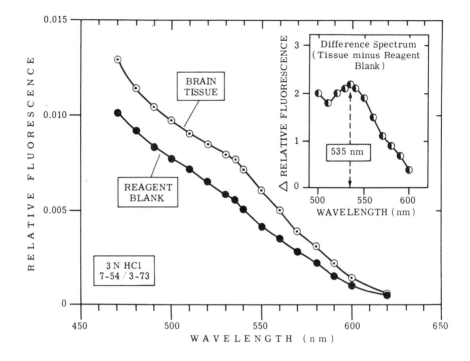

FIG. 3. Fluorescence spectra of the 5-HTP fraction isolated by the procedure described in the METHODS section from rat brain and reagent blank. The difference spectrum is shown in the upper right-hand corner. All readings were made in 3 N HCl with filters 7-54 (excitation wavelength set at 300 nm) and 3-73 (fluorescence).

Lindqvist (1971) reported that it was possible to obtain a fraction containing 5-HTP using ion-exchange procedures. Employing this procedure, we determined the fluorescent spectra of the 5-HTP fraction using conditions similar to those used to obtain the data in Fig. 1. In fact, the spectra shown in Fig. 1 are identical to those obtained for the 5-HTP fraction using the Lindqvist procedure. We then turned to the OPT procedure of Maickel and Miller (1966), which had a higher sensitivity and, hopefully, more specificity.

After finding that we could scale down the OPT method (see EXPERIMENTAL), we confirm that this procedure is more sensitive than the former methods. Standards of 5-HTP in the low picomole range could be measured easily and reproducibly. We used this procedure to measure 5-HTP in tissue samples. In Fig. 4, data are shown that indicate that there is at least one component still present which adds to the fluorescence at 480 nm. In the insert, the difference spectra show that the reagent blank is not an adequate blank in this procedure

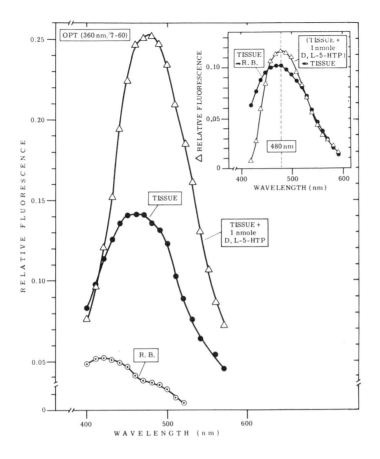

FIG. 4. Fluorescence spectra of the 5-HTP fraction (as described for Fig. 3) subjected to OPT reaction (see EXPERIMENTAL) with and without added 5-HTP (1 nmole). R.B. refers to reagent blank. The difference spectrum are presented in the upper right-hand corner. Excitation was at 360 nm with a 7-60 filter.

(note nongaussian distribution at wavelengths lower than 480 nm), even though the amount of fluorescence in tissue is linear to 1.5 g of tissue (Fig. 5; additional data, which are not shown, indicate that the linearity is also present to 2 g of tissue) and that the line passes through the origin. Since there is apparently present in the 5-HTP fraction one or more components in small amounts in addition to 5-HTP which react with OPT and fluorescence at wavelengths below 480 nm, we tried three different tissue blanks. The tissue blank of Andén and Magnusson (1967) for 5-HT was tried on our samples. It was found that adding ascorbic acid plus ferricyanide and irradiating the samples with ultraviolet light to destroy 5-HTP did not work in the OPT procedure.

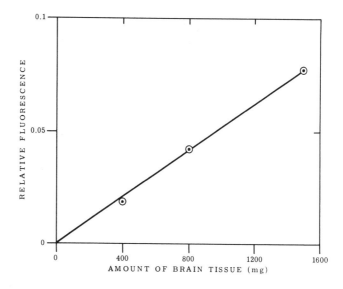

FIG. 5. Relationship between the relative fluorescence yield of the "5-HTP" fraction (as described for Fig. 3) reacted with OPT and amount of brain tissue in the sample (see EXPERIMENTAL).

However, using H_2O_2 with the sample prior to adding OPT (H_2O_2 oxidized tissue blank) or heating the 50-μl sample separately from 100 μl of 10 N HCl plus 5 μl of OPT to 60 min at 85°C and then combining the constituents (heated tissue blank) produced similar amounts of fluorescence which were slightly higher than that of the reagent blank. However, these tissue blanks may not be "perfect" blanks. In the case of the H_2O_2 oxidized tissue blank, we may have destroyed other interfering compounds in addition to the 5-HTP. In the case of the separated and heated tissue blank, one does not correct for the possibility that the interfering compounds in the tissue blank can react with OPT. However, the difference spectra using the separated heated tissue blank produced a near-gaussian distribution pattern. Therefore, if this blank is correct, the 5-HTP content (mean ± SEM) in rat brain is 0.23 ± 0.018 nmole/g. However, this value depends on the tissue blank used in the calculation. Still more information is required—either positively correct for the interfering compounds in the 5-HTP fraction or separate the 5-HTP present in the sample and then measure its level.

We turned to separating 5-HTP from interfering compounds in the 5-HTP fraction by chromatographic methods. With Jepson's procedure (1955), the R_f of authentic 5-HTP, 5-methoxytryptophan (5-MTP), 5-HT, tryptophan, and 5-HIAA were 0.36, 0.55, 0.57, 0.62, and 0.79,

TABLE 1. *Content of 5-hydroxytryptophan in brain*

Species	Authors	Content[a]
Mouse	Wiegand and Scherfling (1962)	<0.45
Mouse	Lindqvist (1971)	<0.14
Rat	Eccleston et al. (1965)	<0.23
Rat	Lindqvist (1971)	<0.32
Rat	Carlsson et al. (1972)	<0.05
Rat	Fischer and Aprison (1972)	1.19 ± 0.11
Rat	Present study[b]; (a) reagent blank	0.32 ± 0.01
	(b) H_2O_2 oxidized tissue blank	0.31 ± 0.01
	(c) heated tissue blank	0.25 ± 0.01

[a]Where known, the mean (nmoles/g of tissue) is given with the SEM. Our values given are based on seven determinations.
[b]See EXPERIMENTAL for description.

respectively. Using the 5-HTP fraction, we found no 5MTP, 5-HT, or 5-HIAA. Positive identification of 5-HTP was obtained based on the R_f value and color formation (slate blue) with Ehrlich reagent. We also noted the presence of two yellow spots and one due to tryptophan. The yellow spots and tryptophan on elution from the chromatogram showed some OPT fluorescence. The identity of each yellow spot is unknown, but they may be kynurenine (Gál, Armstrong, and Ginsberg, 1966) or a hydroxy derivative of kynurenine and anthranilic acid. The 5-HTP area and an area with no compounds was eluted from unstained paper strips; one portion was used with the OPT procedure and another was used in the 3 N HCl procedure. The complete spectrum was determined in each case and was consistent with that noted for pure 5-HTP. Calculations based on a limited number of samples showed that the content of 5-HTP as determined by this method is approximately 0.28 nmole/g, which is very close to the values obtained with the OPT reaction. Thus, it appears that the heated tissue blank and the H_2O_2 blank may be used with the OPT method to yield reliable data (Table 1). However, based on difference spectra, the former blank is preferred at present because it yields a more gaussian curve.

IV. DISCUSSION AND CONCLUSION

Small amounts of endogenous 5-HTP are present in the rat brain; the mean value (± SEM) that we found in preliminary experiments was 0.23 ± 0.018 nmole/g. Although some other authors have stated that they could not measure these levels accurately, the data presented in this chapter suggest that (1) their methods were not sufficiently sensitive and/or (2) the interfering compounds which are usually found in the 5-HTP fraction were not removed or properly accounted for in their analyses. The content of 5-HTP reported previously by Fischer and Aprison (1972) is in error, whereas others who state that there is no detectible endogenous 5-HTP in rat brain appear also to be in error (see Table 1).

A summary of the values reported in the literature to date along with some of our preliminary data are shown in Table 1. Until the problem of the tissue blank in the 5-HTP assay is solved, the best method of measuring 5-HTP appears to be the use of chromatographic methods. If 5-HTP is first separated from other major 5-hydroxyindole compounds by using the extraction procedure as described in this chapter, the one-dimensional chromatographic (39 hr) procedure appears to be adequate; without prior purification steps one would have to resort to two-dimensional chromatographic procedures such as those described by Randerath (1966) and Jepson (1955). Preliminary data using the chromatographic procedure based on that of Jepson (1955) indicate that the heated tissue blank used in the OPT procedure (micro adaptation) can be used since both procedures yield similar data. Since the values obtained by both procedures are similar, it may be possible that the extraction method described is totally adequate and that the net fluorescence values at 480 nm yield reliable data even though the difference spectra of the fluorescence at the lower wavelengths (400 to 440 nm) suggest a contaminant. In other words, the present tissue blank may adequately correct for the fluorescence at 480 nm even though it does not correct fully at the lower wavelengths. However, the tissue blank is still a problem which is under active investigation. Thus, more work is required before the problem is completely resolved.

In our opinion, this appears to be the current status of the "5-HTP story" in brain.

ACKNOWLEDGMENTS

This investigation was supported in part by research grant GB28715X from the National Science Foundation and research grant MHO3225-14 from the National Institute of Mental Health.

REFERENCES

Andén, N.-E., and Magnusson, T. (1967): An improved method for the fluorimetric determination of 5-hydroxytryptamine in tissues. *Acta Physiologica Scandinavica*, 69:87–94.
Carlsson, A., David, J. N., Kehr, W., Lindqvist, M., and Atack, C. V. (1972):

Simultaneous measurement of tyrosine and tryptophan hydroxylase activities in brain *in vivo* using an inhibitor of the aromatic amino acid decarboxylase. *Naunyn-Schmiedeberg's Archives of Pharmacology*, 275:153–168.

Eccleston, D., Ashcroft, G. W., and Crawford, T. B. B. (1965): 5-hydroxyindole metabolism in rat brain. A study of intermediate metabolism using the technique of tryptophan loading-II Applications and drug studies. *Journal of Neurochemistry*, 12:493–503.

Fischer, C. A., and Aprison, M. H. (1972): Determination of nanomole levels of 5-hydroxytryptophan, 5-hydroxytryptamine and 5-hydroxyindoleacetic acid in the same sample. *Analytical Biochemistry*, 46:67–84.

Fischer, C. A., Kariya, T., and Aprison, M. H. (1970): A comparison of the distribution of 5-hydroxyindoleacetic acid and 5-hydroxytryptamine in four specific brain areas of the rat and pigeon. *Comparative and General Pharmacology*, 1:61–68.

Gal, E. M., Armstrong, J. C., and Ginsberg, B. (1966): The nature of *in vitro* hydroxylation of L-tryptophan by brain tissue. *Journal of Neurochemistry*, 13:643–654.

Jepson, J. B. (1955): Paper chromatography of urinary indoles. *The Lancet*, ii:1009–1011.

Lindqvist, M. (1971): Quantitative estimation of 5-hydroxy-3-indoleacetic acid and 5-hydroxytryptophan in the brain following isolation by means of a strong cation exchange column. *Acta Pharmacologica et Toxicologica*, 29:303–313.

Maickel, R. P., and Miller, F. P. (1966): Fluorescent products formed by reaction of indole derivatives and *o*-phthalaldehyde. *Analytical Chemistry*, 38:1937–1938.

Randerath, K. (1966): *Thin-Layer Chromatography*. Verlag Chemie (Academic Press), pp. 101–105.

Takahashi, R., and Aprison, M. H. (1964): Acetylcholine content of discrete areas of the brain obtained by a near-freezing method. *Journal of Neurochemistry*, 11:887–898.

Toru, M., and Aprison, M. H. (1966): Brain acetylcholine studies: A new extraction procedure. *Journal of Neurochemistry*, 13:1533–1544.

Wiegand, R. G., and Scherfling, E. (1962): Determination of 5-hydroxytryptophan and serotonin. *Journal of Neurochemistry*, 9:113–114.

Advances in Biochemical Psychopharmacology, Vol. 11
Raven Press, New York © 1974

Biochemical Characteristics of Mammalian Brain 5-Hydroxytryptophan Decarboxylase Activity

K. L. Sims*

Laboratory of Neuropharmacology, National Institute of Mental Health,
Saint Elizabeths Hospital, Washington, D.C. 20032

The oxidation of 5-hydroxytryptophan (5-HTP) to 5-hydroxytryptamine (5-HT) is catalyzed in mammalian tissues by an enzymic activity termed 5-hydroxytryptophan decarboxylase (5-hydroxy-L-tryptophan carboxylyase; EC 4.1.1.28). The enzymic activity has been purified from kidney (Lovenberg, Weissbach, and Udenfriend, 1962; Christenson, Dairman, and Udenfriend, 1970; Lancaster and Sourkes, 1972), adrenal gland (Lancaster and Sourkes, 1972), and liver (Awapara, Sandman, and Hanly, 1962). The preparations from kidney following the protein purification procedures were able to catalyze the decarboxylation of other aromatic amino acids, particularly 3,4-dihydroxyphenylalanine (DOPA), and the purification of decarboxylase activity toward the other aromatic amino acids was increased to approximately the same extent as that observed from 5-HTP. Therefore the term, "aromatic L-amino acid decarboxylase" was applied to the kidney protein which catalyzes the decarboxylation of 5-HTP (Lovenberg et al., 1962). However, with the exception of brief mention of a sevenfold purification from "crude extract" (not high-speed supernatant) of dog brainstem 5-HTP decarboxylase activity by Lovenberg et al. (1962), few studies have dealt specifically with this brain enzyme.

Previously, the assumption has been made that the enzymic activity present in kidney is identical to the protein present in nervous tissue responsible for 5-HTP decarboxylation. There are three general phenomena that suggest caution in extrapolating the characteristics of the protein isolated from kidney to the activity which is responsible for serotonin production in nervous tissue. First, within the central nervous system (unlike the kidney, liver, and adrenal medulla) there are stores of 5-hydroxytryptamine which potentially may subserve a transmitter

---

*Present address: Department of Psychiatry, University of Kansas Medical Center, Kansas City, Kansas 66103

function; moreover, these stores of 5-hydroxytryptamine in the CNS are located in specialized neurons which represent an exceedingly small fraction of the total number of neurons in the brain (Fuxe, Hökfelt, and Ungerstedt, 1970; Bloom, 1970). This indicates a striking metabolic specialization within subpopulations of cells within a single organ, and emphasizes the possibility that the metabolic regulation of these 5-hydroxytryptamine stores differs from that of other organs. The second reason for caution in maintaining that the same protein catalyzes the same basic reaction for widely different molecules in all tissues is the accumulated experimental evidence that (1) organ-specific proteins exist, e.g., the S-100 protein of brain (Moore and McGregor, 1965); (2) different proteins are known to catalyze the same reaction within different parts of the same cell, e.g., the isocitrate de-hydrogenase group of proteins; (3) different proteins catalyzing the same reaction may determine the flow of energy within a cell, e.g., the α-glycerophosphate dehydrogenase enzymes (Lee and Crane, 1971); and (4) inter- and intraspecies differences can exist in a nonenzymatic protein which carries out the same function role, e.g., the different myosins of cardiac and skeletal muscle cells (Huszar and Elzinga, 1972).

The last and most compelling reason for studying specifically the decarboxylase activity in brain tissue is the finding of a specific histidine decarboxylase in specialized histamine-generating tissues such as the gastric mucosa (Schwartz, Lampart, and Ross, 1970) which exists although the kidney enzyme is capable of catalyzing the same reaction (Lovenberg et al., 1962). In addition, there were fundamental differences in the results reported for the partially purified decarboxylase preparations from liver and kidney tissues (see Awapara et al., 1962, and Lovenberg et al., 1962). For these reasons, our laboratory began a comprehensive study of the 5-HTP decarboxylase *activity* present in the brain.

Three facets of our work with brain 5-HTP decarboxylase activity will be reviewed in this chapter: (1) those assay conditions which affect what is measured as "activity"; (2) the subcellular and regional distribution of 5-HTP decarboxylase in the CNS; and (3) the effect of 6-hydroxydopamine on brain 5-HTP decarboxylase. The data summarized in this report have been published in detail (Sims, Davis, and Bloom, 1973; Sims and Bloom, 1973). DOPA decarboxylase activity was also examined, but neither its activity nor the question of whether one or two proteins in brain are responsible for DOPA and 5-HTP decarboxylase activity is the focus of the present report.

The technical procedures involved in the 5-HTP decarboxylase assay are outlined in Table 1. This microradiometric technique we have developed measures the radioactivity of the liberated $^{14}CO_2$ from the carboxyl-labeled substrate and is basically a manometric procedure

TABLE 1. *Method and specifications of assay for brain 5-hydroxytryptophan decarboxylase (5-HTP-D)*

Method	Specifications
Basic technique	Microradiometric CO_2 trapping using ^{14}C-carboxy substrate
Blank	50 dpm
Ratio Exp/blank	Routinely $\geqslant 5$
Precision (triplicate aliquots)	SD $< 1\%$
Sensitivity of assay (moles reaction product)	1×10^{-12}
Standard assay[a]	5 μl volume 125 μg (wet wt.) brain
Effect of protein conc. and linearity with time	Proportional: $\leqslant 120$ min
Activity whole rat brain homogenate	0.8 μmole/g/hr

[a]*5-HTP-D standard assay:* Reagent concentrations at incubation are 0.5 mM L-[1-^{14}C] 5-hydroxytryptophan, Sp. Act. 1.0 mC/mmole. 0.02 mM D-isomer, Sp. Act. 26.0. 75 mM Tris, pH 8.3. 0.5 mM Pyridoxal-P. 25 μg brain (fresh)/μl (125 μg total). Samples and heat inactivated homogenate blanks were incubated 60 min at 38°C.

similar in principle to the technique first used to study the 5-HTP-decarboxylase reaction (Clark, Weissbach, and Udenfriend, 1954). The method is highly specific, and its high sensitivity (1×10^{-12} mole reaction product) allows multiple determinations on very small quantities of protein or crude tissue extract. This sensitivity was particularly important and indeed necessary for the determination of multiple decarboxylase activities in subcellular fractions.

The experimental conditions employed have a great effect on what is measured as 5-HTP decarboxylase activity in brain preparations. In Table 2 are listed the various experimental conditions which give the maximal 5-HTP decarboxylase activity in homogenates of brain tissue (Sims et al., 1973). The results depicted in Table 2 for both DOPA and 5-HTP decarboxylase do illustrate one important point. The two activities are influenced differently by changes of pH, temperature, concentration of pyridoxal-5-phosphate (pyridoxal-P), levels of substrate, and by the presence of ascorbate. This suggests that the ratio of the two enzymic activities in brain tissue is subject to many variables (not the least of which is the assay incubation time). These findings

TABLE 2. *Summary of optimal experimental conditions for activity and stability of rat brain decarboxylase: Decarboxylation of DOPA and 5-HTP*

	DOPA	5-HTP
pH	6.8 ± 0.1	8.3 ± 0.2
Pyridoxal-P	0.6 mM	0.1 mM
K_m	0.6 mM	0.016 mM
Substrate	3.0 mM	0.2 mM
Temperature	$38°C$	$45°C$
Presence of ascorbate[a]	No effect	(Inhibits)
Freezing brain in liquid N_2	No effect	No effect
Repeated freezing and thawing[b]	Stable	Stable
Lyophilization	↑20%	–
Triton	↑20%	↓25%

[a]5.0 mM ascorbate produced 50% inhibition of 5-HTP decarboxylase.

[b]Storage in 1 mM pyridoxal-5-phosphate is necessary for stability and, after 1 to 2 freeze-thaws, activity of the preparation is 20% > than that of the original unfrozen homogenate.

indicate that the practice of using 5-HTP or DOPA interchangeably as substrates to assay "DOPA-decarboxylase activity" or 5-HTP-decarboxylase activity may well provide misleading information in studies utilizing brain tissue.

Only preliminary studies of possible product and substrate(s) interaction with the 5-HTP-decarboxylase activity in brain were done due to the crude nature of the preparation. No substrate inhibition was noted at 10 mM 5-HTP (highest studied), and the D-isomer of 5-HTP had no effect on enzymic activity in concentration less than or equimolar to the L-isomer. 3-Hydroxytryptamine (dopamine) and 5-hydroxytryptamine did not inhibit brain 5-HTP-decarboxylase activity in concentrations up to 2.0 mM (highest studied). Under standard assay conditions a 40% inhibition of brain 5-HTP-decarboxylase activity was observed when 5 mM DOPA (no ascorbate) was present; however, higher DOPA concentrations could not be reliably evaluated because DOPA markedly increased both the initial (0 time) blank and the progression of the blank during incubation with 5-HTP. This indicates an unappreciated oxidation (nonenzymatic decarboxylation) of 5-HTP via a direct chemical reaction between 5-HTP and DOPA. Moreover, the direct interaction of DOPA and pryidoxal-P, which exhibits second-order kinetics with a rate constant of 2.7 (Schott and Clark, 1952), clouds any interpretation of inhibition of 5-HTP-decarboxylase activity by DOPA because of the substantial nonspecific interaction of DOPA with coenzyme.

The gross regional distribution of brain 5-HTP-decarboxylase using these specific optimal assay conditions has been described (Sims and Bloom, 1973). Because of the relatively large amount of tissue sampled (1 mg fresh weight), the data serve only as a very rough guide to localization of 5-HTP-decarboxylase and will not be reiterated here. More pertinent to both the biochemistry and anatomic localization of brain 5-HTP-decarboxylase were the subcellular localization studies using the tissue fractionation technique of Gray and Whittaker (1962).

The comparative subcellular localization of several brain decarboxylases responsible for the production of several putative transmitters is listed in Table 3. Previous studies have indicated that the 5-HTP-D activity present in kidney is associated predominantly with the soluble fraction (Lovenberg et al., 1962), but in subcellular fractions derived from brain more than half of the 5-HTP-decarboxylase activity was associated with particulate fractions (Table 3). The data for lactate dehydrogenase is included (LDH) as a cytoplasmic marker (Johnson and Whittaker, 1962). Whereas the distribution of DOPA-decarboxylase activity is almost identical to that of LDH, the localization of 5-HTP-decarboxylase activity is quite different from both. For example, 15% of original homogenate 5-HTP-decarboxylase activity (uncorrected for protein loss) appears in the synaptosomal fraction, and only 4% of the homogenate DOPA-decarboxylase activity is present in this fraction. The data in Table 3 does, however, indicate a striking parallel in the patterns of distribution for 5-HTP-decarboxylase and glutamic acid decarboxylase.

The ratio of DOPA-decarboxylase to 5-HTP-decarboxylase in the various subcellular fractions is calculated in Table 3, and these values indicate that significant differences exist in the cellular distribution of the two brain activities. This differential ratio of brain DOPA- to 5-HTP-decarboxylase activity also occurs within individual regions of brain (without fractionation) following the intracisternal administration of 6-hydroxydopamine (Sims and Bloom, 1973). In the latter instance, the change in brain DOPA-decarboxylase activity relative to that for 5-HTP results from the marked fall in DOPA activity (1/3 to 1/2 of control), whereas 5-HTP-decarboxylase activity does not decrease when measured in the same tissue samples (Sims and Bloom, 1973). In fact, there were surprising substantial increases in 5-HTP-decarboxylase activity in hypothalamus, cerebellum, and medulla-pons regions.

The results of all these aforementioned studies indicate that differences exist between the 5-HTP-decarboxylase activity that is

TABLE 3. *Subcellular distribution of brain decarboxylases*

Fraction	Protein (% of H)	Lactic acid dehydrogenase		Glutamic acid decarboxylase		DOPA-decarboxylase		5-HTP-decarboxylase		Ratio: 5-HTP-D RSA / DOPA-D RSA
		RSA	% of H	RSA	% of H	RSA	% of H	RSA	% of H	
Homogenate (H)		1.0	–	1.0	–	1.0	–	1.0	–	1.0
P$_1$ (crude nuclear)	32	0.31	10	0.61	20	0.40	12.8	0.80	23.0	2.0
P$_2$ (crude mitochondrial)	30	0.53	16	1.00	30	0.38	11.9	1.18	35.4	3.1
P$_3$ (microsomal)	14	0.50	7	0.60	8	0.15	2.4	0.52	7.4	3.5
S$_3$ (supernatant)	19	3.20	61	0.80	15	3.10	60.0	1.86	35.5	0.6
Recovery: % of H	95%		94%		73%		87.1%		101.3%	
Sucrose gradient										
A (0.85 M; myelin)	4.5	0.45	2.2	0.20	1	0.10	0.4	0.20	1.0	2.0
B (1.2 M; synaptosomal)	8.3	0.80	6.7	1.18	10	0.34	3.9	1.54	15.4	4.5
C (Pellet; mitochondrial)	7.7	0.29	2.2	0.71	5	0.25	2.5	0.78	6.3	3.1
X (Intermediate zones)	3.2	0.68	2.2	0.96	3	0.54	1.6	0.93	2.9	1.7
Recovery: % of P$_2$	79%		83%		64%		70.5%		72.0%	

Enzymic rates in the initial homogenate were: DOPA-D 5.4 μmoles/g(fresh)/hr and 5-HTP-D 0.9 μmole/g(fresh)/hr; GAD 52.0 μmoles/g(fresh wt)/hr min and LDH 55 O.D. units/mg protein/15 min.

Abbreviations: RSA, relative specific activity; % of H, percent of total activity present in the initial homogenate; % of P$_2$, percent of total activity present in the crude mitochondrial fraction (P$_2$).

Data summarized from Sims, Davis, and Bloom (1973); and Sims and Davis (1973). Relative specific activity (RSA) is calculated by dividing the specific activity of the fraction by the specific activity of the homogenate.

found in kidney and that of brain. There exists the possibility that a specific 5-HTP decarboxylase may have evolved in neuroepithelial tissue in addition to the "aromatic L-amino acid decarboxylase" isolated from peripheral tissues of mesodermal origin which do not store serotonin. We are currently proceeding with purification of the brain 5-HTP-decarboxylase activity in anticipation that the information gained will further help us understand the regulation of 5-hydroxytryptamine metabolism in brain.

REFERENCES

Awapara, J., Sandman, R. P., and Hanly, C. (1962): Activation of DOPA decarboxylase by pyridoxal phosphate. *Archives of Biochemistry and Biophysics*, 98:520–525.

Bloom, F. E. (1970): Serotonin neurons: Localization and possible physiological role. *International Review of Neurobiology*, 13:27–47.

Christenson, J. G., Dairman, W., and Udenfriend, S. (1970): Preparation and properties of a homogenous aromatic L-amino acid decarboxylase from hog kidney. *Archives of Biochemistry and Biophysics*, 141:356–367.

Clark, C. T., Weissbach, H., and Udenfriend, S. (1954): 5-Hydroxytryptophan decarboxylase, preparation and properties. *Journal of Biological Chemistry*, 210:139–148.

Fuxe, K., Hökfelt, T., and Ungerstedt, U. (1970): Morphological and functional aspects of central monoamine neurons. *International Review of Neurobiology*, 13:93–126.

Gray, E. G., and Whittaker, V. P. (1962): The isolation of nerve endings from brain: An electron-microscope study of cell fragments derived by homogenization and centrifugation. *Journal of Anatomy (London)*, 96:79–88.

Huszar, G., and Elzinga, M. (1972): Homologous methylated and nonethylated histidine peptides in skeletal and cardiac myosins. *Journal of Biological Chemistry*, 247:745–753.

Johnson, M. K., and Whittaker, V. P. (1963): Lactate dehydrogenase as a cytoplasmic marker in brain. *Biochemical Journal*, 88:404–409.

Lancaster, G. A., and Sourkes, T. L. (1972): Purification and properties of hog-kidney 3,4-dihydroxyphenylalanine decarboxylase. *Canadian Journal of Biochemistry*, 50:791–797.

Lee, Y., and Crane, J. E. (1971): L-Glycerol 3-phosphate dehydrogenase. *Journal of Biological Chemistry*, 246:7616–7622.

Lovenberg, W., Weissbach, H., and Udenfriend, S. (1962): Aromatic L-amino acid decarboxylase. *Journal of Biological Chemistry*, 237:89–93.

Moore, B. W., and McGregor, D. (1965): Chromatographic and electrophoretic fractionation of soluble proteins of brain and liver. *Journal of Biological Chemistry*, 240:1647–1654.

Schott, H. F., and Clark, C. T. (1952): DOPA decarboxylase inhibition through the interaction of coenzyme and substrate. *Journal of Biological Chemistry*, 196:449–462.

Schwartz, J. C., Lampart, C., and Ross, C. (1970): Properties and regional distribution of histidine decarboxylase in rat brain. *Journal of Neurochemistry*, 17:1527–1534.

Sims, K. L., and Bloom, F. E. (1973): Rat brain L-3,4-dihydroxyphenylalanine and L-5-hydroxytryptophan decarboxylase activities: Differential effect of 6-hydroxydopamine. *Brain Research*, 49:165–175.

Sims, K. L., and Davis, G. A. (1973): Subcellular localization of succinic semialdehyde dehydrogenase. *European Journal of Biochemistry*, 35:450–453.

Sims, K. L., Davis, G. A., and Bloom, F. E. (1973): Activities of 3,4-dihydroxy-L-phenylalanine and 5-hydroxy-L-tryptophan decarboxylases in rat brain: Assay characteristics and distribution. *Journal of Neurochemistry*, 20:449–464.

Advances in Biochemical Psychopharmacology, Vol. 11
Raven Press, New York © 1974

The Metabolism of Indolealkylamines by Type A and B Monoamine Oxidase of Brain

N. H. Neff, H.-Y. T. Yang, C. Goridis,* and D. Bialek

*Laboratory of Preclinical Pharmacology, National Institute of Mental Health,
Saint Elizabeths Hospital, Washington, D.C. 20032
and
*Centre National de la Recherche Scientifique Centre de Neurochimie,
Strasbourg, France*

I. INTRODUCTION

There are apparently two types of monoamine oxidase [mono-amines: O_2 oxidoreductase (deaminating), E.C. 1.4.3.4.] in most mammalian tissues, and they can be characterized by their sensitivity to inhibitor drugs and their specificity for substrates (see Costa and Sandler, 1972). The endogenous indolealkylamines serotonin (5-hydroxytryptamine) and tryptamine (Saavedra and Axelrod, 1972) are substrates for the monoamine oxidases of brain; however, they are not deaminated by the same enzymes when studied *in vitro*. Differential blockage of their deamination can be demonstrated with the irreversible inhibitor drugs clorgyline [N-methyl-N-propargyl-3-(2,4-dichloro-phenoxy) propylamine] and deprenyl (phenylisopropylmethylpropinyl-amine).

II. THE METABOLISM OF SEROTONIN AND TRYPTAMINE
IN VITRO BY THE MONOAMINE OXIDASES OF
RAT BRAIN

The monoamine oxidases have been divided according to their characteristics into a type A and a type B enzyme (Johnston, 1968). Type A enzyme deaminates serotonin and norepinephrine (Goridis and Neff, 1971a); it is inhibited by low concentrations of clorgyline (Johnston, 1968; Hall, Logan, and Parsons, 1969) and it is rather stable when heated at 50°C (Yang, Goridis, and Neff, 1972). In contrast to type A enzyme, type B enzyme deaminates β-phenylethylamine, it is inhibited by low concentrations of deprenyl (Knoll and Magyar, 1972), and it is rapidly inactivated when heated to 50°C (Yang and Neff,

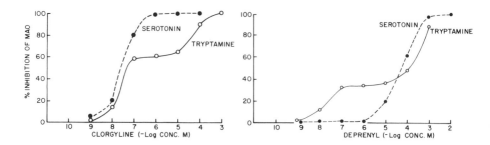

FIG. 1. Inhibition of the monoamine oxidase activity of rat brain by increasing concentrations of clorgyline or deprenyl with serotonin (1.3 mM) or tryptamine (0.1 mM) as substrates. The inhibitor drugs were preincubated with enzyme for 15 min at 22° C prior to adding substrate. The reactions were then continued for an additional 30 min at 37° C.

1973). Some endogenous amines are metabolized by both enzymes, such as dopamine and tyramine (Hall et al., 1969).

When the deamination of tryptamine by a homogenate of rat brain was evaluated in the presence of increasing concentrations of clorgyline, a stepwise inhibition of activity was evident, with a plateau occurring at about 60% inhibition of enzyme activity (Fig. 1). Activity was completely blocked at about 1 mM clorgyline. When the deamination of serotonin was evaluated under similar conditions, a simple sigmoidal curve was observed with complete blockade occurring at about 1 μM clorgyline (Fig. 1). Using the criteria and nomenclature of Johnston (1968) for identifying monoamine oxidases, tryptamine was apparently deaminated by two forms of the enzyme, one which also metabolized serotonin and was readily inhibited by clorgyline (type A enzyme) and one which does not metabolize serotonin and was rather insensitive to clorgyline (type B enzyme).

The drug, deprenyl, in contrast to clorgyline, inhibited the deamination of tryptamine at lower concentrations than were required to inhibit the deamination of serotonin (Fig. 1), and the inhibition of activity increased stepwise, with a plateau occurring at about 35% inhibition. This occurred because deprenyl is a more potent inhibitor of type B enzyme than type A enzyme (Knoll and Magyar, 1972).

The concentration of tryptamine used when testing the inhibitory potential of the two drugs influences the position on the graph where the plateau falls. For example, with the drug clorgyline at 0.5 μM, the plateau occurred at 76, 70, and 64% inhibition of monoamine oxidase when the concentration of tryptamine was 10, 50, and 100 μM, respectively (Table 1). With deprenyl at 0.5 μM, the plateau increased with increasing concentrations of tryptamine (Table 1). One explanation for the shift of the plateau with the concentration of the substrate,

TABLE 1. *Inhibition of rat brain monoamine oxidase activity with changing substrate concentration*

Tryptamine concentration (μM)	Percent inhibition with clorgyline (0.5 μM)	Percent inhibition with deprenyl (0.5 μM)	Total percent inhibition
10	76	20	96
50	70	24	94
100	64	34	98

Clorgyline or deprenyl was incubated at 22°C with the samples for 15 min prior to adding the substrate tryptamine in the concentrations shown. Incubations were continued for 30 min at 37°C.

tryptamine, is that type A enzyme has a lower apparent K_m for tryptamine than type B enzyme. As the concentration of tryptamine is increased, more of it is deaminated by type B enzyme. The presence of type A activity and type B activity, as judged by using specific inhibitor drugs, accounted for about all of the enzyme activity in the homogenate.

III. INDOLEALKYLAMINE METABOLISM IN BRAIN FOLLOWING THE ADMINISTRATION OF CLORGYLINE OR DEPRENYL

When deprenyl or clorgyline were administered intravenously in increasing doses, there was a selective blockade of the monoamine oxidase of brain when tested *in vitro* with the specific substrates serotonin and β-phenylethylamine (Fig. 2) 2 hr after injecting the inhibitor drugs (Yang and Neff, 1973). Clorgyline blocked type A

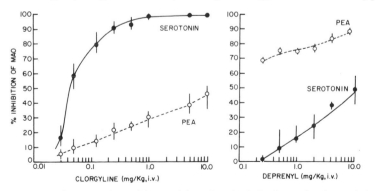

FIG. 2. Blockade of monoamine oxidase activity of rat brain by increasing doses of clorgyline or deprenyl. Animals were killed 2 hr after injection of the drugs and enzyme activity was assayed with serotonin or β-phenylethylamine (PEA) as substrate.

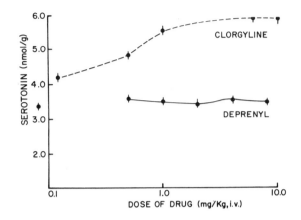

FIG. 3. The concentration of serotonin in rat brain 2 hr after administering clorgyline or deprenyl.

enzyme activity (the deamination of serotonin) preferentially, while deprenyl blocked type B enzyme activity (the deamination of β-phenylethylamine) preferentially. After high doses, the selectivity of the drugs is lost, as would be expected from the *in vitro* studies. Concomitant with the blockade of type A enzyme in brain by clorgyline was the increase of serotonin concentrations in brain (Fig. 3), and we found that serotonin concentrations in brain were still elevated 4 hr after the injections. Deprenyl, in the concentrations studied, had no significant influence on the concentrations of serotonin in brain

TABLE 2. *The concentration of 5-hydroxyindoleacetic in rat brain 2 hr after administering deprenyl or clorgyline*

Treatment[a]	5-Hydroxyindoleacetic acid nmole/g ± SEM (5)
None	2.4 ± 0.2
Clorgyline	0.93 ± 0.02 ($p < 0.01$)
Deprenyl	2.4 ± 0.1

[a]The animals were given 5 mg/kg, i.v., of clorgyline or deprenyl 2 hr before they were killed.

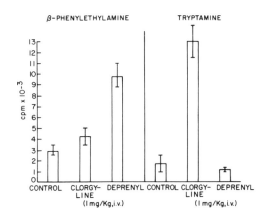

FIG. 4. Radioactive β-phenylethylamine or tryptamine in rat brain 10 min after an intraventricular injection. Rats were treated with clorgyline or deprenyl 2 hr before injecting the radioactive amines.

(Fig. 3). As might be expected, the concentration of the major metabolite of serotonin, 5-hydroxyindoleacetic acid, was decreased following treatment with clorgyline, but not by treatment with deprenyl (Table 2).

To evaluate the *in vivo* metabolism of tryptamine by brain, we injected radioactive tryptamine (70 nmoles) intraventricularly and assayed the quantity of amine remaining in brain 10 min later in rats treated with 1 mg/kg, i.v., of either clorgyline or deprenyl. The metabolism of radioactive β-phenylethylamine (15 nmoles) was evaluated for comparison. Deprenyl treatment depressed the metabolism of β-phenylethylamine, but clorgyline treatment did not (Fig. 4). In contrast, clorgyline treatment depressed the metabolism of tryptamine while deprenyl was ineffective. Fuller and Roush (1972) made similar observations while studying other monoamine oxidase inhibitor drugs.

Our *in vivo* studies and our *in vitro* studies support the hypothesis that there are multiple forms of monoamine oxidase. Tryptamine was apparently metabolized by type A enzyme and type B enzyme when studied *in vitro*. When it is injected into the brain, however, it was almost solely metabolized by type A enzyme. This observation is consistent with the finding that tryptamine was metabolized primarily by type A enzyme when present in low concentrations (Table 1). Although tryptamine is metabolized by both enzymes *in vitro*, there is as yet no conclusive evidence that endogenous tryptamine is metabolized by both enzymes *in vivo*.

IV. THE PINEAL GLAND AND THE OXIDATIVE DEAMINATION OF SEROTONIN

Serotonin is found within the pineal parenchymal cells and within the sympathetic nerves that innervate the gland (Owman, 1964). The serotonin within the nerves probably originates in the parenchymal cells and enters the neurons via the amine reuptake mechanism (Neff, Barrett, and Costa, 1969). Most of the serotonin synthesized within the gland is probably converted to melatonin and its metabolites (Lerner, Case, and Takahashi, 1960). Apparently, only the sympathetic nerves that innervate the pineal contain type A monoamine oxidase, the enzyme which specifically metabolizes serotonin and norepinephrine. We found that, following sympathectomy of the rat pineal gland, the activity toward serotonin was reduced by 70% while only minimal changes toward tyramine (a substrate for both type of enzymes) metabolism were observed (Goridis and Neff, 1971*b*) (Table 3). Moreover, when the monoamine oxidase activities of human sympathetic trunk and pineal gland were evaluated for the presence of type A and type B enzymes using the procedure of Johnston, with tyramine as the substrate, two forms of enzyme were evident (Goridis and Neff, 1972). Sympathetic nerves contained mainly the clorgyline sensitive enzyme (type A enzyme), while the pineal gland contained the clorgyline resistant enzyme (type B enzyme). These observations have two implications. First, the transmitter norepinephrine (a specific substrate for type A enzyme) is preferentially deaminated within the adrenergic nerve endings and not within the pineal cells, since the required enzyme for metabolism is not present in this tissue. This is consistent with the view that intraneuronal monoamine oxidase plays a major role in the

TABLE 3. *Monoamine oxidase activity in rat pineal gland before and after sympathectomy*

Amine substrate	Enzyme activity nmole/pineal/hr ± SEM (N)		Activity loss (%)
	Innervated	Denervated	
Tyramine	9.4 ± 0.5 (4)	8.8 ± 0.8 (5)	6.4
Serotonin	3.4 ± 0.8 (7)	1.0 ± 0.5 (7) $p < 0.01$	70

Mitochondria isolated from the pineal were incubated with 2.1 mM tyramine or 1.2 mM serotonin for 30 min at 37°C. From Goridis and Neff (1971*b*).

deamination of norepinephrine (Kopin and Axelrod, 1963). Moreover, the adrenergic nerve endings within the pineal also have the capacity to metabolize the serotonin that finds its way into the neuron. Second, the virtual absence of type A monoamine oxidase in the pineal cells is consistent with the view that melatonin, and not serotonin, is the end product of tryptophan metabolism in pineal cells (see Wurtman, Axelrod, and Kelly, 1968).

V. SUMMARY

There are apparently two types of monoamine oxidase in mammalian tissues that can be identified *in vivo* and *in vitro*. They may have different active centers or perhaps they may have the same enzyme active centers but properties which have been modified by varying amounts or types of attached membrane material. These enzymes have specific substrates and common substrates and they can be inhibited individually with inhibitor drugs. Serotonin is deaminated *in vivo* and *in vitro* by type A enzyme. Tryptamine is deaminated *in vitro* by type A and B enzyme, but *in vivo* it appears to be deaminated primarily by A enzyme. Therefore, blocking type A enzyme with a drug such as clorgyline should profoundly alter the metabolism of the indolealkylamines of brain, while blocking type B enzyme should have little effect. It appears, however, that serotonin metabolism would be unaltered in pineal gland cells after either drug, since deamination is not a major pathway for serotonin in this tissue.

REFERENCES

Costa, E., and Sandler, M. (1972): *Monoamine Oxidases—New Vistas.* Advances in Biochemical Psychopharmacology, Vol. 5. Raven Press, New York.

Fuller, R. W., and Roush, B. W. (1972): Substrate-selection and tissue-selective inhibition of monoamine oxidase. *Archives Internationales de Pharmacodynamie et de Therapie* (Belgium), 198:207–276.

Goridis, C., and Neff, N. H. (1971a): Monoamine oxidase in sympathetic nerves: A transmitter specific enzyme type. *British Journal of Pharmacology*, 43:814–818.

Goridis, C., and Neff, N. H. (1971b): Evidence for a specific monoamine oxidase associated with sympathetic nerves. *Neuropharmacology*, 10:557–564.

Goridis, C., and Neff, N. H. (1972): Evidence for specific monoamine oxidases in human sympathetic nerve and pineal gland. *Proceedings of the Society for Experimental Biology and Medicine*, 140:573–574.

Hall, D. W. R., Logan, B. W., and Parsons, G. H. (1969): Further studies on the inhibition of monoamine oxidases by M + B 9302 (clorgyline)-1, substrate specificity in various mammalian species. *Biochemical Pharmacology*, 18:1447–1454.

Johnston, J. P. (1968): Some observations upon a new inhibitor of monoamine oxidase in brain tissue. *Biochemical Pharmacology*, 17:1285–1297.

Knoll, J., and Magyar, K. (1972): Some puzzling pharmacological effects of monoamine oxidase inhibitors. *Advances in Biochemical Psychopharmacology*,

5:393–407.

Kopin, I. J., and Axelrod, J. (1963): The role of monoamine oxidase in the release and metabolism of norepinephrine. *Annals of the New York Academy of Sciences*, 107:848–853.

Lerner, A. B., Case, J. D., and Takahashi, Y. (1960): Isolation of melatonin and 5-methoxyindole-3-acetic acid from bovine pineal gland. *Journal of Biological Chemistry*, 235:1992–1997.

Neff, N. H., Barrett, R. E., and Costa, E. (1969): Kinetics and fluorescent histochemical analysis of the serotonin compartments in rat pineal gland. *European Journal of Pharmacology*, 5:348–356.

Owman, C. (1964): Sympathetic nerves probably storing two types of monoamines in the rat pineal gland. *International Journal of Neuropharmacology*, 3:105–112.

Saavedra, J. M., and Axelrod, J. (1972): A specific and sensitive enzymatic assay for tryptamine in tissues. *Journal of Pharmacology and Experimental Therapeutics*, 182:363–369.

Wurtman, R. J., Axelrod, J., and Kelly, D. E. (1968): *The Pineal.* Academic Press, New York.

Yang, H.-Y. T., Goridis, C., and Neff, N. H. (1972): Properties of monoamine oxidases in sympathetic nerve and pineal gland. *Journal of Neurochemistry*, 19:1241–1250.

Yang, H.-Y. T., and Neff, N. H. (1973): β-Phenylethylamine: Specific substrates for type B monoamine oxidase of brain. *Journal of Pharmacology and Experimental Therapeutics*, 187:365–371.

Advances in Biochemical Psychopharmacology, Vol. 11
Raven Press, New York © 1974

Heterogeneity of Rat Brain Mitochondrial Monoamine Oxidase

Moussa B. H. Youdim

MRC Unit of Clinical Pharmacology, University Department of Clinical Pharmacology, Radcliffe Infirmary, Oxford, England

Although biochemical (Youdim and Sourkes, 1965; Johnston, 1968; Jarrott, 1971; Neff and Goridis, 1972), histochemical (Hanker et al., (1973), and electrophoretic studies (Youdim, 1972) have indicated that monoamine oxidase (E.C. 1.4.3.4.) may exist in more than two forms, no adequate explanation has been put forward as to the molecular nature of this multiplicity. The presence of monoamine oxidase "isoenzymes" in different mitochondria has been suggested by Youdim (1972), and recently Kroon and Veldstra (1972) have partially separated subcellular mitochondrial fractions possessing varying mono-amine oxidase activity. Furthermore, phospholipids have been implicated as playing important roles in the activity, inhibitor specificity, and heterogeneity of this enzyme (Tipton, Youdim, and Spires, 1972; Youdim, 1973*a*).

Using histochemical and chemical techniques, a number of investigations have demonstrated that several mitochondrial enzymes have different patterns of distribution within the brain and liver. It has recently been shown that different fractions of mitochondria separated by rate zonal centrifugation do indeed have different enzyme activity with regard to NADH dehydrogenase, succinic dehydrogenase, and α-gylcerolphosphate dehydrogenase (Wilson and Cascarno, 1972). An attempt was therefore made in the present investigation to demonstrate biochemical differences in the brain mitochondrial monoamine oxidase as separated by zonal rate centrifugation. In this study it was considered essential to maintain the prepared brain mitochondria (Gray and Whittaker, 1962) in their physiological state; to avoid artifacts which might influence the enzyme activity, an iso-osmotic gradient of Ficoll-sucrose (Youdim, 1973*b*) was prepared to prevent changes brought about by centrifugation. Fractionation of the brain mitochondria was carried out using a rate zonal rotor. After the run, 40 fractions of 15 ml each were collected and stored at $-20°C$ for future monoamine oxidase activity estimation.

The density of the fractions was measured using an Abbe refractometer, and protein concentration was estimated by the method of Lowry, Rosebrough, Farr, and Randall (1951). The brain mitochondrial protein distribution, in milligrams per milliliter, showed a single broad band spanning 36 fractions, reaching a maximum in fraction 21 and a shoulder at fraction 26, in the lower Ficoll-sucrose gradient.

Using a variety of substrates, kynuramine, benzylamine, dopamine, tryptamine, and tyramine monoamine oxidase activity was measured in the various fractions. Almost 85% of the original activity was recovered using the substrate kynuramine. When the activities were expressed as percent of the highest "tyramine monoamine oxidase" specific activity, the following results were obtained (See Fig. 1). Kynuramine monoamine oxidase remained fairly constant in all fractions (70 to 80%); benzylamine monoamine oxidase (not shown) activity was 90% in fraction 21 and 50% in fraction 32. A comparison with tyramine monoamine oxidase showed that in the fractions with a short elution time, benzylamine and tyramine monoamine oxidase activity are

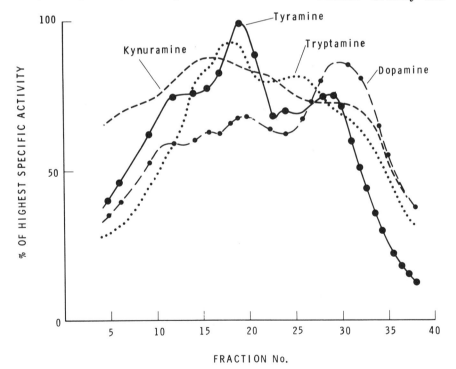

FIG. 1. Distribution of specific activity of monoamine oxidase expressed as percent of highest specific in mitochondrial fractions obtained by rate zonal centrifugation.

similar, but in the fractions with a long elution time (e.g., fraction 32) there is twice as much activity for tyramine. The activity of dopamine monoamine oxidase showed three peaks and increased from 60% in fraction 8 to 93% in fraction 30. In this fraction the activity toward dopamine is greater than that toward tyramine. Tryptamine monoamine oxidase distribution was similar to that of tyramine monoamine oxidase.

Several attempts have been made to characterize different forms of brain mitochondrial monoamine oxidase. These have included separation of different forms of the soluble enzymes by gel electrophoresis (Youdim, Collins, and Sandler, 1969) and the study of several inhibitors on the activity of the mitochondrial enzyme from a number of sources (Youdim and Sourkes, 1965; Neff and Goridis, 1972; Neff, Yang, Goridis, and Bialek, *This Volume*). The existence of different forms of monoamine oxidase has been questioned because of possible artifacts during the solubilization and purification procedure (Collins, 1972). In light of this criticism, special care was taken in this investigation to ensure that the mitochondria were not altered or damaged. To preserve the physiological integrity of mitochondria, the gradient used for their separation consisted of iso-osmotic sucrose with Ficoll added to increase the density. Ficoll is a synthetic polymer of sucrose with negligible osmotic effect since it has a molecular weight of approximately 400,000. The preservation of the basal properties of mitochondria during separation can be tested by studying the permeability of mitochondrial membrane to exogenous NADH (Leninger, 1965). The degree of preservation of the mitochondria used in these experiments was therefore determined by following the rate of oxidation of added NADH at 340 nm. This rate was compared with the rate exhibited by mitochondria of the same fraction which had been made permeable to NADH by ultrasonic disintegration or by freezing and thawing. The ratios between the rate of NADH oxidation in broken mitochondria and that of untreated mitochondria were 5.2, 4.7, 6.1, 4.9, 5.7, and 4.7 for fractions 6, 9, 15, 21, 31, and 36, respectively. These results indicate that the mitochondria were not separated on the gradient according to varying degrees of mitochondrial damage and that the observed differences in monoamine oxidase activities were not artifacts. One may wonder whether the heterogeneity of enzyme activity obtained may result from various degrees of microsomal contamination in the various mitochondria fractions. This is ruled out because extramitochondrial monoamine oxidase has not been detected by electron microscopic studies (Boadle and Bloom, 1969).

The results presented in this report demonstrate that brain mitochondria are heterogeneous with regard to monoamine oxidase; furthermore, one can conclude that in rat brain a monoamine oxidase exists which may preferentially deaminate dopamine and is different from the enzyme system that deaminates tyramine. Kroon and Veldstra (1972) have presented strong evidence for the presence of a "dopamine monoamine oxidase" in synaptosomes. This differs from the synaptosomes that deaminate noradrenaline. Present studies lend further support to the view that multiple forms of monoamine oxidase have a physiological function, and that deamination of nonmethylated biogenic monoamines can take place in neurons. The reuptake mechanism has been thought to be acting as a primary system that terminates the receptor response evoked by amines released extraneuronally during nerve activity (Iversen, 1967). However, with the present knowledge of the multiple forms of monoamine oxidase and their different substrate specificities, the nature of the role played by monoamine oxidase obviously needs reevaluation.

We (Youdim and Benda, 1973, *in preparation*) have also studied the problem of monoamine oxidase multiplicity by inducing antibody to a highly purified rat liver mitochondrial enzyme. Both the liver and brain monoamine oxidase cross-react with this antibody and show two

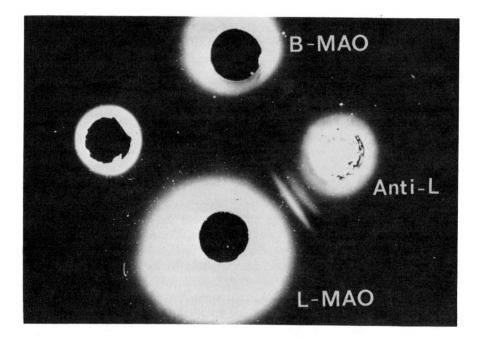

FIG. 2. Antibody (Anti-L) to a highly purified rat liver mitochondrial enzyme cross-reacted with both liver (L-MAO) and brain (B-MAO) monoamine oxidase and showed two enzymatically active immunoprecipitin lines in double diffusion (ouchterlony) using 1.2% Agarose.

enzymatically active immunoprecipitin lines in double diffusion (ouchterlony) using 1.2% Agarose (Fig. 2). These results may suggest the presence of two distinct monoamine oxidase proteins.

REFERENCES

Boadle, M. C., and Bloom, F. E. (1969): A method for the fine structural localization of monoamine oxidase. *Journal of Histochemistry and Cytochemistry*, 17:331–340.

Collins, G. G. S. (1972): Summary of Section I. *Advances in Biochemical Psychopharmacology*, Vol. 5, pp. 129–132. Raven Press, New York.

Gray, E. G., and Whittaker, V. P. (1962): The isolation of nerve endings from brain: An electron-microscopic study of cell fragments derived by homogenization and centrifugation. *Journal of Anatomy*, 96:79–88.

Hanker, J. S., Kusyk, C. J., Bloom, F. E., and Pearse, A. G. E. (1973): The demonstration of dehydrogenases and monoamine oxidase by the formation of osmium block at the site of the Hatchett's Brown. *Histochemie*, 33:205–230.

Iversen, L. L. (1967): *Uptake and Storage of Noradrenaline in Sympathetic Nerves.* Cambridge University Press, Cambridge.

Jarrott, B. (1971): Occurrence and properties of monoamine oxidase in adrenergic neurons. *Journal of Neurochemistry*, 18:7–16.

Johnston, J. P. (1968): Some observation upon a neuro inhibitor of monoamine oxidase in brain tissue. *Biochemical Pharmacology*, 17:1285–1297.

Kroon, M. C., and Veldstra, H. (1972): Multiple forms of rat brain mitochondrial monoamine oxidase subcellular localization. *Federation of European Biochemical Societies Letters*, 24:173–176.

Leninger, A. L. (1965): *The Mitochondria*, p. 147. W. A. Benjamin, New York.

Lowry, O. H., Rosebrough, M. J., Farr, A. L., and Randall, R. J. (1951): Protein measurement with the folin phenol reagent. *Journal of Biological Chemistry*, 193:265–275.

Neff, N. H., and Goridis, C. (1972): Neuronal monoamine oxidase: Specific enzyme types and their rates of formation. *Advances in Biochemical Psychopharmacology*, Vol. 5, pp. 307–324. Raven Press, New York.

Tipton, K. F., Youdim, M. B. H., Spires, I. P. C. (1972): Beef adrenal medulla monoamine oxidase. *Biochemical Pharmacology*, 21:2197–2204.

Wilson, M. A., and Cascarno, J. (1972): Biochemical heterogeneity of rat liver mitochondria separated by rate zonal centrifugation. *Biochemical Journal*, 129:209–218.

Youdim, M. B. H. (1972): Multiple forms of monoamine oxidase and their properties. *Advances in Biochemical Psychopharmacology*, Vol. 5, pp. 67–78. Raven Press, New York.

Youdim, M. B. H. (1973a): Multiple forms of mitochondrial monoamine oxidase. *British Medical Bulletin*, 29:120–123.

Youdim, M. B. H. (1973b): Heterogeneity of rat brain and liver mitochondrial monoamine oxidase: Subcellular fractionation. *Biochemical Society Transactions*, 1:1126–1127.

Youdim, M. B. H., Collins, G. G. S., and Sandler, M. (1969): Multiple forms of rat brain monoamine oxidase. *Nature* (London), 223:626–628.

Youdim, M. B. H., and Sourkes, T. L. (1965): The effect of heat, inhibitors and riboflavin deficiency on monoamine oxidase. *Canadian Journal of Biochemistry*, 43:1305–1318.

Advances in Biochemical Psychopharmacology, Vol. 11
Raven Press, New York © 1974

N-Methylation of Indolealkylamines in the Brain with a New Methyl Donor

Josée Leysen and Pierre Laduron

Department of Neurobiochemistry, Janssen Pharmaceutica, Beerse, Belgium

I. INTRODUCTION

It has been suggested by several investigators, as discussed by Himwich (1970), that anomalous methylation of biogenic amines may somehow be involved in schizophrenia. In this connection, bufotenin and N,N-dimethyltryptamine have been detected in the urine of certain schizophrenic patients (Fischer, Lagravere, Vazquez, and Di Stefano, 1961; Narasimhachari and Himwich, 1973); these compounds are known to produce psychotomimetic reactions (Himwich, 1970). However, up to now, possible mechanisms by which these dimethylated amines can originate in the brain have not been clearly identified.

Very recently, Morgan and Mandell (1969) and Mandell and Morgan (1971) reported the *in vitro* formation of bufotenin during the incubation of a brain enzyme with S-adenosylmethionine (SAM) and serotonin. Furthermore, Saavedra and Axelrod (1972) described both *in vivo* and *in vitro* formation of N-methylated tryptamine in the brain; in the former experiment, ^{14}C-tryptamine was injected intracisternally in the rat and thereafter N-methyltryptamine and N,N-dimethyltryptamine were identified in the brain; in the latter experiment, tryptamine was incubated with a brain enzyme preparation with SAM as methyl donor. However, all enzymatic assays using SAM have yielded very low activities, which have not always been reproducible and which have been associated with many methodological difficulties.

In our laboratory, another methyl donor, 5-methyltetrahydrofolic acid (5-MTHF), was used successfully (Laduron, 1972); when this compound was incubated with catecholamines and an enzyme preparation from rat brain, considerable activity was observed and dopamine (DA) was proved to be converted into epinine, first by TLC (Laduron, Gommeren, and Leysen, 1973; Laduron, 1973) and later by mass fragmentography (*unpublished results*). As this method had successfully been applied to catecholamines, we checked other potential substrates

TABLE 1. *Substrate for N-methyltransferase with 5-MTHF in rat brain*

Active	Inactive
Dopamine	N,N'-Dimethyldopamine
Epinine	3,4-Dihydroxyfenyl acetic acid
3-Hydro-4-methoxyphenylethylamine	Bufotenin
3-Methoxy-4-hydroxyphenylethylamine	N,N-Dimethyltryptamine
3,4-Methoxyphenylethylamine	Tryptophol
Tryptamine	5-Methoxytryptophol
N-Methyltryptamine	Melatonin
N_1-Methyltryptamine	Histamine
5-Methoxytryptamine	Dimethylamphetamine
Serotonin	
N-Methylserotonin	
Amphetamine	
N-Methylamphetamine	
Norepinephrine	
Epinephrine	
Normetanephrine	
Mescaline	

for their ability to be N-methylated with rat brain enzyme using 5-MTHF. Among the compounds listed in Table 1, nearly all the primary and secondary amines appeared to be substrates. Owing to their possible importance in psychotic disorders, the indolealkylamines were studied further. The purpose of the present chapter is to report on a more detailed investigation of the N-methylation of these substrates.

II. PROPERTIES OF THE RAT BRAIN N-METHYLTRANSFERASE ENZYME

At first, a choice of various better-known indoles with different side chains was made, to find out whether these substrates were methylated on the nitrogen or on the oxygen, using 5-MTHF. The results (Table 2) show that under the experimental conditions employed (i.e., high substrate concentration, 1.6×10^{-5} M methyl-^{14}C 5-MTHF, sodium phosphate buffer, pH 6.4, 2 hr incubation at 37°C, fixation with borate or acid, and extraction of the radioactive product with an appropriate organic solvent), high activity was observed using 5-hydroxytryptamine (5-HT) and tryptamine as substrate. When the secondary amines were tested enzyme activities were decreased, and they were absent completely with the tertiary amine, bufotenin. On the other hand, 5-methoxytryptamine and N_1-methyltryptamine were found to be

TABLE 2. *N-Methyltransferase with various indole derivatives*

Substrate	nmoles/hr/mg protein	Substrate	nmoles/hr/mg protein
HO—[indole]—$CH_2CH_2NH_2$ (N-H) 5-HT	0.186	[indole]—$CH_2CH_2NH_2$ (N_1–CH_3) N_1-MT	0.058
HO—[indole]—$CH_2CH_2NHCH_3$ (N-H) N-MHT	0.060	[indole]—CH_2CH_2OH (N-H) Tryptophol	0
HO—[indole]—$CH_2CH_2N(CH_3)_2$ (N-H) N-DMHT	0	CH_3O—[indole]—CH_2CH_2OH (N-H) 5-MTryptophol	0
CH_3O—[indole]—$CH_2CH_2NH_2$ (N-H) 5-MT	0.112	CH_3O—[indole]—$CH_2CH_2NHCOCH_3$ (N-H) Melatonin	0
[indole]—$CH_2CH_2NH_2$ (N-H) T	0.080	HO—[indole] (N-H) 5-HI	0
[indole]—$CH_2CH_2NHCH_3$ (N-H) N-MT	0.028		

highly N-methylated, while other indoles, such as tryptophol, 5-methoxytryptophol, melatonin, and 5-hydroxyindole, did not act as substrates. From these results it seems very likely that, using brain enzyme and 5-MTHF as methyl donor, the indolealkylamines are methylated on the nitrogen of the side chain.

The physical properties of the N-methylation reaction with indolealkylamines are very similar to those of the reaction with the catecholamines. For both kinds of substrate, the pH-optimum value was 6.4, as is shown in Fig. 1, and both displayed linear kinetics over a fairly well-defined time and enzyme concentration range.

The substrate saturation curves showed a rather complex behavior pattern, as illustrated in Fig. 2. With 3-hydroxy-4-methoxyphenylethylamine (4-MDA) as well as with tryptamine, incubated at varying concentrations, a clearly intermediary plateau region was obtained in the v vs. S plots. Since no rectangular hyperbole was found for these reactions, S/v vs. S plots were of course not linear, but showed several linear sections instead. The same type of S/v vs. S plot was obtained when the concentration of the methyl donor 5-MTHF varied while the concentration of either 4-MDA or tryptamine was kept fixed. This

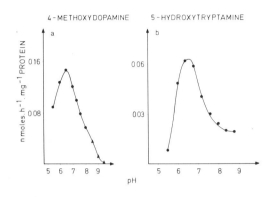

FIG. 1. pH activity curves of brain N-methyltransferase using 5-MTHF as methyl donor (a) with 4-MDA and (b) with 5-HT as substrate.

kinetic pattern seemed unusual at first sight, but in recent papers by Teipel and Koshland (1969) and Engel and Ferdinand (1973) several enzymes have been reported to behave in that way. In a theoretical approach to this kinetic problem, Teipel and Koshland (1969) calculated that such enzymes must possess a total of more than two substrate binding sites, and the relative magnitude of the intrinsic catalytic or binding constants of these sites first decrease, then increase

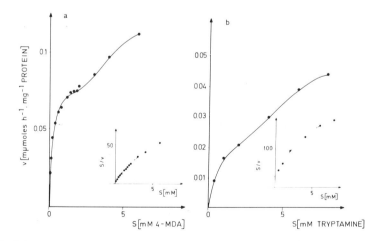

FIG. 2. Substrate saturation curves with brain N-methyltransferase using varying concentrations of (a) 4-MDA and (b) tryptamine, both with a constant concentration of 1.6×10^{-5}M 5-MTHF.

as the enzyme is saturated. The reaction mechanism of the brain N-methyltransferase using 5-MTHF as methyl donor is far from being elucidated. However, the foregoing results suggest that the reaction is a highly complex one, and that the enzyme probably has a regulatory function.

III. SPECIFICITY OF THE METHYL DONOR

Various methylating enzymes were further isolated from a brain preparation obtained by ammonium sulfate precipitation either by sucrose gradient centrifugation or gel filtration through a Sephadex G-200 column (Leysen, Gommeren, and Laduron, 1973). These two approaches yielded quite comparable results: a higher molecular weight enzyme peak could be located in the elution fractions assayed using 5-MTHF and 4-MDA; a lower molecular weight enzyme peak could be detected when SAM was incubated together with 3,4-dihydroxyphenyl-acetic acid (DOPAC) under conditions for O-methylation, that is, at pH 7.9 in the presence of Mg^{2+}. This latter peak overlapped the histamine-N-methyltransferase enzyme peak which also required SAM as methyl donor. A typical elution pattern of the first two enzymes is presented in Fig. 3a.

After pooling and concentrating, the two enzyme preparations were tested for their ability to methylate various catechol-like substrates (Leysen et al., 1973). It was found that the higher molecular weight enzyme was highly specific toward 5-MTHF as methyl donor and catalyzed methyl transfer on the nitrogen only. The second enzyme peak required SAM exclusively and appeared to be catechol-O-methyl-transferase (COMT) (EC 2.1.1.6).

When the eluted fractions of an identical column experiment were tested with 5-HT and SAM, only very slight activity was obtained corresponding to the position of COMT. However, similar activity was found when substrate was omitted from the incubation mixture (Fig. 3a). Similarly, in fractions provided by gradient centrifugation, no activity could be found with indolealkylamines using SAM as methyl donor. On the other hand, when 5-HT was incubated with 5-MTHF, considerable activity was obtained in fractions corresponding to the N-methyltransferase enzyme, as shown in Fig. 3b. The failure to detect N-methyltransferase activity in the brain using SAM is in disagreement with the findings reported by Morgan and Mandell (1969), Mandell and Morgan (1971), and Saavedra and Axelrod (1972). This led us to perform a detailed investigation of the specificity of the methyl donor with indolealkylamines as a substrate. The method described by Saavedra and Axelrod (1972) using SAM was slightly modified in that three different blanks were prepared, one using boiled enzyme, a

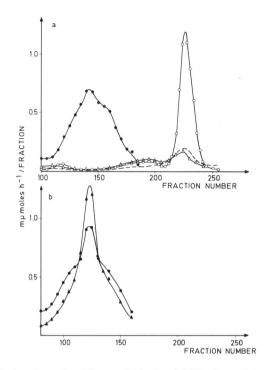

FIG. 3. Activity in fractions eluted from a Sephadex G-200 column determined with ●———● 5-MTHF and 4-MDA, ▲———▲ 5-MTHF and 5-HT, ○———○ SAM and DOPAC, △———△ SAM and 5-HT, ‑‑‑‑ SAM without substrate.

second to which substrate was added only after fixation of the incubation mixture, and the third without any substrate. Four different substrates, serotonin, tryptamine, N-methyltryptamine, and bufotenin, were tested in this manner with two different enzyme preparations, a supernatant and a P_2 ammonium sulfate fraction. The results are summarized in Table 3. It is obvious that the activities were completely different, depending on the blank which was substracted. The activities which remained after evaporation of the organic extract were extremely low and variable. To avoid evaporation, a somewhat drastic method, the radioactive methanol, formed enzymatically (Axelrod and Daly, 1965), was removed in another way, by lyophilizing the reaction mixture immediately after incubation. This method also yielded very low and irreproducible activities. However, when 5-MTHF was used as methyl donor, much higher activities were obtained in a very reproducible way for three different enzyme preparations, as shown in Table 3.

TABLE 3. *N-Methyltransferase activity (nmoles h^{-1} mg protein^{-1}) with brain enzyme and different methyl donors*

| Substrate | Substracted blank | Methyl-^{14}C SAM Supernatant | | Methyl-^{3}H SAM P_2 $(NH_4)_2SO_4$ fraction | | Methyl-^{14}C 5-MTHF | | |
		Before evaporation	After evaporation	Before evaporation	After evaporation	Supernatant	P_2 fraction	Peak I Seph. G-200
Serotonin	Boiled enzyme	0.151	0	0.005	0.0006			
	Substr. after fix.	0	0.007	0.0002	0.0002			
	Without substr.	0	0	0.00008	0.0003	0.083	0.146	0.199
Tryptamine	Boiled enzyme	0.156	0	0.006	0.0004			
	Substr. after fix.	0	0.009	0.0006	0.00008			
	Without substr.	0	0.0009	0.0004	0.00008	0.045	0.090	0.109
N-CH$_3$tryptamine	Boiled enzyme	0.164	0					
	Substr. after fix.	0	0.003					
	Without substr.	0	0			0.024	0.043	0.071
Bufotenin	Boiled enzyme	0.157	0	0.006	0.0002			
	Substr. after fix.	0	0	0	0			
	Without substr.	0	0.004	0	0	0.0005	0.004	0.008

IV. DISCUSSION

The N-methyltransferase that we found in the brain displays a very high specificity toward 5-MTHF as methyl donor, but seems to be less specific toward the substrates which become N-methylated. According to the results of our investigation, we may assume that the same enzyme N-methylates catecholamines as well as indolealkylamines and other nonbiogenic amines. Indeed, the identical pH optimum was noted for all substrates, each of which gave rise to the same typical kinetic behavior. Furthermore, using two different purification procedures, only one methylating enzyme specific to 5-MTHF was detected. It should be pointed out that this enzyme, which can be named 5-methyltetrahydrofolate-N-methyltransferase (5-MT-NMT), exclusively catalyzes methylations on the nitrogen. Methylation of the amine function could not be demonstrated using a brain enzyme preparation and SAM. Apart from the fact that SAM was required by the specific histamine-N-methyltransferase (EC 2.1.1.8), in the brain this latter methyl donor appeared to be reserved for O-methylation processes, at least in our experimental conditions. This statement is in agreement with some results of Mandel, Ho Sam Ahn, Vanden Heuvel, and Walker (1972), who were not able to observe any N-methyltransferase activity using SAM and a human brain enzyme preparation.

The assumption that 5-MTHF plays a direct role as methyl donor for the methylation of indolealkylamines is certainly not incompatible with the *in vivo* determinations previously described. The experiment of Saavedra and Axelrod (1972), in which radioactive tryptamine was injected intracisternally, proves that N-methylation of this substrate does indeed occur in the brain, but it does not indicate which methyl donor is involved. Considering the much higher activity obtained *in vitro* using 5-MTHF compared with SAM, we may suppose that 5-MTHF is also active *in vivo* as methyl donor in the N-methylation reaction. This assumption is further supported by the finding that in the cerebrospinal fluid the concentration of 5-MTHF is much higher than in the serum (Herbert and Zalusky, 1961; Weckman and Lehtovora, 1969). Moreover, 5-MTHF is frequently taken up into neural tissue after being formed from folate congeners in the liver (Allen and Klipstein, 1970; Levitt, Nixon, Pincus, and Bertino, 1971). Since these data from different research areas agree so well, and even amplify each other, we suggest that the psychotomimetic amines such as bufotenin and N,N-dimethyltryptamine detected in the urine of schizophrenics might originate in the brain itself by an N-methylation reaction employing 5-MTHF. Nevertheless, this does not exclude the possibility that these compounds may be formed in other organs, such as the lung, where Axelrod (1962) has found an appropriate enzyme.

However, in the latter case, many problems arise with regard to penetration of such psychotomimetic compounds through the blood-brain barrier in active form.

However, it seems more likely to assume that, if these compounds are elaborated in excess in schizophrenic disorders, they are formed directly in the brain; and it seems more than likely that 5-MTHF rather than SAM is involved in the N-methylation reactions which give rise to them.

REFERENCES

Allen, C. D., and Klipstein, E. A. (1970): Brain folate concentration in rats receiving diphenylhydantoin. *Neurology*. 4:403 (Abstr.).

Axelrod, J. (1962): The enzymatic N-methylation of serotonin and other amines. *Journal of Pharmacology and Experimental Therapeutics*, 138:28—33.

Axelrod, J., and Daly, J. (1965): Pituitary gland: Enzymatic formation of methanol from S-adenosyl-methionine. *Science*, 150:892—893.

Engel, P. C., and Ferdinand, W. (1973): The significance of abrupt transitions in Lineweaver—Burk plots with particular reference to glutamate dehydrogenase: Negative and positive co-operativity in catalytic rate constants. *Biochemical Journal*, 131:97—105.

Fischer, E., Fernandez Lagravere, T. A., Vazquez, A. J., and Di Stefano, A. O. (1961): A bufotenin-like substance in the urine of schizophrenics. *Journal of Nervous and Mental Disease*, 133:441—444.

Herbert, V., and Zalusky, R. (1961): Selective concentration of folic acid activity in cerebrospinal fluid. *Proceedings of the American Society of Experimental Biology*, 20:453—458.

Himwich, H. E. (1970): Indoleamines and the schizophrenias. In: *Biochemistry, Schizophrenias and Affective Illnesses*, edited by H. E. Himwich, pp. 79—122. Williams & Wilkins, Baltimore.

Laduron, P. (1972): N-methylation of dopamine to epinine in brain tissue using N-methyltetrahydrofolic acid as the methyl donor. *Nature New Biology*, 238:212—213.

Laduron, P. (1973): New concepts on the N-methylation reactions of biogenic amines in adrenal medulla and brain. *Abstracts of III International Catecholamine Symposium* (Strasbourg).

Laduron, P., Gommeren, W., and Leysen, J. (1973): N-methylation of biogenic amines. I. Characterization and properties of an N-methyltransferase in rat brain using 5-methyltetrahydrofolic acid as the methyl donor. *Biochemical Pharmacology (submitted for publication)*.

Levitt, M., Nixon, P. F., Pincus, J. H., and Bertino, J. R. (1971): Transport characteristics of folates in cerebrospinal fluid: A study utilizing doubly labeled 5-methyltetrahydrofolate and 5-formyltetrahydrofolate. *Journal of Clinical Investigations*, 50:1301—1308.

Leysen, J., Gommeren, W., and Laduron, P. (1973): N-methylation of biogenic amines. II. Specificity of the methyl donor for O- and N-methyltransferase in rat brain. *Biochemical Pharmacology (submitted for publication)*.

Mandell, A., and Morgan, M. (1971): Indole (ethyl) amine N-methyltransferase in human brain. *Nature New Biology*, 230:85—87.

Mandel, L. R., Ho Sam Ahn, Vanden Heuvel, W. J. A., and Walker, R. W. (1972): Indoleamine-N-methyltransferase in human lung. *Biochemical Pharmacology*, 21:1197—2000.

Morgan, M., and Mandell, A. (1969): Indole (ethyl) amine N-methyltransferase in the brain. *Science*, 165:492—494.

Narasimhachari, N., and Himwich, H. E. (1973): GC-MS identification of bufotenin in urine samples from patients with schizophrenia or infantile autism. *Life Science*, 12:475—478.

Saavedra, J., and Axelrod, J. (1972): Psychotomimetic N-methylated tryptamines: Formation in brain *in vivo* and *in vitro*. *Science*, 175:1365—1366.

Teipel, J., and Koshland, D. E., Jr. (1969): The significance of intermediary plateau regions in enzyme saturation curves. *Biochemistry*, 8:4656—4663.

Weckman, N., and Lehtovora, R. (1969): Logarithmic-normal distribution of cerebrospinal fluid folate concentrations. *Experientia* (Basel), 25:585—586.

Advances in Biochemical Psychopharmacology, Vol. 11
Raven Press, New York © 1974

Multiple N-Methyltransferases for Aromatic Alkylamines in Brain

Louise L. Hsu and Arnold J. Mandell

Department of Psychiatry, University of California, San Diego, La Jolla, California 92037

I. INTRODUCTION

Over a decade ago Axelrod (1961) reported the presence of a nonspecific N-methyltransferase for aromatic amines in mammalian tissue. Subsequently, enzymatic activity capable of N-methylating both serotonin and tryptamine in brains of chicks, sheep, and humans was reported from our laboratories (Morgan and Mandell, 1969; Mandell and Morgan, 1971; Mandell, Buckingham, and Segal, 1971). The presence of enzymatic activity capable of N-methylating aromatic amines in rat brain and human brain, as well as in other organs, has been confirmed (Saavedra and Axelrod, 1972; Saavedra, Coyle, and Axelrod, 1973). In all the foregoing experiments S-adenosyl-L-methionine (SAM) was used as the methyl donor. Laduron (1972) recently reported using 5-methyltetrahydrofolic acid (5-MTHF) as a methyl donor in his studies of the formation of epinine by N-methylation of dopamine. We concluded that the same donor could be used in studies of other N-methylation processes in brain. The following is a preliminary report on the enzymatic N-methylation of two specific aromatic alkylamines, tryptamine and β-phenylethylamine, in the presence of 5-MTHF and of SAM.

II. MATERIALS AND METHODS

5-MTHF-^{14}C, barium salt (51 mC/mmole) was obtained from Amersham/Searle. S-Adenosyl-L-methionine-methyl-^{3}H (SAM-^{3}H) (4.55 C/mM) was obtained from New England Nuclear Corporation. Tryptamine was obtained from Cal Biochem Company, and β-phenylethylamine from Sigma Chemical Company. Silica Gel IB flexible sheets for thin-layer chromatography (20 × 20 cm) were purchased from J. T. Baker Chemical Company. Sephadex G-100 was obtained from Pharmacia Fine Chemicals, Inc. Other chemicals were purchased from standard sources in maximal obtainable purity. Sprague-Dawley strain male rats (100 to 150 g) from Carworth were used in these studies.

Whole rat brains were homogenized in five volumes of distilled water or various buffer solutions at $4°C$ immediately after the rats were sacrificed by decapitation. Whole brain homogenates (WH) were centrifuged at 40 K rpm for 40 min to yield a 100,000 $\times$ g supernate (S_0). The supernate was then partially purified with ammonium sulfate. Aliquots of WH, S_0, and the pellets obtained from precipitation at 0 to 20% (Pp_1), 20 to 40% (Pp_2), 40 to 50% (Pp_3), and 50 to 60% (Pp_4) ammonium sulfate saturation, suspended in 0.02 M K-phosphate (pH 6.8), were dialyzed overnight against the same buffer, and enzyme activity was assayed in all fractions. Portions of fresh WH and supernate were also used for the assay of enzyme activity. Undialyzed Pp_3 [40 to 50% $(NH_4)_2SO_4$] was applied to a Sephadex G-100 column (1.5 $\times$ 20 cm) which was equilibrated with 0.05 M K-phosphate buffer (pH 6.5). Protein was eluted with the same buffer and collected in fractions of 1 ml for 2 hr. Blue Dextran was used as marker to determine the void volume of the column, which was estimated at 12.5 ml. Enzyme activity was subsequently assayed in fractions containing appreciable concentrations of protein. Protein concentrations in all our experiments were assayed according to Lowry, Rosebrough, Farr, and Randall (1951).

N-Methyltransferase activity was assayed with modifications of the methods of Laduron (1972) or Morgan and Mandell (Morgan and Mandell, 1969; Mandell and Morgan, 1971). Each reaction mixture contained 150 μmoles of K-phosphate buffer at various ph values, 0.25 μC of 5-MTHF-^{14}C or 0.15 μC of SAM-^{3}H, 10 μmoles of tryptamine or β-phenylethylamine, 0.5 μmoles of sodium metabisulfite, 2.5 μmoles of ethylenediamine tetraacetic acid (EDTA), and an enzyme aliquot. The total volume of each incubation mixture was 0.5 ml. After incubation at $37°C$ for 2 hr (if not otherwise indicated), the reaction was stopped by the addition of 1 ml of saturated sodium borate solution (pH 11), and the radioactive product was extracted with 7 ml of toluene containing 3% isoamyl alcohol. The extracts were shaken for 40 sec and centrifuged for 5 min at 2 K rpm. Portions (4 ml each) of the organic phase were transferred to counting vials, and the solvent in each vial was evaporated to dryness at $80°C$ in a chromatography oven. The residue was taken up with 2 ml of ethanol and 10 ml of Aquasol, and radioactivity was measured in a Beckman (LS-233) liquid scintillation spectrometer. Boiled enzyme samples were used as controls in all the assays.

We analyzed the radioactive products in the Pp_{3D} fraction by thin-layer chromatography. The extracting solvent was evaporated at room temperature under nitrogen atmosphere. The residue was taken up with 0.1 ml of ethanol, applied in strips 4.5 cm long to a silica gel sheet, and developed in a solvent system of isopropanol:NH_4OH:H_2O (100:2.8:10). Twenty-five micrograms each of N-monomethyl-tryptamine (MMT) and N,N-dimethyltryptamine (DMT) were used as carriers. A standard mixture of 25 μg of MMT, 25 μg of DMT, and 50 μg of tryptamine was applied for comparison. A solution of 1 g of 4(p)-dimethylaminobenzaldehyde in 50 ml of concentrated HCl and 50 ml of ethanol was used as the indicator spray. After spraying and drying, visible bands of silica gel (including the origin and the solvent front) were scraped from the plastic sheet and triturated in 1 ml of ethanol, 2 ml of distilled H_2O, and 10 ml of Aquasol preparatory to measurement of radioactivity.

III. RESULTS

A. The effects of pH and Na^+ and K^+ ions on N-methyltransferase activity in brain supernates with 5-MTHF-^{14}C as methyl donor and tryptamine as substrate: We assayed enzyme activity of whole-brain

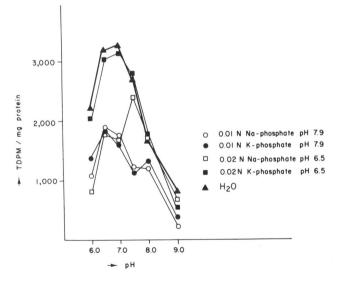

FIG. 1. The effects of pH and Na^+ and K^+ ions on N-methyltransferase activity in brain supernates with 5-MTHF-^{14}C as methyl donor and tryptamine as substrate. Each individual rat brain was homogenized in one of the five media. Homogenates were centrifuged at 40 K rpm to yield 100,000 $\times$ g supernates. The N-methyltransferase activity was assayed in the supernates at pH levels ranging from 6.0 to 9.0 (See Methods). The incubation time in this experiment was 90 min. Each point represents the average (± 10%) net TDPM/mg protein/90 min of duplicate samples.

homogenates in various media as described by several authors. Specifically, Mandell et al. (1971) used 0.1 N K-phosphate buffer (pH 7.9); Saavedra and Axelrod (1972) used 0.01 N Na-phosphate buffer (pH 7.9); and Laduron (1972) used distilled water. To determine the best medium for demonstrating N-methylating enzymatic activity in brain supernates when 5-MTHF was the methyl donor and tryptamine the substrate, we measured N-methyltransferase activity at various pH levels in supernates obtained from rat brain homogenized in five different media: 0.01 N sodium phosphate, pH 7.9; 0.01 N potassium phosphate, pH 7.9; 0.02 N sodium phosphate, pH 6.5; 0.02 N potassium phosphate, pH 6.5; and double distilled water, pH 5.5. The incubation period was 90 min. We measured enzymatic activity in the incubation mixtures at pH levels ranging from 6.0 to 9.0. The optimal pH for enzymatic activity in the supernates was between 6.5 and 7.0, regardless of the original medium (Fig. 1). The supernate in distilled water yielded the highest enzyme activity, and that in 0.02 N potassium phosphate buffer, pH 6.5, yielded the second highest enzymatic activity. Henceforth, we chose distilled water as the homogenizing medium.

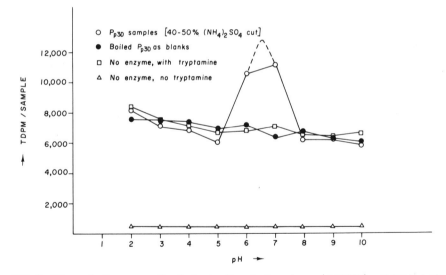

FIG. 2. Effect of pH on aromatic alkylamine-N-methyltransferase (AANMT) activity in brain with 5-MTHF as methyl donor and tryptamine as substrate. Pp_{3D} was obtained from protein precipitation when 100,000 X g supernate (in water) was saturated 40 to 50% with ammonium sulfate. The precipitate was picked up with a minimal amount of 0.02 N K-phosphate (pH 6.8) and dialyzed overnight against the same buffer. The protein content in each sample was 0.25 mg in assays of enzyme activity (see Methods). All enzyme samples and blanks were assayed at pH levels ranging from 2.0 to 10.0. Each point represents the average of determinations in triplicate. Note the contrast between the activity in blanks with and without substrate (see text).

B. The effect of pH on N-methyltransferase activity in partially purified fractions obtained from $(NH_4)_2SO_4$ precipitation, with 5-MTHF as methyl donor and either tryptamine or β-phenylethylamine as substrate: There is a discrepancy between the present finding of a pH optimum ranging from 6.5 to 7.0 and the optimum of 7.9 reported elsewhere (Morgan and Mandell, 1969; Mandell and Morgan, 1971; Mandell et al., 1971; Saavedra and Axelrod, 1972; Saavedra et al., 1973). This might have been caused by the change in methyl donor. However, it is difficult to explain the discrepancy between our results and Laduron's (1972), since in both cases the same donor, 5-MTHF-^{14}C, was used. Laduron reported optimal enzyme activity in the partially purified ammonium sulfate precipitate between pH 8.2 and 8.4, with dopamine as substrate. We explored the effect of pH on N-methyltransferase in the partially purified ammonium sulfate precipitate as well, with 5-MTHF-^{14}C as methyl donor and either tryptamine or β-phenylethylamine as substrate. We assayed N-methyltransferase activity in both Pp_{3D} (dialyzed) and Pp_{4D} (dialyzed) at pH levels ranging from 5.5 to 9.0. Optimal enzymatic activity was observed between pH 6.5 and 7.0 for both Pp_{3D} (Fig. 2) and Pp_{4D}. Similar results indicating the pH dependency of the enzyme were observed when β-phenylethylamine was used as substrate, again with

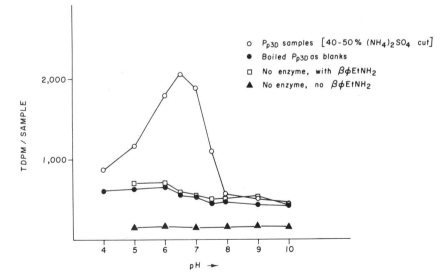

FIG. 3. Effect of pH on AAMNT activity in brain with 5-MTHF as methyl donor and β-phenylethylamine (βφEtNH$_2$) as substrate (see legend for Fig. 2). Each point represents the average of determinations made in triplicate, except ▲ represents the average of duplicate determinations.

5-MTHF-^{14}C as methyl donor (Fig. 3). Optimal enzyme activity with β-phenylethylamine as substrate was also observed between pH 6.5 and 7.0. Since in all these experiments we had found the optimal enzyme activity within a pH range of 6.5 to 7.0, we chose a pH of 6.5 for our further studies of the enzyme or enzymes.

Blanks with neither enzyme nor tryptamine yielded almost no background counts. Blanks with tryptamine and boiled enzyme gave unexpectedly high counts, which were comparable to those (also unexpected) of blanks with tryptamine but no enzyme. These phenomena suggested to us the possibility of a nonenzymatic reaction between methyl donor and substrate. Consequently, blanks with boiled enzyme *and* substrate are required for precise characterization of enzymatic activity.

C. *Time course and enzyme concentration with 5-MTHF-^{14}C as methyl donor at pH 6.5:* The enzyme activity in Pp$_3$ increased linearly for the first hour and leveled off gradually over the next 2 hr in the presence of either tryptamine or β-phenylethylamine. Enzyme activity was linear with protein concentration (0.1 to 1.0 mg) when tryptamine was the substrate and Pp$_{4D}$ was the enzyme source, and when β-phenylethylamine was the substrate and Pp$_{3D}$ was the enzyme source.

D. *Michaelis-Menten constants (K_m values) for tryptamine and for β-phenylethylamine with 5-MTHF-^{14}C as methyl donor:* K_m values of the enzyme for both substrates were estimated by double reciprocal

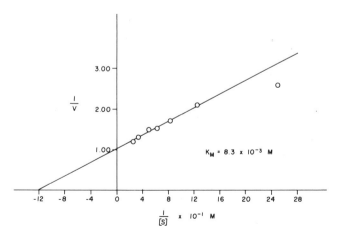

FIG. 4. Lineweaver-Burk double reciprocal plot showing substrate kinetics from AANMT (Pp$_{3D}$) at pH 6.5 with 5-MTHF-^{14}C as methyl donor and tryptamine as substrate. The protein content in each sample was 0.25 mg. Enzyme activity was measured at various substrate concentrations and expressed as nanomoles per milligram of protein per hour (V). Each point represents the average of duplicates. Blanks contained substrate and boiled enzyme.

plots. As shown in Fig. 4, the K_m for tryptamine was about 8.3×10^{-3} M. For β-phenylethylamine the K_m was approximately 6.25×10^{-3} M.

E. *Product identification by thin-layer chromatography:* High radioactivity, isographic with N-monomethyltryptamine, and smaller amounts of radioactivity, isographic with N, N-dimethyltryptamine, were found in Pp_{3D} *and* in the blanks containing tryptamine and boiled enzyme. However, both radioactive products gave much higher counts in Pp_{3D} than in the blanks (Fig. 5).

F. *Partial purification of N-methyltransferase with 5-MTHF-^{14}C or SAM-^{3}H as methyl donor and tryptamine or β-phenylethylamine as substrate:* Table 1 summarizes the specific enzyme activity in dialyzed and undialyzed samples of whole homogenate, $100,000 \times g$ supernate, and fractions from ammonium sulfate precipitation Pp_2 (20 to 40%) and Pp_3 (40 to 50%). Either 5-MTHF-^{14}C or SAM-^{3}H was used as methyl donor. Either tryptamine or β-phenylethylamine was used as substrate. Pp_3 demonstrated the highest enzyme activity for both substrates, in both dialyzed and fresh undialyzed samples, when compared to other fractions. Activity in Pp_3 was about fivefold greater than in whole homogenate. Dialysis enhanced the enzyme activity in all fractions studied, except for WH when SAM-^{3}H was the methyl donor. Pp_2 had the second highest enzyme activity.

Enzyme activity was assayed in various fractions (1 ml each) collected after Pp_3 was applied to a Sephadex G-100 column. Four assays were carried out, that is, in the presence of each donor in

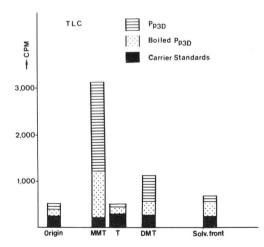

FIG. 5. Thin-layer chromatograms for analysis of products from the enzyme reaction (see Methods). The solvent front was 14.5 cm from the origin. R_f values for N-monomethyltryptamine (MMT), tryptamine, and N,N-dimethyltryptamine (DMT) were 0.31, 0.45, and 0.62, respectively. This experiment was repeated several times with similar results.

TABLE 1. *Specific Activity of N-methyltransferase in rat brain (TDPM/mg protein)*

Methyl donor	5-MTHF-^{14}C		SAM-^{3}H		Protein concentration (mg/ml)
Methyl acceptor	Tryptamine	$\beta\phi$EtNH$_2$	Tryptamine	$\beta\phi$EtNH$_2$	
Enzyme source					
WH	602	618	401	453	10.0
WH$_D$	1,810	1,537	–	68	6.4
100,000g S$_0$	2,137	1,632	243	409	6.2
100,000g S$_{0D}$	4,393	3,814	1,315	516	6.8
(NH$_4$)$_2$SO$_4$P$_{p_{2D}}$	5,393	4,111	2,136	601	8.3
(NH$_4$)$_2$SO$_4$P$_{p_{3D}}$	8,899	7,909	11,140	1,759	4.1
Sephadex G-100 [(NH$_4$)$_2$SO$_4$Pp$_3$]					
Fraction 7–10	3,396	7,338	1,419	545	0.64
11–12	22,271	–	11,799	–	0.88
13–18	2,342	814	–	3,754	0.23

Protein content in each assay of Sephadex fractions was 0.08 mg and 0.2 mg, while in other fractions it was from 0.2 to 0.5 mg. Each number on the table represents the average ($\pm$ 10%) net TDPM/mg protein of duplicates. Blanks contained substrate and boiled enzyme.

combination with each substrate, respectively (Table 1). When 5-MTHF-^{14}C was used as methyl donor, fractions 7 to 10 showed more enzyme activity for β-phenylethylamine than for tryptamine. Fractions 11 and 12 showed high enzyme activity specifically for tryptamine and no enzyme activity for β-phenylethylamine. Fractions 13 to 18 showed no enzyme activity for tryptamine but appreciably high activity for β-phenylethylamine. These results suggest the possible existence of two N-methyltransferases: one specific for tryptamine, the other non-specific for aromatic amines.

IV. DISCUSSION

This work confirms and extends some previous findings. Although Saavedra et al. (1973) reported that a blank without enzyme or substrate was equivalent to a blank containing boiled enzyme, we found, with either SAM or 5-MTHF as the methyl donor, that boiled blanks consistently gave higher activity counts than blanks without enzyme or substrate. In addition, extractable radioactivity from the blank containing boiled enzyme shows on chromatograms as mono- or dimethyltryptamine (Fig. 5). We invariably controlled for nonenzymatic N-methylation by using boiled enzyme blanks. Laduron (1972) also found evidence of a small degree of nonenzymatic methylation of dopamine to epinine, and therefore he used boiled enzyme blanks.

Whereas Laduron (1972) reported a pH optimum between 8.2 and 8.4 for reactions involving 5-MTHF and dopamine, we found the pH optimum to be between 6.5 and 7.0 in reactions involving 5-MTHF and either β-phenylethylamine or tryptamine. Reactions involving SAM have a pH optimum of 7.9, regardless of substrate, as we have previously reported (Mandell et al., 1971).

5-MTHF has low affinity for either β-phenylethylamine or tryptamine as a substrate. The K_m is in the range of 1×10^{-3} M, two orders of magnitude higher than the K_m we reported for SAM in relation to tryptamine. It is difficult to interpret the functional significance of these kinetic constants without knowing the concentrations of substrate in subcellular pools. However, it would seem that an enzyme using 5-MTHF might function only when concentrations of biogenic amines in the brain were high, as they are, for example, when monoamine oxidase is inhibited. Recent reports of genetically low levels of monoamine oxidase activity in the platelets of schizophrenics may be relevant to this speculation (Murphy and Wyatt, 1972).

N-Methyltransferase activity in fractions 11 and 12 (Table 1), which appears to be specific for an idoleamine substrate regardless of whether the methyl donor is 5-MTHF (at pH 6.5) or SAM (at pH 7.9), was a surprise. N-Monomethyl-β-phenylethylamine has a partition coefficient between the aqueous and organic phases of extraction comparable to that of N-monomethyltryptamine. This comparability allows us to define the specificity of the enzyme.

Activity in the Sephadex G-100 fractions was consistently enriched between 15- and 25-fold over that in whole homogenate. A portion of this enrichment is apparently the result of loss of a dialyzable inhibitor of the enzyme(s); this can be seen by comparing the activities in $100{,}000 \times g\,S_0$ and $100{,}000 \times g\,S_{0D}$ in Table 1.

Fractions 7 to 10 from the Sephadex column appear to manifest N-methylation of both β-phenylethylamine and indoleamine substrates. The activity in these particular fractions may well be that reported by Saavedra et al. (1973).

V. CONCLUSION

There appear to be at least two N-methylating systems for aromatic alkylamines in brain. One is specific to an indoleamine substrate and may correspond to what we have previously called indole(ethyl)-amine-N-methyltransferase (Morgan and Mandell, 1969; Mandell and

Morgan, 1971; Mandell et al., 1971). The other is nonspecific and may correspond to the enzyme called aromatic-amine-N-methyltransferase by Saavedra et al., 1973). Both can use either SAM or 5-MTHF as a methyl donor, but apparently when SAM is the methyl donor, there is greater affinity for the indoleamine substrate than when 5-MTHF is the methyl donor. These enzymes will be of interest to those pursuing a theory of psychotic disorder that involves the biotransformation of neurotransmitters.

ACKNOWLEDGMENTS

This work was supported by National Institute of Mental Health grant 14360. We are grateful to Mark Crissey for performing the assays.

REFERENCES

Axelrod, J. (1961): Enzymatic formation of psychotomimetic metabolites from normally occurring compounds. *Science*, 134:343.

Laduron, P. (1972): N-Methylation of dopamine to epinine in brain tissue using N-methyltetrahydrofolic acid as the methyl donor. *Nature New Biology*, 238:212–213.

Lowry, O. H., Rosebrough, N. J., Farr, A. L., and Randall, R. J. (1951): Protein measurement with the Folin phenol reagent. *Journal of Biological Chemistry*, 193:265–275.

Mandell, A. J., Buckingham, B., and Segal, D. S. (1971): Behavioral, metabolic and enzymatic studies of a brain indole(ethyl)amine N-methylating system. In: *Brain Chemistry and Mental Disease*, edited by B. T. Ho and W. M. McIsaac. Plenum, New York.

Mandell, A. J., and Morgan, M. (1971): Indole(ethyl)amine N-methyltransferase in human brain. *Nature New Biology*, 230:85–87.

Morgan, M., and Mandell, A. J. (1969): Indole(ethyl)amine N-methyltransferase in the brain. *Science*, 165:492–493.

Murphy, D. L., and Wyatt, R. J. (1972): Reduced monoamine oxidase activity in blood platelets from schizophrenic patients. *Nature* 238:225–226.

Saavedra, J. M., and Axelrod, J. (1972): Psychotomimetic N-methylated tryptamines: Formation in brain *in vivo* and *in vitro*. *Science*, 172:1365–1366.

Saavedra, J. M., Coyle, J. T., and Axelrod, J. (1973): The distribution and properties of the nonspecific N-methyltransferase in brain. *Journal of Neurochemistry*, 20:743–752.

Advances in Biochemical Psychopharmacology, Vol. 11
Raven Press, New York © 1974

N-Methyltetrahydrofolic Acid: The Physiological Methyl Donor in Indoleamine N- And O-Methylation

Shailesh P. Banerjee and Solomon H. Snyder

Departments of Pharmacology and Experimental Therapeutics and of Psychiatry and Behavioral Sciences, Johns Hopkins University School of Medicine, Baltimore, Maryland 21205

The N-methylation of biogenic amines has been proposed by numerous authors for a major role in certain mental illnesses (Szara, 1956; Pollin, Cardon, and Kety, 1961; Tanimukai, Gunther, Spaide, Bueno, and Himwich, 1970; Mandell and Spooner, 1968). Axelrod described an enzyme which actively N-methylated a variety of biogenic amines, including indoleamines, but which could not be demonstrated in tissues other than the rabbit lung (Axelrod, 1961, 1962). More recently, N-methylation of indoleamines has been reported in chick and human brain (Morgan and Mandell, 1969; Mandell and Morgan, 1971) and in a variety of mammalian tissues after dialysis (Saavedra and Axelrod, 1972). All these enzymatic activities utilize S-adenosylmethionine (AMe) as a methyl donor and, except for the rabbit lung enzyme, their activity is relatively feeble (Axelrod, 1961, 1962; Morgan and Mandell, 1969; Mandell and Morgan, 1971). Laduron (1972) recently observed that dopamine can be N-methylated to epinine using methyltetrahydrofolic acid (MTHF) as the methyl donor. We now report the methylation of several biogenic amines by mammalian and avian tissues by an enzyme for which MTHF appears to be the primary methyl donor. Both N- and O-methylation is mediated by MTHF.

Tissues from male rats (150 to 200 g), male rabbits (1.5 kg), and 5-day-old Leghorn chicks (mixed sex) were homogenized in 10 volumes of 0.005 M sodium phosphate buffer (pH 7.9) and the homogenates centrifuged at 100,000 × *g* for 60 min. Supernatant preparations were assayed with or without dialysis for 12 hr against 100 to 200 volumes of the same buffer. Incubation mixtures contained, at final concentrations, sodium phosphate buffer, pH 7.9 (0.005 M), amine substrate (5 mM), ^{14}C-MTHF (50 mC/mmole; New England Nuclear Corp.) (1 μM) or ^{14}C-AMe (50 mC/mmole; New England Nuclear Corp.) (1 μM) together with enzyme (0.5 to 2.0 mg protein/ml) in a final volume of 0.5 ml. After incubating for 30 to 60 min at 37°C, 1 ml of 0.5 M borate buffer, pH 10, was added and the radioactive product was extracted into 6 ml of an organic solvent, selected on the basis of the substrate utilized. After shaking for 10 min, the extract was centrifuged, a 5-ml portion of the organic phase was transferred to a counting vial, and the solvent evaporated to dryness at 80°C in a chromatography oven. After dissolving the residue in 2 ml of ethanol, 10 ml of toluene phosphor were added and the radioactivity was counted. Blank values, consisting of incubation mixtures in which the substrate amine was omitted, were subtracted routinely and gave values similar to blanks obtained by heating the enzyme preparation. Enzyme activity was linear for at least 60 min and with tissue concentration in the range utilized. Methylated products were identified by thin-layer chromatography in three solvent systems after evaporating the organic residue to dryness. In all cases the radioactive peaks coincided with methylated forms of the amines studied as substrates. The identity of the various methylated metabolites will be discussed below.

As reported previously (Axelrod, 1961, 1962), both undialyzed and dialyzed supernatant preparations of rabbit lung methylated serotonin and tyramine vigorously with AMe as the methyl donor (Table 1). By contrast, undialyzed rat lung, rat brain, and rabbit brain failed to methylate significantly either substrate with AMe, similar to previous results (Saavedra, Coyle, and Axelrod, 1973), while undialyzed chick brain did methylate amines with AMe, confirming the results of Morgan and Mandell (1969). However, except for rabbit lung, which methylated amines as efficiently when dialyzed as when not dialyzed, methylation by other tissues was greatly enhanced by dialysis with both AMe and MTHF as methyl donors.

Rabbit lung differed from other tissues in its methylating properties. Dialyzed supernatant preparations of rabbit lung methylated serotonin and tyramine four to six times more efficiently with AMe than with MTHF. By contrast, in all other tissues methylation of tyramine with MTHF was considerably greater than with AMe and no serotonin methylation could be detected with AMe. In all tissues examined, serotonin was methylated with MTHF 1.5 to 3 times more efficiently than tyramine.

In both rat and chick, the heart had the highest enzyme activity with MTHF. In rabbit, chick, and rat, brain had virtually the least activity of the tissues examined. Thus amine methylation with MTHF as the methyl donor is a ubiquitous enzymatic activity, with less activity in brain than in some other tissues. The rabbit lung enzyme differs from other tissues in methyl donor preference, but shows a similar amine substrate preference.

TABLE 1. Species andt tissue distribution of methyltransferase activity

Methyltransferase activity[a]

Tissue	AMe as methyl donor			MTHF as methyl donor			Ratio of enzyme activity with AMe to activity with MTHF	
	Serotonin	Tyramine	Ratio serotonin/ tyramine	Serotonin	Tyramine	Ratio serotonin/ tyramine	Serotonin	Tyramine
Rabbit lung	32.0	24.0	1.33	8.0	4.0	2.0	4.0	6.0
Brain	0	0.05	0	2.6	1.4	1.9	0	0.04
Liver	0	0.2	0	2.1	1.2	1.7	0	0.08
Rat brain	0	0.4	0	3.0	1.3	2.3	0	0.31
Liver	0	1.2	0	4.3	1.4	3.1	0	0.90
Lung	0	0.35	0	4.0	1.3	3.1	0	0.27
Heart	0	0.4	0	8.0	3.0	2.7	0	0.14
Chick brain	0	1.0	0	6.0	4.0	1.5	0	0.25
Heart	0	2.0	0	26.0	11.0	2.3	0	0.18

Tissues were homogenized in 10 volumes of 5 μM Na-phosphate buffer, pH 7.9, and enzyme activity was assayed in the 100,000 g supernatant fraction after dialysis with serotonin (5 mM) or tyramine (5 mM) as substrates and S-adenosylmethionine (AMe) (1 μM) or 5-methyltetrahydrofolic acid (MTHF) (1 μM) as methyl donors. Data are the mean of three experiments whose results varied less than 20%.

[a] nmoles/mg protein/hr.

The MTHF-requiring N-methylating enzyme was partially purified from rat brain (Snyder and Banerjee, 1974). In an initial fractionation with ammonium sulfate, the specific activity of fractions precipitating at 30 to 45% and 45 to 60% saturation was similar with AMe and MTHF as methyl donors and with tyramine and tryptamine as substrates. The fraction precipitating at 75% saturation and the resulting supernatant fraction failed to methylate tyramine or tryptamine with AMe, but methylated these amines vigorously with MTHF. The specific activity of the final supernatant fraction after 75% ammonium sulfate saturation was enhanced four- to fivefold over the original homogenate with tyramine and tryptamine as amine substrates and MTHF as methyl donor. This fraction, retaining 40% of the enzyme activity of the original tissue homogenate, was further purified by a negative absorption on alumina-C γgel, resulting in a further four- to fivefold enhancement of specific activity for methylation of tyramine and tryptamine with MTHF, but still with no methylating activity in the presence of AMe. Thus this two-step procedure resulted in a 20-fold purification of the enzyme with 25% recovery of total activity.

The K_m of the partially purified enzyme for MTHF was 10 μM, and

TABLE 2. *Substrate specificity of partially purified rat brain methyltransferase with MTHF as methyl donor*

Substrate	Specific Activity	Extraction procedure
Tyramine	4.4	1
Tryptamine	4.3	2
Serotonin	8.8	3
β-Phenylethylamine	2.0	4
Octopamine	2.7	1
N-Methyltryptamine	3.4	2
N-Methylserotonin	7.2	3
Desmethylimipramine	0.0	3
5-Methoxytryptamine	2.0	2
Bufotenin (N,N-dimethylserotonin)	8.4	3

The enzyme preparation used was the supernatant fraction after 75% ammonium sulfate precipitation. Enzymatic activity is expressed as nmoles/mg protein/hr. The following specific extraction procedures were used for the optimal isolation of a given product: (1) extracted with a mixture of toluene and isoamyl alcohol (3:2) and dried overnight at 80°C in a chromatography oven; (2) extracted with a toluene isoamyl alcohol mixture (97:3) and dried in oven; (3) extracted with isoamyl alcohol and dried in oven; (4) extracted as in (2) but counted with no previous drying procedure because of the volatility of the product.

its K_m for serotonin was 0.2 mM. The affinity constant for tyramine was about 1 mM whether MTHF or AMe served as methyl donor (Snyder and Banerjee, 1974). The purified enzyme was most active toward serotonin, bufotenin, and N-methylserotonin, which were methylated about twice as efficiently as tyramine or tryptamine (Table 2). Octopamine, 5-methoxytryptamine, and β-phenylethylamine were somewhat poorer substrates than tyramine and tryptamine. The relative activity of amine substrates was the same for the purified enzyme as for the original supernatant of rat brain homogenate.

Thin-layer chromatography revealed two distinct types of methylated metabolite after incubation with partially purified or initial supernatant preparations. With tryptamine as substrate, the labeled product was predominantly N-methyltryptamine, with only small amounts of material migrating in a similar manner to N,N-dimethyl-tryptamine (Fig. 1). By contrast, N-methylation of serotonin appeared to represent only a minor pathway, with the bulk of radiolabeled product present as 5-methoxytryptamine (Fig. 2). To confirm the existence of O-methylation of 5-hydroxyindoles, a reaction heretofore described only for melatonin synthesis in the pineal gland (Axelrod and Weissbach, 1961), we tested bufotenin, as a possible substrate for methylation. Bufotenin, the N,N-dimethylated derivative of serotonin, is not likely to be further N-methylated. However, bufotenin was almost as active a substrate as serotonin (Table 2). Moreover, thin-layer chromatographic analysis revealed that the product of bufotenin methylation with MTHF was N,N-dimethyl-5-methoxytryptamine (Fig. 3). The small amounts of radioactivity migrating in the vicinity of bufotenin might conceivably represent methyl exchange with ^{14}C-MTHF.

Whether N-methylating and O-methylating activities are mediated by the same enzyme is not clear. We compared the relative ability of four different ammonium sulfate fractions of the partially purified enzyme from rat liver to O-methylate bufotenin and to N-methylated 5-methoxytryptamine. The ratio of methylation of these two amines was the same in 0 to 30%, 30 to 45%, 45 to 60%, and 60% ammonium sulfate fractions. This suggests that the two different types of methylation might involve the same enzyme.

Our studies of the influence of AMe- on MTHF-mediated methylation argue in favor of distinct enzymes for methyl donation by AMe and MTHF, respectively. AMe in concentrations ranging from 5×10^{-6} M to 10^{-4} M did not alter MTHF-mediated methylation of serotonin by supernatant preparations of rat liver. The failure of serotonin to function as a substrate with AMe also suggests the existence of two enzymes with different methyl donors. O-methylation might require MTHF, while MTHF and AMe could both serve as donors for N-methylation.

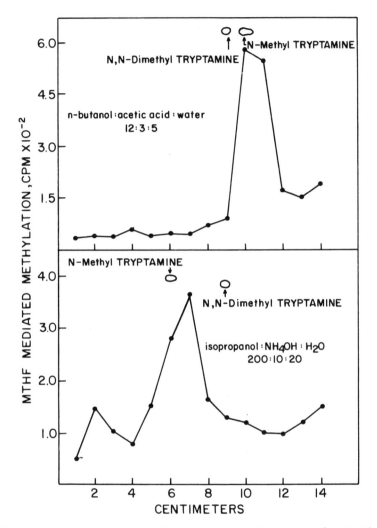

FIG. 1. Thin-layer chromatographic analysis of products of incubation of tryptamine with ^{14}C-MTHF and a partially purified methyltransferase from rat liver. The incubation condition is described in the text and was for a period of 60 min. The radioactive methylated product was extracted into water-saturated isoamyl alcohol. After shaking and centrifuging, the organic phase was evaporated to dryness. Five samples were pooled by redissolving in 2 ml of ethanol. One hundred microliters of the ethanol extract was employed for thin-layer chromatography in the indicated solvent system.

In most tissues the enzyme is much more active with MTHF than with AMe, suggesting that MTHF may be the normal methyl donor. AMe has been generally regarded as a universal methyl donor, especially in methylations of biogenic amines. Our findings and those of Laduron emphasize that MTHF may be an important biological methyl donor. In

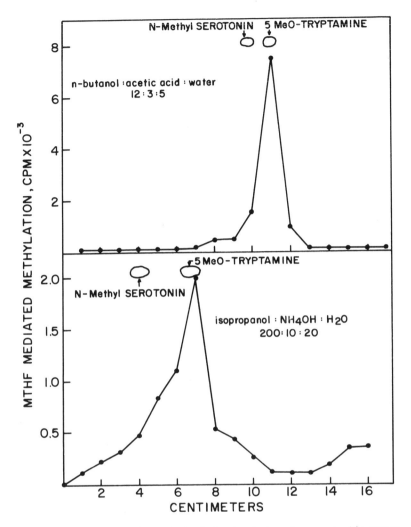

FIG. 2. Identification of the products obtained after incubation of serotonin, ^{14}C-MTHF, and partially purified rat liver methyltransferase by thin-layer chromatography. The procedure is described in Fig. 1. 5-Methoxytryptamine is designated MeO-tryptamine in the figure.

other experiments, we have found that MTHF can serve as the methyl donor for histamine methyltransferase from rat brain and hydroxyindole-N-methyltransferase from bovine pineal, although in both cases AMe is about 50 times as efficient a methyl donor.

Our demonstration of methylation of biogenic amines by MTHF may have bearing upon amine methylation theories of mental illness. Although the enzyme has broad substrate specificity, serotonin, which

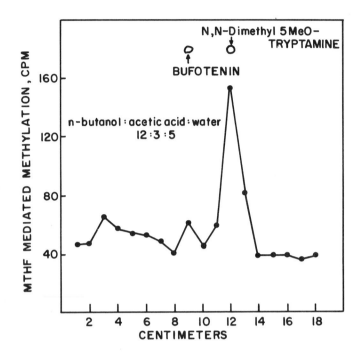

FIG. 3. Methylation of bufotenin by rat liver methyltransferase in the presence of ^{14}C-MTHF. The incubation and thin-layer chromatographic procedures are described in Fig. 1.

is widely implicated in mental illness, is the best substrate. Interestingly, enzymatic activity is greatest in tissues other than the brain. A physiological role for this enzyme is suggested by the recent demonstration of significant amounts of 5-methoxytryptamine in mammalian hypothalamus (Green, Koslow, and Costa, 1973). The enzymatic formation of 5-methoxy-N, N-dimethyltryptamine is of interest, because this compound appears to be a potent psychedelic agent (Holmsted and Lindgren, 1967).

ACKNOWLEDGMENTS

Supported by U.S. Public Health Service grants MH-18501, NS-07275, DA-00246, grants of the John A. Hartford Foundation, the Scottish Rite Foundation, Research Scientist Development Award MH-33128 to S.H.S., and a Canadian Medical Research Council Fellowship to S.P.B.

REFERENCES

Axelrod, J. (1961): Enzymatic formation of psychotomimetic metabolites from normally occurring compounds. *Science,* 134:343–344.

Axelrod, J. (1962): The enzymatic N-methylation of serotonin and other amines. *Journal of Pharmacology and Experimental Therapeutics,* 138:28–33.

Axelrod, J., and Weissbach, H. (1961): Purification and properties of hydroxyindole-O-methyltransferase. *Journal of Biological Chemistry,* 236:211–213.

Green, A. R., Koslow, S. H., and Costa, E. (1973): Identification and quantitation of a new indolealkylamine in rat hypothalamus. *Brain Research,* 51:371–374.

Holmstedt, B., and Lindgren, J.-E. (1967): Chemical constituents and pharmacology of South American snuffs. In: *Ethnopharmacologic Search for Psychoactive Drugs* edited by D. H. Efron, pp. 339–373. U.S. Public Health Service, Washington, D.C.

Laduron, P. (1972): N-Methylation of dopamine to epinine in brain tissue using N-methyltetrahydrofolic acid as the methyl donor. *Nature New Biology,* 238:212–213.

Mandell, A. J., and Morgan, M. (1971): Indole(ethyl)amine N-methyltransferase in human brain. *Nature New Biology,* 230:85–87.

Mandell, A. J., and Spooner, C. E. (1968): Psychochemical research studies in man. *Science,* 162:1442–1453.

Morgan, M., and Mandell, A. J. (1969): Indole(ethyl)amine N-methyltransferase in the brain. *Science,* 165:492–493.

Pollin, W., Cardon, P. V., Jr., Kety, S. S. (1961): Effects of amino acid feedings in schizophrenic patients treated with iproniazid. *Science,* 133:104–105.

Saavedra, J. M., and Axelrod, J. (1972): Formation of psychotomimetic N-methylated tryptamines in brain *in vivo* and *in vitro. Science,* 175:1365–1366.

Saavedra, J. M., Coyle, J. T., and Axelrod, J. (1973): The distribution and properties of the non-specific N-methyltransferase in brain. *Journal of Neurochemistry,* 20:743–752.

Snyder, S. H., and Banerjee, S. P. (1974): Amine in schizophrenia. *Journal of Neurochemistry (in press).*

Szara, S. (1956): Dimethyltryptamine: Its metabolism in man; the relation of its psychotic effect to the serotonin metabolism. *Experientia,* 12:441–442.

Tanimukai, H., Gunther, R., Spaide, J., Bueno, J. R., and Himwich, H. E. (1970): Detection of psychotomimetic N,N-dimethylated indoleamines in the urine of four schizophrenic patients. *British Journal of Psychiatry,* 117:421–430.

Advances in Biochemical Psychopharmacology, Vol. 11
Raven Press, New York © 1974

5-Methoxytryptamine: A Possible Central Nervous System Transmitter

Stephen H. Koslow

Laboratory of Preclinical Pharmacology, National Institute of Mental Health, Saint Elizabeths Hospital, Washington, D.C. 20032

I. INTRODUCTION

In the past few years, evidence has been published suggesting that serotonin is not the only centrally active indolealkylamine. Microspectrofluorometic studies in the rat have shown that there is a new intraneuronal monoamine with cell bodies in the raphe region having fibers descending to the spinal cord (Bjorklund, Falck, and Stenevi, 1970) and ascending through the mesencephalon and hypothalamus (Bjorklund, Falck, and Stenevi, 1971a). These neurons were termed B-type and were differentiated from A-type (catecholamine-containing) and C-type (serotonin-containing) on the basis of their fluorescent excitation and emission maxima and rate of photodecomposition. The B-type fluorophor has spectral and histochemical properties of an indolylethylamine, similar to 6-hydroxytryptamine, 5,6-dihydroxytryptamine, N-methyl-5-hydroxytryptamine, and 5-methoxytryptamine (Bjorklund, Falck, and Stenevi, 1971b).

Although the B-type fluorophor appears to be an indole, its biosynthetic pathway and storage mechanism seem to be different from that of serotonin. With *p*-chlorophenylalanine *(p*-CPA), brain serotonin levels are markedly decreased by inhibiting tryptophan-5-hydroxylase, the first step in the synthesis of serotonin from tryptophan. This treatment does not notably affect the B-type fluorophor in the cells of the raphe nucleus (Bjorklund et al., 1971a). Similarly, Aghajanian and Asher (1971) have shown that following L-tryptophan loading, there is a biochemically measurable increase in whole-brain serotonin, but the only fluorescent increase is seen in the axons of the raphe neurons. If *p*-CPA was given prior to the L-tryptophan, then whole-brain serotonin levels significantly decreased as expected; there was, however, no change in the L-tryptophan induced intensity of the fluorescence of the raphe cells. Drugs such as norfenfluramine and p-chloramphetamine have also been shown to reduce serotonin concentrations in several brain areas, but they are relatively ineffective in lowering the serotonin concentrations of the hypothalamus (Costa and Revuelta, 1972a,b), the area to which the raphe system has its ascending projection. Although there are many possible explanations for some of these findings, one plausible suggestion is that the substance being measured as serotonin in the hypothalamus is not serotonin but some other compound.

To test the hypothesis of the presence of a new indole in the central nervous system, we analyzed the hypothalamus for various indolealkyl-amines (Green, Koslow, and Costa, 1973). The hypothalamus was chosen since it is easily and reproducibly dissected from the rat brain and is known to contain B-type neurons as described by Bjorklund et al. (1971a). In these studies, it is of utmost importance that the method utilized for quantitation have a high degree of specificity. A highly specific method, having a sensitivity greater than 1 pmole, is obtained by the use of gas chromatography-mass spectrometry (GC-MS). This technique has been applied to the measurement of catecholamines (Koslow, Cattabeni, and Costa, 1972) and indolealkylamines (Catta-beni, Koslow, and Costa, 1972). The specificity of the measurement is based on the GC retention time of the compound and structural verification of the compound by multiple ion detection (MID).

II. METHODOLOGY

The compounds measured were serotonin (S), N-acetylserotonin (NAS), 5-methoxytryptamine (5-MT), and melatonin (M). All of these compounds are known to be localized in the pineal gland, whereas S is known to be present in the central nervous system and gastrointestinal tract as well. The indoles are extracted by homogenizing the tissue with 0.1 M $ZnSO_4$ followed by neutralization with 0.1 M $Ba(OH)_2$. In brief, an aliquot of the supernatant is dried down in the presence of internal standards (α-methylserotonin, α-MS; N-acetyltryptamine, NAT) and reacted for 3 hr with pentafluoropropionic anhydride. After drying, the residue is reconstituted in ethyl acetate and an aliquot injected into the GC port of an LKB 9000 GC-MS. The indoles are analyzed as the pentafluoropropionyl (PFP) derivatives, thereby giving all of these compounds adequate vapor pressure for GC. The internal standards are used to give reproducible quantitation; α-MS is the internal standard for S and 5-MT, while NAT is the internal standard for NAS and M.

For quantitation, the intensity of the base peak (most abundant fragment, 100%) is recorded at the compound's GC retention time (Table 1). The specificity of the measurement is based on a partial structural reconstruction of the compound being measured. This is

TABLE 1. *Gas chromatographic (GC) retention times and positive ions monitored for quantitation of the five substituted indole alkylamine pentafluoropropionyl (PFP) derivatives*

Indole PFP derivative	GC retention time (min)	Positive ion recorded for quantitation (m/e)[a]
α-Methylserotonin	2.7	465
Serotonin	3.4	451
N-Acetylserotonin	4.5	492
N-Acetyltryptamine	5.8	330
5-Methoxytryptamine	8.0	306
Melatonin	12.3	360

[a]m/e, mass-to-charge.

TABLE 2. *Specificity of GC-MS measurement as determined by multiple ion detection (MID)*

Indole PFP derivative	m/e Positive ions monitored		Authentic compound	Hypothalamus extract	Blood extract
Serotonin	451	(100)	2.3	2.3	2.3
	438	(44)			
5-Methoxytryptamine	306	(100)	1.0	1.0	1.0
	319	(96)			
N-Acetylserotonin	492	(100)	14.3	—	—
	473	(96)			
Melatonin	213	(40)	2.1	—[a]	—[a]
	198	(19)			

[a]Because of the low concentrations of melatonin present, MID could not verify the structure of melatonin. Thus specificity is based only on the detection of the molecular ion (m/e 360) at the GC retention time of authentic melatonin. MID was not done for the internal standards. Tissue samples were, however, analyzed in the absence of internal standards and showed no "biological background" at the retention times and m/e of the internal standards.

accomplished by measuring the relative abundance of two or more characteristic fragments (m/e) at the compound's GC retention time (Table 2). The ratios of the ion density of two or more fragments measured in the tissue extract are compared to the fragment ratio obtained from authentic standards; if there is agreement, then the measurement is taken as being specific. A more complete description of the methodology has been published (Koslow, Cattabeni, and Costa, 1973; Koslow and Green, 1973).

III. RESULTS

A. Hypothalamic Indolealkylamines

Analysis of hypothalamic extracts by MID verified the presence of S, and also for the first time demonstrated the existence of 5-MT and M in this area (Table 2) (Green et al., 1973). MID analysis for S and 5-MT revealed the presence of fragment ratios typical of these compounds at their specific retention times (2.3 and 1.0, respectively.) A fragment ratio was not obtained for M because of the low concentrations present, and thus its identification is only tentative and based on the recording of the molecular ion for melatonin (PFP) (m/e 360) at the GC retention time for authentic melatonin. These compounds have also been identified in a similar manner in whole rat blood.

Since it has previously been believed that these compounds (5-MT and M) are localized only in the pineal gland, it is possible that their presence in the hypothalamus merely reflects the uptake of these compounds either from the blood or CSF. To test this possibility, animals were pinealectomized and the indole content of the hypothalamus measured 1 month later. As shown in Table 3, pinealectomy does not significantly alter the hypothalamic concentration of these indoles. These results cannot be taken as conclusive, since it has been suggested by Bjorklund, Owman, and West (1972) that following pinealectomy there may be functionally significant structures remaining in the region of the lamina intercalaris, which may still be a source of these indoles.

Both the origin and the functional significance of these compounds in the hypothalamus remain to be established. It is interesting to speculate that these compounds might function as neurotransmitters. Supporting this is the finding that 5-MT, when applied iontophoretically to brainstem neurons, mimics the depolarizing action of S (Bradley, 1972). Also, the microspectrofluorometric studies of Bjork-

TABLE 3. *Concentration of hypothalamic indoles in normal and pinealectomized rats*

	Control (8)	Pinealectomized (3)
Serotonin	3.96 ± 0.23	3.63 ± 0.51
N-Acetylserotonin	—[a]	—[a]
5-Methoxytryptamine	0.62 ± 0.02	0.46 ± 0.07
Melatonin	$1.5 \ \pm 0.16$	$1.4 \ \pm 0.20$

Results are expressed in nanamoles per gram (Mean ± SE). Number of determinations shown in brackets. Pinealectomy was performed 21 to 28 days before analysis.

[a]Not detectable.

lund, Falck, and Stenevi (1971*b*) indicate that the unknown mesencephalic indole may be 5-MT.

IV. CONCLUSIONS

Although there are many questions to be answered about these compounds in the brain, the fact that they are present must be considered when analyzing S content. Particular caution should be used if S is assayed by either the OPT method (Maickel, Cox, Saillant, and Miller, 1968) or the radioimmunoassay described by Peskar and Spector (1973). With the OPT method all five substituted indoles have identical spectral characteristics, and in the case of 5-MT there is significant contribution to the S value (Green et al., 1973). In using the radioimmunoassay, 5-MT is almost as effective as S in inhibiting ^{3}H-serotonin binding. In view of the lack of knowledge we have regarding the presence or absence of compounds in the central nervous system, it is necessary to pursue many of the problems involving brain function with techniques that offer a high degree of specificity to the measurement.

In summary, 5-MT is present in the rat hypothalamus. M has also been identified in this area, but analysis by MID must be accomplished to say this with any high degree of certainty. The 5-MT may be the same interneuronal amine described by Bjorklund et al. (1971*a*), but before this can be claimed with any certainty, a subcellular distribution study must be done on hypothalamic 5-MT and the indole content of the raphe nucleus must be determined.

REFERENCES

Aghajanian, G. K., and Asher, I. M. (1971): Histochemical fluorescence of raphe neurons: Selective enhancement by tryptophan. *Science*, 172:1159–1161.

Bjorklund, A., Falck, B., and Stenevi, V. (1970): On the possible existence of a new intraneuronal monoamine in the spinal cord of the rat. *Journal of Pharmacology and Experimental Therapeutics*, 175:525–532.

Bjorklund, A., Falck, B., and Stenevi, V. (1971*a*): Classification of monoamine neurons in the rat mesencephalon: Distribution of a new monoamine neuronal system. *Brain Research*, 32:269–285.

Bjorklund, A., Falck, B., and Stenevi, V. (1971*b*): Microspectrofluorimetric characterization of monoamines in the central nervous system: Evidence for a new neuronal monoamine-like compound. In: *Histochemistry of Nervous Transmission, Progress in Brain Research*, Vol. 34, pp. 63–73. Elsevier, Amsterdam.

Bjorklund, A., Owman, C. H., and West, K. K. (1972): Peripheral sympathetic innervation and serotonin cells in the habenular regions of the rat brain. *Zeitschrift für Zellforschung und Mikroskopische Anatomie*, 127:570–579.

Bradley, P. B. (1972): Mechanism of action of psychotomimetic drugs. *Fifth International Congress on Pharmacology*, San Francisco, Abstract 167.

Cattabeni, F., Koslow, S. H., and Costa, E. (1972): Gas chromatography-mass spectrometric assay of four indolealkylamines of rat pineal. *Science*, 178:166–168.

Costa, E., and Revuelta, A. (1972a): Norfenfluramine and serotonin turnover rate in the rat brain. *Biochemical Pharmacology*, 21:2385–2393.

Costa, E., and Revuelta, A. (1972b): (−)-p-Chloroamphetamine and serotonin turnover in rat brain. *Neuropharmacology*, 11:291–295.

Green, A. R., Koslow, S. H., and Costa, E. (1973): Identification and quantitation of a new indole alkylamine in rat hypothalamus. *Brain Research*, 51:371–374.

Koslow, S. H., Cattabeni, F., and Costa, E. (1972): Norepinephrine and dopamine: Assay by mass fragmentography in the picomole range. *Science*, 176:177–180.

Koslow, S. H., Cattabeni, F., and Costa, E. (1973): Quantitative mass fragmentography of some indole alkylamines of the rat pineal gland. In: *Pineal Gland: Proceedings of a Workshop*, edited by D. Klein. Spectrum Press, New York.

Koslow, S. H., and Green, A. R. (1973): Analysis of pineal and brain indole alkylamines by gas chromatography-mass spectrometry. In: *Advances in Biochemical Psychopharmacology*, Vol. 7, edited by E. Costa and B. Holmstedt, pp. 33–43. Raven Press, New York.

Maickel, R. P., Cox, R. H., Saillant, J., and Miller, F. P. (1968): A method for the determination of serotonin and norepinephrine in discrete areas of rat brain. *International Journal of Neuropharmacology*, 7:275–281.

Peskar, B., and Spector, S. (1973): Serotonin; radioimmunoassay. *Science*, 179:1340–1341.

Advances in Biochemical Psychopharmacology, Vol. 11
Raven Press, New York © 1974

Serotonin-Binding Proteins Isolated by Affinity Chromatography

Jean Chen Shih, Samuel Eiduson, Edward Geller*, and E. Costa**

*The Neuropsychiatric Institute and Brain Research Institute, Departments of Psychiatry and Biological Chemistry, School of Medicine, UCLA, Los Angeles, California 90024, *Neurobiochemistry Laboratory, Veterans Administration Brentwood Hospital, Los Angeles, California 90073, and **Laboratory of Preclinical Pharmacology, Saint Elizabeths Hospital, Washington, D.C. 20032*

Transmitters released into the synaptic junction are presumed to exert their effects at specific receptor sites on the postsynaptic membrane. In addition, some transmitters are known to be inactivated by reuptake through the presynaptic membrane, presumably also on specific receptor sites. It is widely believed that there should be such specific membrane receptor sites for serotonin within those areas of the nervous system where serotonin may be acting. Two of these areas are the hypothalamus and the spinal cord.

Techniques for isolating pinched-off nerve endings, or synaptosomes, are now well developed and offer several advantages in attempting to isolate these membrane receptors. Mainly, these advantages are (1) the relative enrichment of such sites by the elimination of the bulk of the perikaryal proteins, and (2) when suitable conditions are employed, the postsynaptic membrane generally remains attached to the isolated synaptosome. This chapter describes our attempts to isolate the serotonin receptor site using the technique of affinity chromatography and elution with two drugs which pharmacological evidence suggests compete with serotonin for its receptor site.

Synaptosomal membranes were prepared from hypothalamus and spinal cord as described by Whittaker and Barker (1972). Synaptosomes were lysed in water, centrifuged at 12,000 X g for 30 min, and the supernatant suspension put on a discontinuous gradient of 0.4, 0.6, 0.8, 1.0, and 1.2 M sucrose. The three distinct bands that separated at the 0.4 M to 0.6 M, 0.6 M to 0.8 M, and 0.8 to 1.0 M interfaces were pooled and used as the source of membranes.

These membranes were incubated at pH 7.4 in the presence or absence of 1% Triton X100 for 40 min at 20°C with ^{14}C-serotonin Sepharose (5 mg membrane protein to 5 ml 8.5 X 10^{-4} M ^{14}C-serotonin Sepharose) prepared via a diazonium derivative as described by

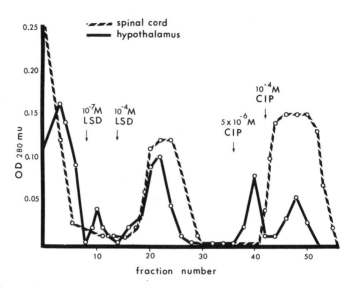

FIG. 1. The protein separation pattern of synaptosomal membrane obtained with drug elution from a serotonin-affinity column. Each protein peak was eluted by various concentrations of drugs as indicated. See text for details.

Cuatrecasas (1970). At the end of the incubation period the mixture was loaded into a 1.2 × 5 cm column and washed with 30 ml of 0.05 M phosphate buffer, pH 7.4, until the absorbance of the eluate at 280 nm was zero. The column was eluted first with either lysergic acid diethylamide (LSD) at concentrations of 10^{-7} to 10^{-4} M or with chlorimipramine (CIP) at the same concentrations. When no further protein was removed by the first drug, the column was then eluted with the other drug.

One such elution pattern is shown in Fig. 1. Proteins that did not bind to the Sepharose appeared in the void volume and in the buffer wash. It is clear that certain proteins were eluted with the different concentrations of the drugs. The CIP-eluted protein fraction (derived either from hypothalamus or spinal cord) migrated to the same area during gel electrophoresis when the CIP elution was used either before or after the LSD. In the hypothalamus proteins were seen following each of the elution concentrations, but in the spinal cord the protein was eluted apparently only by the higher concentration (10^{-4} M) of drugs. In relation to the total protein applied to the Sepharose, a higher percentage of bound protein was seen in the hypothalamus.

The protein fractions eluted from the column by 10^{-4} M LSD and 10^{-4} M CIP were pooled, concentrated by lyophilization, and run on disc gel electrophoresis at pH 4.5 (Reisfeld, Lewis, and Williams, 1962) and pH 8.3 (Davis, 1964). A single protein band was observed in each case (Fig. 2). The protein eluted by LSD migrated differently from the protein eluted by CIP at both pH values. Identical results were obtained when membranes were incubated in the presence of Triton X100.

To characterize further the specificities of these proteins, adult rats were injected with 6-OH-dopamine in the lateral ventricle as described by Bloom, Algeri, Groppetti, Revuelta, and Costa (1969) and the serotonin-binding proteins were isolated by the same methods reported above. In this experiment, the concentration of brain norepinephrine in the drug-treated animals was 50% lower than in the controls, while the concentration of serotonin was the same in both. No differences were seen in elution patterns or in gel electrophoresis of the proteins derived from 6-hydroxydopamine-treated animals compared to control animals. These observations suggest that the serotonin-binding proteins may not be located in the membranal system destroyed by 6-OH-dopamine.

In summary, we have isolated two serotonin-binding proteins using a serotonin-affinity column. One of these proteins was eluted by 10^{-4} M LSD and the other by 10^{-4} M CIP. Since these two proteins have

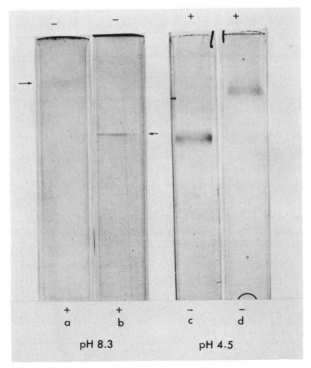

FIG. 2. The gel electrophoretic pattern of serotonin-binding proteins eluted from the affinity column by 10^{-4} M CIP (a,c) and by 10^{-4} M LSD (b,d). The application of sample for a and b is at the cathode while the application for c and d is at the anode.

different electrophoretic mobilities, and since the protein patterns were not changed by the order of elution, it strongly suggests to us that they are two different proteins. It is tempting to speculate also that perhaps the protein eluted by CIP may be presynaptic while the protein eluted by LSD may be located postsynaptically. These data by themselves, of course, cannot resolve this question.

Since 6-OH-dopamine did not alter the concentration of these proteins, it suggests that the serotonin-binding proteins may not be located in the membranal system which is destroyed by 6-OH-dopamine.

ACKNOWLEDGMENTS

We are grateful to Professor Peter Lomax (Department of Pharmacology, School of Medicine, The Center of the Health Sciences, UCLA) for help with the injections of 6-OH-dopamine. The skillful technical assistance of Stephen Huyett, Janet Seager, Catherine Campbell, and Carol J. Siporin is also gratefully acknowledged.

This work was supported in part by National Institute of Mental Health grant MH19734, National Institutes of Health grant HD04364, and The Grant Foundation.

REFERENCES

Bloom, F. E., Algeri, S., Groppetti, A., Revuelta, A., and Costa, E. (1969): Lesions of central norepinephrine terminals with 6-OH-dopamine: Biochemistry and fine structure. *Science*, 166:1284–1286.

Cuatrecasas, P. (1970): Protein purification by affinity chromatography, derivatives of agarose and polyacrylamide beads. *Journal of Biological Chemistry*, 245:3059–3065.

Davis, B. J. (1964): Disc electrophoresis—II. Method and application to human serum proteins. *Annals of the New York Academy of Sciences*, 121:404–427.

Reisfeld, R. A., Lewis, U. J., and Williams, D. E. (1962): Disk electrophoresis of basic proteins and peptides on polyacrylamide gels. *Nature*, 195:281–283.

Whittaker, V. P., and Barker, L. A. (1972): The subcellular fractionation of brain tissue with special reference to the preparation of synaptosomes and their component organelles. In: *Methods of Neurochemistry*, Vol. 2, edited by Rainer Fried. Marcel Dekker, New York.

Advances in Biochemical Psychopharmacology, Vol. 11
Raven Press, New York © 1974

Affinity Chromatography for Subfractionation of 5-Hydroxytryptamine-, LSD-Binding Proteins from Cebrebral and Nerve-Ending Membranes

E. Mehl and L. Weber

*Max-Planck-Institute of Psychiatry, Division of Neurochemistry,
Munich, West Germany*

In this volume many examples have been presented of synergistic and antagonistic effects of LSD on 5-hydroxytryptamine (5-HT) systems. We have obtained evidence that LSD and 5-HT exert their action on identical macromolecules. Six fractions, isolated by affinity chromatography, bound 5-HT as well as LSD. The affinity gel contained a covalently coupled LSD analogue. Fiszer and De Robertis (1969) purified a 5-HT-binding proteolipid fraction by the method of organic solvent extraction and gel permeation chromatography.

I. EXPERIMENTAL

The ligand of the affinity gel was synthesized according to the method of Garbrecht (1959). The yield was 0.45 g of *iso*-D- and D-lysergic acid diaminohexane amide (LDA), when using 3.5 mmoles of lithium D-lysergate and 25 mmoles of 1,6-diamino-*n*-hexane. LDA was purified on a 120-ml column of neutral Al_2O_3 with *n*-butanol/ethanol/water, 16:3:1 (by volume). Impurities were not detectable on thin-layer chromatography (silica gel; ethyl acetate/*n*-propanol/25% NH_3, 9:7:4, by volume). As determined by both ninhydrin reaction and fluorimetry with LSD as reference, the LDA contained 1 mole of free amino group per mole of lysergic acid amide.

The LDA was coupled via the free amino group to the succinylated N,N'-diaminodipropylamine derivative of Sepharose 4B prepared by the method of Cuatrecasas (1970) using N-cyclohexyl-N'-[β-(N-methyl-morpholinio)-ethyl]-carbodiimide *p*-toluene sulfonate. To establish the extent of coupling by a modified Salkowski reaction, samples of 50 mg of settled gel, or succinylated gel blank, were mixed with 70 μl of 60% perchloric acid and 2 μl of 0.5 M $FeCl_3$, incubated for 30 min at 45°C, and assayed by absorbance at 475 nm. LDA concentration in the settled gel was 0.4 mM.

The cerebral cortex of mini pigs was dissected out for isolation of nerve-ending membranes (NM) by zonal rotor centrifugation (Cotman, Mahler, and Anderson, 1968). For comparison, a crude particulate fraction (CM) was prepared by homogenizing the cortex tissue in ten volumes water followed by centrifugation at $100,000 \times g$ for 2 hr. The proteins were extracted at $22°C$ while being stirred. One milliliter of 0.2% sodium dodecyl sulfate (SDS) in 50 mM sodium phosphate, pH 7.3, was used per 2.5 mg of protein. After centrifugation at $100,000 \times g$ for 2 hr, supernatants were first filtered through blank gels, containing the succinylated gel derivative without the LDA ligand, and then passed through the affinity columns (1.5 g of LDA gel) at 5 ml/hr. Further conditions are described under Results. The eluted 2.7-ml fractions were dialyzed against seven 20 l-changes of buffer (0.02% NaN_3, 0.1% SDS, 0.1 M NaCl in 66 mM sodium phosphate, pH 7.3) for binding assay.

II. RESULTS AND DISCUSSION

Since desoxycholate and water were ineffective in extracting the bulk of 5-HT-binding proteins from nerve-ending membranes, SDS was used. With this detergent, a 50% solubilization was achieved and the total binding capacity of the membranes preserved. When 40 mg of NM proteins were applied, the affinity gel initially retained 0.4 mg of protein. The eluents (a–h) contained standard buffer (150 mM NaCl, 50 mM Tris-HCl, pH 7.4, and 0.1% SDS) plus various additions: 0.6 M NaCl (eluent b, 100 ml), 2.0 M urea (eluent c, 100 ml), 5 mM D,L-tryptophan (eluent d, 220 ml). The eluents (a–d) eluted 0.2 mg of nonspecifically adsorbed proteins from the column. After elution with the specific eluents (e–h) and dialysis, the binding capacity of the fractions was verified (Fig. 1). The binding fractions obtained by the highly specific eluents (indicated by arrows) were six to seven times more concentrated in synaptic membranes (NM) than in CM. Fulfilling the condition of pharmacologic 5-HT, LSD-antagonism, all the binding fractions bound 5-HT as well as LSD (Table 1) in a reversible manner. From 20 g of cortex, 1 μg of the LSD-eluted synaptic protein was obtained with subunits of 70,000 and 80,000 daltons (by SDS gel electrophoresis). Before designating the two synaptic 5-HT, LSD-binding proteins as "5-HT receptors," biophysical bilayer techniques must be applied.

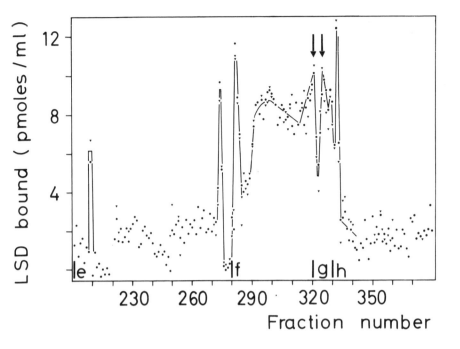

FIG. 1. Profile of ³H-LSD-binding proteins after affinity chromatography of extracts from nerve-ending membranes. The fractions, 201 to 380, specifically eluted by a linear gradient of 0.1 to 30 mM 5-HT (eluent e), 1 mM 1-methyl-D-lysergic acid butanolamide (eluent f), 0.7 mM LSD (eluent g), and 2 M urea, pH 10 (eluent h), were exhaustively dialyzed and assayed in 400-μl ultrafiltration cells (Paulus, 1969) in the presence of 0.012% SDS and 7 × 10⁻⁹ M ³H-LSD. (The sample contained 4000 cpm; maximal binding was 1100 cpm; filter blank was 80 to 120 cpm.)

TABLE 1. *5-HT-binding capacity of LSD-binding proteins*

| Fraction number | Nerve-ending membranes | | Crude membranes |
	5-HT-binding (pmoles/ml)	Binding ratio[a] (5-HT/LSD)	Binding ratio (5-HT/LSD)
274	8.9 ± 3.1	1.0 ± 0.2	1.4 ± 0.1
282	10.5 ± 3.0	1.0 ± 0.3	1.9 ± 0.2
298	11.3 ± 2.7	1.3 ± 0.3	0.9 ± 0.1
319	12.8 ± 0.5	1.2 ± 0.1	(0.5 ± 0.1)
325	14.8 ± 0.7	1.5 ± 0.1	(0.8 ± 0.2)
332	19.2 ± 3.9	1.5 ± 0.3	0.9 ± 0.1

[a] Binding ratio: pmoles of ³H-5-HT/ml to pmoles of ³H-LSD/ml; 7.0 × 10⁻⁹ M LSD and 2.0 × 10⁻⁸ M 5-HT, respectively, were used. Assay was done in duplicate or triplicate.

ACKNOWLEDGMENTS

We gratefully acknowledge the technical assistance of Lisa Guiard, Jutta Redemann, and Christian Suchanek, and the generosity of the Sandoz AG in supplying LSD and methysergide. The work was sponsored by Deutsche Forschungsgemeinschaft (Sonderforschungs-bereich 51).

Part of this work will be presented in the doctoral dissertation of L. Weber.

REFERENCES

Cotman, C. W., Mahler, H. R., and Anderson, N. G. (1968): Isolation of a membrane fraction enriched in nerve-end membranes from rat brain by zonal centrifugation. *Biochimica et Biophysica Acta,* 163:272–275.

Cuatrecasas, P. (1970): Protein purification by affinity chromatography. *Journal of Biological Chemistry,* 245:3059–3065.

Fiszer, S., and De Robertis, E. (1969): Subcellular distribution and chemical nature of the receptor for 5-hydroxytryptamine in the central nervous system. *Journal of Neurochemistry,* 16:1201–1209.

Garbrecht, W. L. (1959): Synthesis of amides of lysergic acid. *Journal of Organic Chemistry,* 24:368–72.

Paulus, H. (1969): A rapid and sensitive method for measuring the binding of radioactive ligands to proteins. *Analytical Biochemistry,* 32:91–100.

Advances in Biochemical Psychopharmacology, Vol. 11
Raven Press, New York © 1974

Comparative Studies of Brain 5-Hydroxytryptamine and Tryptamine

P. J. Knott, C. A. Marsden, and G. Curzon

*Department of Neurochemistry, Institute of Neurology,
Queen Square, London, England*

Physiological changes of tryptophan concentration influence brain 5-hydroxytryptamine (5-HT) metabolism (Curzon, Joseph, and Knott, 1972; Perez-Cruet, Tagliamonte, Tagliamonte, and Gessa, 1972). In addition, it is now known that the brain contains other indolealkylamines as well as 5-HT—tryptamine in particular (Saavedra and Axelrod, 1972). These findings stimulated us to make comparative studies of brain 5-HT and tryptamine metabolism, paying particular attention to regional distribution and to effects of altered brain tryptophan concentration and brain lesions.

I. METHODS

Male Sprague-Dawley rats (weight 180 to 220 g) were killed by guillotine and their brains divided into regions (Glowinski and Iversen, 1966). Tryptophan, 5-HT, and 5-hydroxyindoleacetic acid (5-HIAA) were determined as described by Curzon et al. (1972) and tryptamine by the method of Saavedra and Axelrod (1972). Electrolytic lesions in dorsal and medial raphe nuclei were made stereotaxically using coordinates given by Pellegrino and Cushman (1967); frontal 0.4 mm, sagittal 0.0 mm, horizontal 0.6 and 2.5 mm.

II. RESULTS

A. *Regional Distribution of Tryptophan and Indole Metabolites*

Regional distributions found for tryptophan, 5-HT, 5-HIAA and tryptamine are shown in Fig. 1. Tryptophan (Fig. 1a) was distributed essentially uniformly except for the hypothalamus, where its concentra-

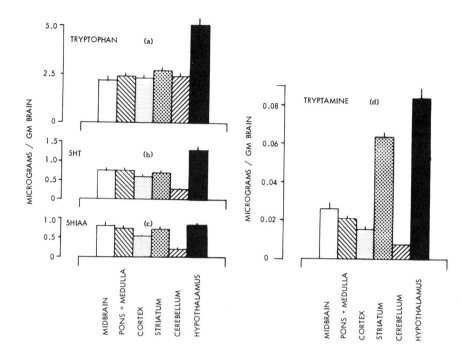

FIG. 1. Histograms (a), (b), and (c) show the regional distribution of tryptophan, 5-HT and 5-HIAA, respectively, in the brains of 17 untreated rats. All three determinations were made on each individual region. Histogram (d) shows the regional distribution of tryptamine, which was determined on a separate group of 25 rats by pooling regions from five animals for each determination so that five determinations were made for each region (except for the cerebellum, where only two determinations were made). Values are given in micrograms per gram wet weight and are means ± SE.

tion was found to be consistently higher than in other brain regions. 5-HT (Fig. 1b) and tryptamine (Fig. 1d) showed similarities of distribution, with highest levels in the hypothalamus and lowest in the cerebellum. Concentrations of the two amines in the midbrain and pons + medulla were not markedly different. The striatum showed a striking difference between amines; here tryptamine but not 5-HT concentration was higher than in all other regions except hypothalamus. 5-HIAA distribution (Fig. 1c) followed that of 5-HT except that hypothalamic 5-HIAA concentration was not outstandingly high.

Since distribution patterns of 5-HT and tryptamine were not identical, the results suggested that the latter was not synthesized specifically within serotoninergic neurons. To study this question further, electrolytic lesions were placed in the dorsal and medial raphe nuclei of 14 rats to destroy ascending serotoninergic fibers. Ten days later, although brain 5-HT was reduced about 70%, tryptamine was reduced only 30% relative to sham-operated control rats. Both lesioned and sham-operated rats were treated with pargyline (75 mg/kg) 2 hr before killing.

B. Effect of Altered Cerebral Tryptophan Concentration on Brain Indoles

Food deprivation increases whole-brain tryptophan concentration in the rat (Curzon et al., 1972; Perez-Cruet et al., 1972) and led to significant percentage increases in all brain regions studied. These changes were least marked in the hypothalamus (Table 1). Food deprivation for 24 hr increases 5-HIAA and in some regions also 5-HT. Combined increases of 5-HT and 5-HIAA were in decreasing order, as follows: midbrain, pons + medulla, cortex, striatum, cerebellum, hypothalamus.

Two hours after i.p. injection of 50 mg/kg of tryptophan, percentage increases of tryptophan in all regions studied were comparable. However, in the hypothalamus, increases of the 5-hydroxyindoles were relatively less than in other brain regions. Furthermore, both the percentage increases and their rank order were very similar to those in the food deprivation group (Table 1).

The refractory behavior of the hypothalamus was studied further by

TABLE 1. *Effect of altered brain tryptophan concentration on 5-hydroxyindoles*

Region	Food deprivation (% increases)		Tryptophan administration (% increases)	
	Tryptophan	5-HT + 5-HIAA	Tryptophan	5-HT + 5-HIAA
Midbrain	69[a]	41[a]	126[a]	50[b]
Pons-medulla	76[c]	35[c]	106[a]	33[c]
Cortex	57[c]	27[c]	124[a]	27[b]
Striatum	57[b]	20[c]	92[a]	15 ns
Cerebellum	61[a]	17 ns	103[b]	16[c]
Hypothalamus	39[c]	14 ns	144[b]	9 ns
n	(6)	(6)	(7)	(8)

Food deprivation: food was withdrawn from rats at 1600 hr and they were killed 24 hr later together with a control group. Tryptophan administration: tryptophan (50 mg/kg) was given i.p. and the rats were killed 2 hr later together with control animals previously injected with saline. Percentage increases of 5-hydroxyindoles are given by combining the increases of 5-HT and 5-HIAA. n = number of determinations.

[a] $p < 0.001$.
[b] $p < 0.01$.
[c] $p < 0.05$.

using probenecid (200 mg/kg, i.p., 2 hr) to block the egress of 5-HIAA. Somewhat surprisingly, although it led to a marked (+80%) and significant increase of 5-HIAA in the rest of the brain, there was only a small (+19%) and nonsignificant increase in the hypothalamus. It is not known whether this difference is due to special characteristics of 5-HIAA transport or of probenecid transport in the hypothalamus.

Relative unresponsiveness of tryptamine synthesis to altered brain tryptophan concentration was manifest in work on whole brain. Thus, although 24-hr fasting consistently increases brain tryptophan and 5-HIAA concentrations, that of tryptamine did not alter significantly (control 37 ± 7.9 (SE) ng/g, food deprived 47 ± 12.7 ng/g). Since it was possible that both tryptamine synthesis and degradation were increasing at the same rate, groups of fed and food-deprived rats were treated with pargyline (75 mg/kg, i.p., 2 hr) to inhibit degradation of tryptamine to indoleacetic acid. Even under these conditions, cerebral tryptamine was not significantly raised by fasting (130 ± 22 ng/g) when compared with the pargyline-treated, fed control group (128 ± 17 ng/g).

III. DISCUSSION

Our results suggest that not only is indole metabolism particularly active in the hypothalamus with high concentration of tryptophan, tryptamine, and 5-HT, but that it also has unusual characteristics at this site. The relatively high tryptophan concentration is consistent with the findings of Pujol, Sordet, Petitjean, Germain, and Jouvet (1972) in the cat, but does not agree with earlier work in the dog (Eccleston, Ashcroft, Moir, Parker-Rhodes, Lutz, and O'Mahoney, 1968). Results (Knott, Marsden, and Curzon, *to be published*) suggest that it is not a methodological artefact; neither is it due to contamination by blood tryptophan nor to the presence of tryptamine or 5-methoxytryptamine. It may be significant that whole-brain tryptophan concentrations found in our work are at the lower limit of the range of values reported in the literature. Therefore, it might be that our findings reflect a hypothalamic mechanism not directly detectable at higher tryptophan concentrations.

Fernstrom and Wurtman (1971) point out that the influence of physiological plasma tryptophan changes on brain 5-HT synthesis appears to contradict the concept of the resistance of critical brain metabolism to peripheral changes. However, the 5-HT and 5-HIAA changes shown in Table 1 suggest that the above influence may not be uniformly present throughout the brain and that it is minimal in the hypothalamus. This is consistent with the finding (Curzon and Green, 1971) that the increase of hypothalamic 5-HIAA on restraint stress was less than in other brain regions. In the case of tryptamine, results obtained do suggest that its metabolism in the brain may be resistant to physiological changes in peripheral tryptophan concentration.

More detailed study of the apparently special characteristics of 5-HT metabolism in the hypothalamus is needed—especially in view of the results with probenecid, which suggest that the relationship between 5-HIAA concentrations and 5-HT turnover may itself not be the same in the hypothalamus as in other regions.

A completely parallel regional distribution of tryptamine and 5-HT would be consistent with tryptamine occurring exclusively within serotoninergic neurons. However, the relatively high striatal tryptamine and the less marked reduction of tryptamine than of 5-HT after raphe lesions suggest that a large fraction of brain tryptamine synthesis occurs outside the raphe 5-HT neuronal system. Possibly, tryptamine is also synthesized by aromatic amino acid decarboxylase in neurons of other types. The high striatal tryptamine could result from the very high decarboxylase activity found there by Sims, Davis, and Bloom (1973), while the high hypothalamic tryptamine may be related both to the high decarboxylase activity and high tryptophan concentration in this region. It would be of interest to determine whether nigral lesions or treatment with 6-hydroxydopamine also causes a fall of brain tryptamine.

ACKNOWLEDGMENTS

We thank the MRC for financial support of this work.

REFERENCES

Curzon, G., and Green, A. R. (1971): Regional and subcellular changes in the concentration of 5-hydroxytryptamine and 5-hydroxyindoleacetic acid in the rat brain caused by hydrocortisone, DL-α-methyltryptophan, L-kynurenine and immobilization. *British Journal of Pharmacology*, 43:39—52.

Curzon, G., Joseph, M. H., and Knott, P. J. (1972): Effects of immobilization and food deprivation on rat brain tryptophan metabolism. *Journal of Neurochemistry*, 19:1967—1974.

Eccleston, D., Ashcroft, G. W., Moir, A. T. B., Parker-Rhodes, A., Lutz, W., and O'Mahoney, D. P. (1968): A comparison of 5-hydroxyindoles in various regions of dog brain and cerebrospinal fluid. *Journal of Neurochemistry*, 15:947—957.

Fernstrom, J. D., and Wurtman, R. J. (1971): Brain serotonin content: Physiological dependence on plasma tryptophan levels. *Science*, 173:149—152.

Glowinski, J., and Iversen, L. L. (1966): Regional studies of catecholeamines in the brain—1. The disposition of (^{3}H) norepinephrine, (^{3}H) dopamine and (^{3}H) DOPA in various regions of the brain. *Journal of Neurochemistry*, 13:655—669.

Pellegrino, L. J., and Cushman, A. J. (1967): In: *A Stereotaxic Atlas of the Rat Brain*. Appleton-Century-Crofts, New York.

Perez-Cruet, J., Tagliamonte, A., Tagliamonte, P., and Gessa, G. L. (1972): Changes in brain serotonin metabolism associated with fasting and satiation in rats. *Life Sciences* 11, Part 2:31—39.

Pujol, J. F., Sordet, F., Petitjean, F., Germain, D., and Jouvet, M. (1972): Insomnie et métabolisme cérébral de la sérotonine chez le chat: Etude de la synthese et de la libération de la sérotonine mésurees *in vitro* 18 hr aprés destruction du système du raphe. *Brain Research*, 39:137—149.

Saavedra, J. M., and Axelrod, J. (1972): A specific and sensitive enzymatic assay for tryptamine in tissue. *Journal of Pharmacology and Experimental Therapeutics*, 182:363—369.

Sims, K. L., Davis, G. A., and Bloom, F. E. (1973): Activities of 3,4-dihydroxy-L-phenylalanine and 5-hydroxy-L-tryptophan decarboxylase in rat brain: Assay characteristics and distribution. *Journal of Neurochemistry*, 20:449—464.

Advances in Biochemical Psychopharmacology, Vol. 11
Raven Press, New York © 1974

Viral Infections in the Central Nervous System and 5-Hydroxytryptamine Metabolism

Björn-Erik Roos and Erik Lycke

*Departments of Pharmacology and Virology, University of
Göteborg, Göteborg, Sweden*

In contrast to many bacterial infections, virus infections of the mouse brain will affect the levels of monoaminergic transmitter substances. For virus infections such as vaccinia, increased levels of 5-hydroxyindoleacetic acid (5-HIAA) are observed, in rabies an increase of homovanillic acid (HVA) concentration, and in herpes simplex infection changes in concentration of both 5-HIAA and HVA may be seen (Lycke, Modigh, and Roos, 1970). In herpes simplex-infected mice, the increase of acid monoamine metabolites depends on the severity of the virus infection (Lycke and Roos, 1968). Inhibitors of virus multiplication will cause brain concentrations of the acids to return to normal.

The high levels of 5-HIAA and HVA are probably not due to a direct and specific interaction of virus-coded proteins with the formation of amines. Histologically, there seems to be little or no cytopathic change in the transmitter-synthesizing neurons of the brainstem. On the other hand, the degree of neuronal destruction of postsynaptic neurons in the hemispheres seems to correlate with the increase in acids, indicating that changes in receptor structures of postsynaptic neurons might be of considerable importance.

Because viral infections produce very little change in levels of 5-hydroxytryptamine (5-HT) and dopamine (DA) in the brain, increases in 5-HIAA and HVA might reflect an increased release of the corresponding transmitter from the nerve endings, and thus increased monoamine synthesis and turnover. However, high concentrations of acid metabolites might also be due to impaired efflux from the brain. We have spent much effort to clarify which of these two possibilities is responsible for our findings. All our experiments suggest that the viral infection in some way causes an increased release of transmitter and thus an increased synthesis and metabolism. After monoamine oxidase inhibitor treatment, efflux of the acids is equal in infected and uninfected animals, and the disappearance of DA and norepinephine (NE) after pretreatment with a catecholamine synthesis blocker is more rapid in infected than in uninfected animals. The same holds true when blocking the synthesis of 5-HT. Administration of precursors such as 5-HTP and L-DOPA causes a more vigorous turnover of their corresponding amines in infected animals than in uninfected, but the disappearance of acid monoamine metabolites from the brain is equal in both cases. Pretreatment with probenecid, which blocks active transport of acid monoamine metabolites in the rat brain, causes a more rapid accumulation of acids in infected than in uninfected mice (Lycke et al., 1970; Lycke and Roos, 1972a, b). But again, the increase in release of 5-HT and also of DA and NE is most probably due to disturbances in the postsynaptic area, and this might explain variations in reaction pattern we see after infections with different types of virus.

As a possible hypothesis for the pathogenesis of Parkinson's disease, it might be suggested that an intense turnover of transmitter substances during infection is followed by a secondary exhaustion of monoamine metabolism. It might be important to remember that although the main changes in monoaminergic transmission in parkinsonism are in the nigrostriatal pathway and thus in synthesis and release of DA, appreciable changes are also seen in levels of 5-HT and 5-HIAA. When newborn mice were infected intraperitoneally with an attenuated strain of Coxsackie B4, a mortality rate of about 35% was observed. Thirty percent of the mice survived the infection without demonstrating any signs of disease, whereas about 35% survived after having showed various encephalitic symptoms. The most prominent of these were spastic paralyses, rigidity, and gait abnormalities. According to the progress of the disease and their recover powers, animals were divided into four groups and observed for about 2 months. The first of the groups contained animals without any signs of disease; in the second were those which seemed to have recovered totally after less severe signs; the third still had abnormalities of gait; and the fourth group of mice showed spasticity of varying severity.

At birth the dopaminergic system of the mouse brain is not fully developed. It does not become mature before the end of the first month of life. In Coxsackie B4-infected animals, in the beginning there were no major differences in DA level from that of normal animals. After about 2 months, however, low concentrations of amine tended to be present, particularly in animals with severe disease. This tendency was more obvious for NE values. A slower maturation of NE-synthesizing neurons in infected animals was apparent, and only those mice which had undergone infection without any symptoms of disease seemed capable of compensating for this deficit later in life.

It thus seems possible that as a result of a virus encephalitis in mice, a dysfunction of catecholamine metabolism can occur later in life. Preliminary data show that the same happens to 5-HT metabolism. An increased synthesis during active encephalitis might thus be followed by a state of exhaustion of amine-producing nerve cells. The possible implications of these observations for Parkinson's disease or endogenous depression, for example, can as yet be only speculative, however.

ACKNOWLEDGMENTS

This work was supported by a grant from the Swedish Medical Research Council (B73-04X-2728-05B).

REFERENCES

Lycke, E., Modigh, K., and Roos, B.-E. (1970): The monoamine metabolism in viral encephalitides of the mouse. I. Virological and biochemical results. *Brain Research*, 23:235–246.

Lycke, E., and Roos, B.-E. (1968): Effect on the monoamine-metabolism of the mouse brain by experimental herpes simplex infection. *Experientia*, 24:687–689.

Lycke, E., and Roos, B.-E. (1972a): The monoamine metabolism in viral encephalitides of the mouse. II. Turnover of monoamines in mice infected with herpes simplex virus. *Brain Research*, 44:603–613.

Lycke, E., and Roos, B.-E. (1972b): Studies on increased turnover of brain monoamines induced by experimental herpes simplex infection. *Acta Pathologica Microbiologica Scandinavica*, Section B. 80:596–701.

Advances in Biochemical Psychopharmacology, Vol. 11
Raven Press, New York © 1974

Possible Role of Free Serum Tryptophan in the Control of Brain Tryptophan Level and Serotonin Synthesis

G. L. Gessa and A. Tagliamonte

Institute of Pharmacology, University of Cagliari, Cagliari, Italy

I. INTRODUCTION

The present study provides evidence that the rate-limiting step in brain serotonin synthesis is not the activity of tryptophan hydroxylase, but the concentration of tryptophan at the sites of serotonin. This concentration, in turn, depends mainly, but not exclusively, on the concentration of free tryptophan in serum.

II. RELATIONSHIP BETWEEN BRAIN TRYPTOPHAN CONCENTRATION AND SEROTONIN SYNTHESIS

A peculiar characteristic of tryptophan hydroxylase, the enzyme which converts tryptophan to 5-hydroxytryptophan, is that this enzyme, unlike tyrosine hydroxylase, has a Michaelis constant (K_m) for its substrate much higher than the concentration of tryptophan present in the whole mammalian brain (Lovenberg, Jéquier, and Sjoerdsma, 1968).

Tryptophan hydroxylase also seems not to be saturated *in vivo*, since the administration of exogenous tryptophan has been shown to increase the formation of brain serotonin and 5-hydroxyindoleacetic acid (5-HIAA) (Eccleston, Ashcroft, and Crawford, 1965; Fernstrom and Wurtman, 1971a).

These considerations indicate that serotonin synthesis should depend on the concentration of tryptophan in the serotonin neuron.

Accordingly, we found that all treatments so far tested which had been shown to increase serotonin synthesis also increased brain tryptophan concentration. Conversely, all treatments so far tested which were found to increase brain tryptophan also increased 5-HIAA levels, except phenelzine, a monoamine oxidase inhibitor which inhibits

5-HIAA formation, and fenfluramine, which has been reported to release serotonin protected from monoamine oxidase inactivation (Le Duarec, Schmitt, and Laubie, 1966).

The concentrations of brain tryptophan and 5-HIAA were influenced neither by cold exposure for 2 hr nor by many other psychotropic drugs such as *a*-methyltyrosine, morphine, apomorphine, chlorpromazine, desipramine, and haloperidol (Tagliamonte, Tagliamonte, Perez-Cruet, Stern, and Gessa, 1971*a*). These results are summarized in Fig. 1 and the experimental conditions in Table 1.

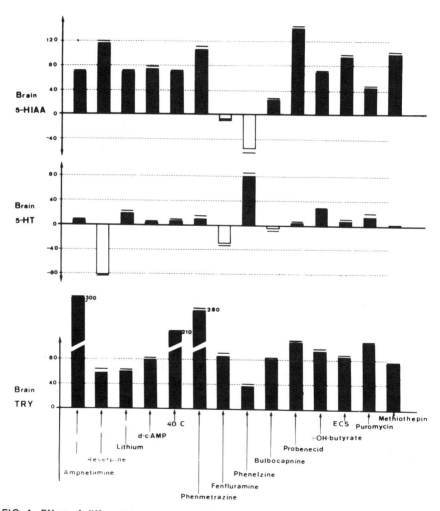

FIG. 1. Effect of different treatments on the level of tryptophan, 5-HT, and 5-HIAA in the rat brain. Columns represent percent changes from normal. Dosages, time schedules, and references are reported in Table 1.

TABLE 1. *Dosages, time schedules, and references of the data reported in Fig. 1*

Treatment (no. of experiments)	mg/kg, i.p. (except where indicated)	Interval between last treatment and sacrifice (min)	Increase in 5-HT turnover reported by
D-Amphetamine (8)	10	90	Reid, 1970
Reserpine (9)	5	90	Tozer et al., 1966
Lithium carbonate (26)	60 mg/kg, twice daily for 6 days	360	Perez-Cruet et al., 1971
Dibutyryl-cyclic-AMP (13)	100 μg/animal intraventricularly	90	Forn et al., 1972
40°C environment (12)	Animals were kept at this temperature for 120 min.	–	Reid et al., 1968
Phenmetrazine (8)	20	120	–
DL-fenfluramine (14)	10	120	–
Phenelzine (9)	100	120	–
Bulbocapnine (14)	60	120	Tagliamonte et al., 1971c
Probenecid (18)	300	120	
γ-hydroxybutyrate (8)	1500	120	Spano and Przegalinski, 1973
ECS (21)	–	60	Tagliamonte et al., 1972
Puromycin (18)	50 μg/animal intraventricularly	120	Biggio et al., 1972
Methiothepin (6)	50	60	Monachon et al., 1972

III. RELATIONSHIP BETWEEN SERUM TRYPTOPHAN
AND BRAIN TRYPTOPHAN

The results reported above prompted us to study the mechanism controlling tryptophan transport from blood into brain.

We observed that among treatments found to increase brain tryptophan concentrations, some enhanced, some left unaltered, and others decreased the concentration of plasma tryptophan. Therefore we concluded that brain tryptophan levels are independent of tryptophan concentration in plasma (Tagliamonte et al., 1971a).

However, the fact that tryptophan is the only amino acid which circulates in plasma combined to serum proteins, and that only its free fraction can cross the blood-brain barrier, led us to investigate whether the changes of free tryptophan concentration in serum might reflect the change of brain tryptophan content and, therefore, serotonin turnover.

In order to test this hypothesis, we studied:

1. the effect of the administration of L-tryptophan on the concentrations of free and total serum tryptophan, brain tryptophan, and 5-HIAA;

2. the effect on brain tryptophan content of drugs capable of displacing serum tryptophan from its protein binding; and

3. the effect of different treatments, which increase brain serotonin synthesis, on free and total tryptophan concentrations in serum.

A. *Effect of L-Tryptophan Administration on Free and Total*
Serum Tryptophan and Brain Tryptophan

Figure 2 shows the time course of the changes in free and total serum tryptophan and brain tryptophan following oral or intraperitoneal administration of L-tryptophan. It appears that the changes in brain tryptophan were time-related and proportional to changes in free serum tryptophan and not in total serum tryptophan.

B. *Effect of Tryptophan Displacement from Protein Binding on*
Brain Tryptophan

The results reported in Table 2 show that salicylate, probenecid, and chlofibrate, which are able to release tryptophan from its binding on serum proteins (McArthur and Dawkins, 1969; Tagliamonte, Biggio, and Gessa, 1971b; Gessa, *unpublished observations*), are also capable of

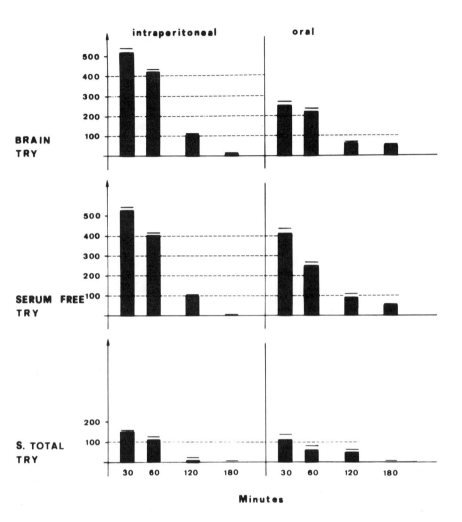

FIG. 2. Percent changes in plasma and brain tryptophan (TRY) induced by intraperitoneal or oral administration of L-TRY (50 mg/kg) in rats. Experimental details are reported in Tagliamonte et al., 1973b.

increasing brain tryptophan and 5-HIAA. A release of serum trypto-phan from its protein binding would explain both the decrease in the absolute level of the amino acid in serum and its increase in the CNS.

We have found that these drugs also stimulate brain serotonin synthesis (Tagliamonte, Biggio, Vargiu, and Gessa, 1973a; Tagliamonte, Tagliamonte, Gessa, Duce, Maffei, and Gessa, 1971c).

TABLE 2. *Effect of sodium salicylate, probenecid, and chlofibrate on free and total serum tryptophan, brain tryptophan, 5-HT, and 5-HIAA*

Treatment (no. of experiments	mg/kg	Serum tryptophan (µg/g)		Brain (µg/g)		
		Total	Free	Trypto-phan	5-HT	5-HIAA
None (50)	–	24.61 ± 0.66	2.46 ± 0.09	3.31 ± 0.09	0.66 ± 0.06	0.73 ± 0.07
Na-salicylate (24)	50 i.p.	11.75 ± 0.19^a	3.08 ± 0.11^a	4.15 ± 0.07^a	0.78 ± 0.09	0.93 ± 0.05^a
Probenecid (18)	150 i.p.	16.20 ± 0.25^a	3.40 ± 0.15^a	5.30 ± 0.30^a	0.70 ± 0.08	1.35 ± 0.11^a
Chlofibrate (16)	200 o.s.	6.85 ± 0.24^a	4.85 ± 0.14^a	4.37 ± 0.50^a	0.68 ± 0.03	1.11 ± 0.08^a

Na-salicylate, probenecid, and chlofibrate were injected 60, 120, and 180 min before killing, respectively. Each value is the average ± SE of the number of experiments reported in parentheses.

[a] $p < 0.01$ with respect to control value.

C. Effect of Treatments Which Increase Serotonin Synthesis on Free Serum Tryptophan

As Fig. 3 shows, both D-amphetamine and LiCl, which stimulate serotonin synthesis, also significantly increase free serum tryptophan. These drugs do not release tryptophan from its protein binding when they are added *in vitro* at the maximal concentration which can be reached *in vivo* after their administration.

On the contrary, electroconvulsive shock (ECS) and intracerebral administration of puromycin, which stimulate serotonin synthesis (Tagliamonte, Tagliamonte, Di Chiara, Gessa, and Gessa, 1972; Biggio, Mereu, Vargiu, and Tagliamonte, 1972), increase brain tryptophan and 5-HIAA levels but influence neither total nor free serum tryptophan levels.

The latter results indicate that, under certain experimental conditions, the level of brain tryptophan may increase independently of changes in its free concentrations in serum.

IV. ROLE OF TRYPTOPHAN IN THE CONTROL OF SEROTONIN SYNTHESIS UNDER PHYSIOLOGICAL CONDITIONS

We have previously shown that brain tryptophan level and serotonin turnover are higher in fasted than in fed rats (Perez-Cruet, Tagliamonte, Tagliamonte, and Gessa, 1972).

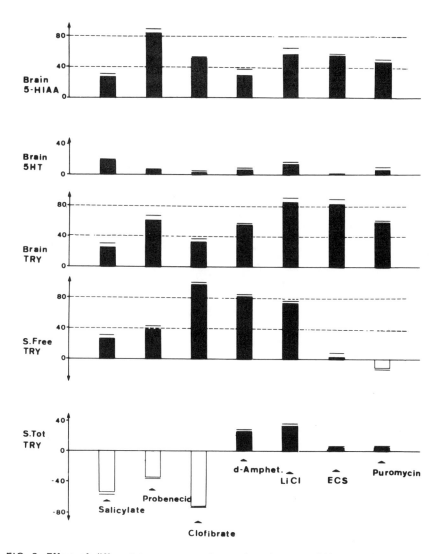

FIG. 3. Effect of different treatments on free and total serum TRY, brain TRY, 5-HT, and 5-HIAA in rats. Columns represent percent changes from normal. Each value is the mean ± SE of at least eight determinations. Values for salicylate, probenecid, and chlofibrate are the same as those reported in Table 2. D-Amphetamine was given at the dose of 5 mg/kg, i.p., 60 min prior to sacrifice. LiCl was given at the dose of 100 mg/kg, i.p., twice daily for 5 days; animals were killed 120 min after the last treatment. ECS were produced as described in Tagliamonte et al., 1972. Puromycin was given intraventricularly at the dose of 200 μg/ animal 180 min prior to sacrifice.

As Table 3 shows, the concentration of total tryptophan in serum is lower in fasted than in fed rats, indicating also that under psychological conditions brain tryptophan and serotonin turnover are independent of the level of total serum tryptophan.

TABLE 3. *Free and total tryptophan in serum tryptophan, serotonin, and 5-HIAA in brain of fasted and fed rats*

Condition[a]	Time of sacrifice[b]	Serum tryptophan		Brain tryptophan (μg/g)	Brain serotonin (μg/g)	Brain 5-HIAA (μg/g)
		Free (μg/ml)	Total (μg/ml)			
Fasted	12:00	4.56 ± 0.13	18.04 ± 0.16	5.67 ± 0.06	0.66 ± 0.03	0.61 ± 0.04
Fed	12:00	2.78 ± 0.11[c]	28.21 ± 0.19[c]	3.57 ± 0.04[c]	0.64 ± 0.04	0.43 ± 0.02[c]
Fasted	24:00	4.79 ± 0.09	17.52 ± 0.13	5.83 ± 0.03	0.62 ± 0.04	0.64 ± 0.03
Fed	24:00	3.01 ± 0.12[c]	27.81 ± 0.21[c]	3.48 ± 0.06[c]	0.63 ± 0.02	0.41 ± 0.05[c]

Each value is the average ± SE of at least 20 determinations. Experimental details are reported in Tagliamonte et al., 1973*b*.

[a] Rats were fasted for 24 hr or fed for 2 hr.

[b] Rats killed at noon were rats trained to eat their daily food from 10:00 to 12:00; rats killed at 24:00 were trained to eat their food from 22:00 to 24:00.

[c] $p < 0.001$ with respect to fasted rats.

On the contrary, the concentrations of free serum tryptophan are much higher in fasted than in fed rats. These changes are due neither to circadian variations nor to stress (Tagliamonte, Biggio, Vargiu, and Gessa, 1973*b*).

V. AMINO ACIDS COMPETING WITH TRYPTOPHAN FOR THE SAME TRANSPORT INTO THE CNS

Studies on passage of amino acids from blood to brain and competition studies on uptake of amino acids by brain slices (Cohen and Lajtha, 1972) or synaptosomes (Grahame-Smith and Parfitt, 1970) have shown that many of the aromatic amino acids share the same transport mechanism, so that a high concentration of one can lower the passage or the uptake of others.

These conditions suggest that tryptophan concentration in brain may depend not only on the concentration of free serum tryptophan but also, as suggested by Fernstrom and Wurtman (Fernstrom and Wurtman, 1971*b*), on the concentrations of other amino acids sharing the same transport system with tryptophan.

To test this hypothesis, we studied the effect of the intragastric administration of different amino acid mixtures not containing tryptophan on serum and brain tryptophan levels and serotonin metabolism.

Different rats were treated orally with the tryptophan-free amino acid mixture reported in Table 4: that is, the first group received a mixture of six amino acids sharing with Try the same transport system from blood to brain (Oldendorf, 1971) (mixture A); the second group received nine of the ten essential amino acids, excluding tryptophan, plus the semiindispensable glycine [this mixture (B) includes five of the amino acids contained in mixture A]; the third group received ten amino acids, none of which compete with Try for its transport into the brain [this mixture comprises only three essential amino acids (mixture C)];

TABLE 4. *Effect of the oral administration of different amino acid mixtures on tryptophan (TRY), serotonin (5-HT), and 5-hydroxyindoleacetic acid (5-HIAA) in brain*

Amino acid mixture	TRY (μg/g)	5-HT (μg/g)	5-HIAA (μg/g)
Control	5.30 ± 0.06	0.62 ± 0.03	0.55 ± 0.04
Mixture A	4.30 ± 0.04[a]	0.59 ± 0.01 N.S.	0.43 ± 0.03[a]
Mixture B	2.27 ± 0.01[b]	0.46 ± 0.04[b]	0.32 ± 0.02[b]
Mixture C	5.26 ± 0.04 N.S.	0.63 ± 0.03 N.S.	0.53 ± 0.05 N.S.

Each value is the average $\pm$ S.E. of 12 determinations. Rats were given by gavage 10 ml/kg of one of the following amino acid mixtures in distilled H_2O:

Mixture A: L-phenylalanine 9, L-leucine 4, L-isoleucine 8, L-tyrosine 5, L-methionine 8, L-histidine 4 mg/ml, respectively.

Mixture B: L-phenylalanine 9, L-leucine 4, L-isoleucine 8, L-methionine 8, L-histidine 4, L-lysine 20, L-threonine 7, L-valine 10, L-arginine 0.8, glycine 60 mg/ml, respectively.

Mixture C: L-lysine 20, L-threonine 7, L-arginine 0.8, glycine 60, L-serine 3.25, Ac. glutamic 6.25, aspartic 6.25, L-proline 7, 22, L-hydroxyproline 7, L-alanine 12.50 mg/ml, respectively.

Rats were sacrificed 2 hr after treatment.

[a] $p < 0.01$.

[b] $p < 0.001$.

the fourth group received a corresponding volume of distilled H_2O. Each amino acid was given at a dose corresponding to one-fourth the amount present in the diet the animal normally consumes in 24 hr.

The results of the experiments are reported in Tables 4 and 5. They show that:

1. The administration of the mixture containing the six amino acids which share with tryptophan the same transport cause a modest but significant reduction in brain tryptophan and 5-HIAA levels but change neither free nor total serum Try levels.

TABLE 5. *Effect of the amino acid mixtures listed in Table 1 on free and total tryptophan (Try) levels in serum*

Treatment	Serum tryptophan (μg/ml)	
	Free	Total
Control	2.53 ± 0.02	15.45 ± 0.10
Mixture A	2.61 ± 0.05 N.S.	16.32 ± 0.14 N.S.
Mixture B	0.86 ± 0.03^a	5.12 ± 0.27^a
Mixture C	2.51 ± 0.02 N.S.	15.21 ± 0.12 N.S.

The values reported in this table and in Table 1 were obtained from the same animals.

[a]$p < 0.001$.

To obtain the same degree of reduction in brain tryptophan, much higher, nonphysiological amounts of each individual amino acid should have been administered, if given alone (Yuwiler and Geller, 1965). These results suggest that the concentrations of each individual amino acid sum up in competing with tryptophan for the same carrier.

2. The administration of the mixture of essential amino acids (mixture B) dramatically reduced tryptophan, serotonin, and 5-HIAA in brain and free and total tryptophan in serum.

The dramatic decline in free and bound serum tryptophan produced by the mixture of the essential amino acids may be due to the removal of tryptophan from serum by the liver, allowing protein synthesis to proceed. This hypothesis is supported by the fact that mixture C, which contained only three essential amino acids, was ineffective.

However, the results do not allow us to clarify whether the reduction in brain tryptophan produced by mixture B is due to the decreased concentration of free serum tryptophan per se or to the resultant imbalance between free tryptophan concentration and that of the other competing amino acids.

The finding that the decrease in brain tryptophan induced by either amino acid mixtures was accompanied by a decline in 5-HIAA levels indicates that not only increases but also decreases in brain tryptophan reflect parallel changes in brain serotonin turnover.

VI. CONCLUSIONS

The principal conclusions that we can draw from the data presented above are the following:

1. The changes in the level of brain tryptophan were always associated with parallel changes in serotonin synthesis and vice versa. This applies to pharmacological or physiological conditions, such as fasting and satiation. These findings suggest that the concentration of tryptophan into the whole brain reflects that into the sites of serotonin synthesis.

2. The concentration of tryptophan into the serotonin neuron seems to be the rate-limiting step of brain serotonin synthesis.

3. The results reported do not allow us to decide whether tryptophan transport into the brain is controlled, under physiological conditions, either by the ratio between free tryptophan to other competing amino acids or by the absolute concentration of free tryptophan in serum. An imbalance between free tryptophan and other amino acids circulating in plasma may actually occur under different conditions, since the ratio of free to bound tryptophan in serum is not constant: It is changed by drugs, it increases during fasting, and, in humans, it has a circadian rhythm, with a zenith at midnight (Gessa et al., *Life Sciences, in press*).

The data reported raise some basic questions:

1. Are there any conditions in which changes in serotonin synthesis may occur without parallel changes in brain tryptophan concentrations?

2. The finding that methiothepin increases brain tryptophan concentrations is particularly interesting, since this drug is considered to stimulate serotonin synthesis through a transsynaptic feedback mechanism secondary to the blockade of the central serotonin receptors (Monachon, Burkard, Jalfre, and Haefely, 1972). Does this finding mean that blockade of serotonin receptors can activate tryptophan transport into the CNS?

3. Is serotonin synthesis related to nerve activity or, in general, does an increase of tryptophan and 5-HIAA concentration in the CNS reflect changes of brain function?

REFERENCES

Biggio, G., Mereu, G., Vargiu, L., and Tagliamonte, A. (1972): Effect of protein synthesis inhibition on tryptophan level and serotonin turnover in the rat brain. *Rivista di Farmacologia e Terapia*, III:229–236.

Cohen, S. R., and Lajtha, A. (1972): Amino acid transport. In: *Handbook of Neurochemistry*, edited by Abel Lajtha, vol. 7, pp. 543–572. Plenum Press, New York.

Eccleston, D., Ashcroft, G. W., and Crawford, T. B. B. (1965): 5-Hydroxyindole metabolism in rat brain. A study of intermediate metabolism using the technique of tryptophan loading—II. *Journal of Neurochemistry,* 12:493–503.

Fernstrom, J. D., and Wurtman, R. J. (1971a):   Brain serotonin content: Physiological dependence on plasma tryptophan levels. *Science,* 173:149–152.

Fernstrom, J. D., and Wurtman, R. J. (1971b): Brain serotonin content: Increase following ingestion of carbohydrate diet. *Science,* 174:1023.

Forn, J., Tagliamonte, A., Tagliamonte, P., and Gessa, G. L. (1972): Stimulation by dibutyryl cyclic AMP of serotonin synthesis and tryptophan transport in brain slices. *Nature New Biology,* 237:245–247.

Grahame-Smith, D. G., and Parfitt, A. G. (1970): Tryptophan transport across the synaptosomal membrane. *Journal of Neurochemistry,* 17:1339–1353.

Le Douarec, J. C., Schmitt, H., and Laubie, M. (1966): Etude pharmacologique de la fenfluramine et de ses isoméres optiques. *Archives of International Pharmacodynamics,* 161:206–232.

Lovenberg, W., Jéquier, E., and Sjoerdsma, A. (1968): Tryptophan hydroxylation in mammalian systems. *Advances in Pharmacology,* 6A:21–36.

McArthur, J. N., and Dawkins, P. D. (1969): The effect of sodium salicylate on the binding of L-tryptophan to serum proteins. *Journal of Pharmacy and Pharmacology,* 21:744–750.

Monachon, M. A., Burkard, W. P., Jalfre, M., and Haefely, W. (1972): Blockade of central 5-hydroxytryptamine receptors by methiothepin. *Naunyn-Schmiedeberg's Archiv für Pharmakologie,* 274:192–197.

Oldendorf, W. H. (1971): Uptake of radiolabeled essential amino acids by brain following arterial injection. *Proceedings of the Society for Experimental Biology and Medicine,* 136:385–386.

Perez-Cruet, J., Tagliamonte, A., Tagliamonte, P., and Gessa, G. L. (1971): Stimulation of serotonin synthesis by lithium. *Journal of Pharmacology and Experimental Therapeutics,* 178:325–330.

Perez-Cruet, J., Tagliamonte, A., Tagliamonte, P., and Gessa, G. L. (1972): Changes in brain serotonin metabolism associated with fasting and satiation in rats. *Life Sciences,* 11:31–39.

Reid, W. D. (1970): Hyperthermia and increased turnover of brain serotonin produced by amphetamine. *Federation Proceedings,* 29:747.

Reid, W. D., Volicer, L., Smookler, H., Beaven, M. A., and Brodie, B. B. (1968): Brain amines and temperature regulation. *Pharmacology,* 1:329–344.

Spano, P. F., and Przegalinski, E. (1973): Stimulation of serotonin synthesis by anesthetic and non-anesthetic doses of gamma-hydroxybutyrate. *Pharmacological Research Communications,* 5:55–69.

Tagliamonte, A., Biggio, G., and Gessa, G. L. (1971b): Possible role of "frée" plasma tryptophan in controlling brain tryptophan concentrations. *Rivista di Farmacologia e Terapia,* II:251–255.

Tagliamonte, A., Biggio, G., Vargiu, L., and Gessa, G. L. (1973a): Increase of brain tryptophan and stimulation of serotonin synthesis by salicylate. *Journal of Neurochemistry,* 20:909–912.

Tagliamonte, A., Biggio, G., Vargiu, L., and Gessa, G. L. (1973b): Free tryptophan in serum controls brain tryptophan level and serotonin synthesis. *Life Sciences,* 12:277–287.

Tagliamonte, A., Tagliamonte, P., Di Chiara, G., Gessa, R., and Gessa, G. L. (1972): Increase of brain tryptophan by electroconvulsive shock in rats. *Journal of Neurochemistry,* 19:1509–1512.

Tagliamonte, A., Tagliamonte, P., Gessa, R., Duce, M., Maffei, C., and Gessa, G. L. (1971c): Increase of brain tryptophan by probenecid. *Rivista di Farmacologia e Terapia,* II:207–213.

Tagliamonte, A., Tagliamonte, P., Perez-Cruet, J., Stern, S., and Gessa, G. L. (1971a): Effect of psychotropic drugs on tryptophan concentration in the rat brain. *Journal of Pharmacology and Experimental Therapeutics*, 177:475–480.

Tozer, T. N., Neff, N. H., and Brodie, B. B. (1966): Application of steady-state kinetics to the synthesis rate and turnover time of serotonin in the brain of normal and reserpine-treated rats. *Journal of Pharmacology and Experimental Therapeutics*, 153:177–182.

Yuwiler, A., and Geller, E. (1965): Serotonin depletion by dietary leucine. *Nature*, 208:83–84.

Advances in Biochemical Psychopharmacology, Vol. 11
Raven Press, New York © 1974

Control of Brain Serotonin Levels by the Diet

John D. Fernstrom and Richard J. Wurtman

*Laboratory of Neuroendocrine Regulation, Department of Nutrition and Food
Science, Massachusetts Institute of Technology, Cambridge, Massachusetts 02139*

Brain serotonin concentrations are rapidly influenced by changes in brain tryptophan (Moir and Eccleston, 1968; Fernstrom and Wurtman, 1971*a*). Brain tryptophan levels respond to acute variations in the plasma concentrations of tryptophan and of other neutral amino acids that compete with tryptophan for uptake into the brain (tyrosine, phenylalanine, leucine, isoleucine, and valine) (Blasberg and Lajtha, 1965; Fernstrom and Wurtman, 1972*b*). The ingestion of food alters the plasma concentrations of all these amino acids, and consequently, brain tryptophan and serotonin levels (Fernstrom and Wurtman, 1971*b*, 1972*a*, 1974). Individual food constituents (e.g., protein, carbohydrate) differ in their effects on the plasma amino acid pattern, and thus on brain tryptophan and serotonin (Fernstrom and Wurtman, 1971*b*, 1974).

The concentrations of tryptophan and of most other amino acids in human and rat plasma fluctuate characteristically during each 24-hr period (Wurtman, 1970; Wurtman, Rose, Chou, and Larin, 1968; Fig. 1). Among human subjects who eat three meals a day, tryptophan levels are lowest at 2 to 4 a.m. and rise by 50 to 80%, attaining a plateau in the late morning or early afternoon. In rats, the daily nadir and peak occur 8 to 10 hr later, a shift consistent with their nocturnal feeding habits. The plasma amino acid rhythms are not generated simply by the cyclic ingestion of dietary protein, inasmuch as they persist in human volunteers who eat essentially no protein for 2 weeks (Wurtman et al., 1968). They do disappear in subjects placed on a total fast (Marliss, Aoki, Unger, Soeldner, and Cahill, 1970), which suggests that they are *not* truly circadian, but of nutritional origin (perhaps resulting from the postprandial release of insulin and other hormones which modify tissue uptake of amino acids).

The existence of plasma amino acid rhythms suggested that the quantities of these compounds available to the brain and other tissues for synthesis of proteins and low-molecular-weight derivatives might

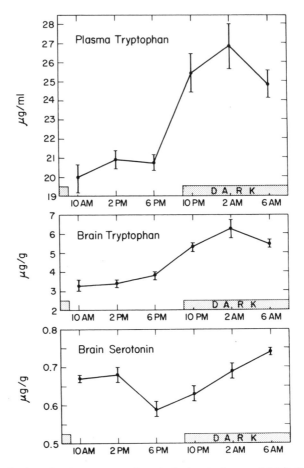

FIG. 1. Daily rhythms in plasma tryptophan, brain tryptophan, and brain 5-hydroxytrypt-amine. Groups of ten rats kept in darkness from 9 p.m. to 9 a.m. were killed at intervals of 4-hr. Vertical bars indicate standard errors of the mean. In all cases values obtained at 6 a.m differ significantly from those at 6 p.m. ($p < 0.001$).

also change diurnally, possibly in response to food consumption. To explore the possible significance of plasma amino acid rhythms, we attempted to determine whether experimentally induced fluctuations of the same amplitude as those occurring diurnally could cause parallel changes in the rate at which a particular amino acid was incorporated into proteins or converted to a particular low-molecular-weight compound. The amino acid whose plasma concentration seemed most likely to influence its metabolic fate was tryptophan, the least abundant amino acid in most tissues and foods (Wurtman and Fernstrom, 1972). Daily rhythms in the ingestion of tryptophan-containing proteins (and, presumably, in the concentration of tryptophan delivered to the liver via the portal venous circulation) were already known to control protein synthetic mechanisms in rat liver; that is, the nocturnal increase in food consumption was followed by increases in the proportion of hepatic RNA aggregated to polysomes (Fishman, Wurtman, and Munro, 1969) and in the quantity of the enzyme tyrosine transaminase (Wurtman, 1970).

As the dependent variable in our studies, we looked for changes in brain serotonin content among rats with a spontaneous daily rhythm or treatment-induced fluctuation in plasma tryptophan. Three lines of evidence had suggested that the amount of tryptophan available to the brain might control serotonin synthesis: (1) a daily rhythm in the brain concentrations of both tryptophan and serotonin (Albrecht, Visscher, Bittner, and Halberg, 1956; Wurtman and Fernstrom, 1972; Fig. 1); (2) the reportedly high K_m for tryptophan of tryptophan hydroxylase, relative to whole-brain tryptophan concentrations (Lovenberg, Jéquier, and Sjoerdsma, 1968); and (3) the great increase of brain concentrations of serotonin and its chief metabolite, 5-hydroxyindole-acetic acid (5-HIAA) caused by large doses of tryptophan (50 to 1600 mg/kg, i.p.) (Moir and Eccleston, 1968).

Initial experiments were designed to determine whether brain serotonin concentrations could be increased by raising brain tryptophan from daily nadir values to just below peak nocturnal levels. The administration of L-tryptophan (12.5 mg/kg, i.p.; less than 5% of the tryptophan that a 200-g rat would consume daily in 10 to 20 g of rat chow) at noon produced peak elevations in plasma and brain tryptophan that were within the nocturnal range of untreated rats (Figs. 1 and 2) and caused brain serotonin levels to rise by 20 to 30% ($p < 0.01$) within 1 hr of treatment (Fernstrom and Wurtman, 1971a). Doses of 25 mg/kg caused proportionately greater increases in both brain tryptophan and brain serotonin. Larger doses of tryptophan, which caused brain tryptophan concentration to rise well beyond its physiological range, produced no further increments in brain serotonin (Fig. 2).

The increase in brain serotonin produced by the very small doses of tryptophan showed that a normal daily change in plasma and brain tryptophan can influence brain serotonin synthesis. However, a substrate-induced rhythm in serotonin synthesis is probably not the *only* factor causing daily rhythms in brain serotonin; brain serotonin levels might also reflect rhythms in serotonin release or intraneuronal metabolism.

Now that small *increases* in plasma tryptophan had been shown to cause parallel changes in brain serotonin, we next attempted to determine whether physiological *decreases* in the plasma amino acid could depress the serotonin content.

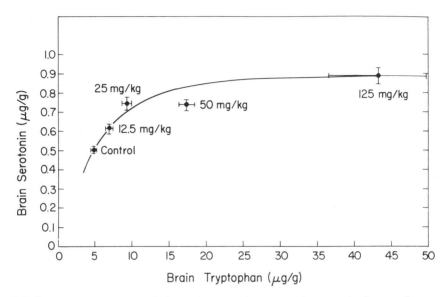

FIG. 2. Dose-response curve relating brain tryptophan and brain serotonin. Groups of ten rats received L-tryptophan (12.5, 25, 50, or 125 mg/kg, i.p.) at noon, and were killed 1 hr later. Horizontal bars represent standard errors of the mean for brain tryptophan; vertical bars represent standard errors of the mean for brain serotonin. All brain tryptophan levels were significantly higher than control values ($p < 0.001$). All brain serotonin levels were significantly higher than control values ($p < 0.01$). Plasma tryptophan rose 22% over control levels in rats injected with the 12.5 mg/kg dose ($p < 0.02$) (Fernstrom and Wurtman, 1971a).

Rats similar to those used in the previous experiments were fasted overnight and then received a dose of insulin (2 I.U./kg, i.p.) known to lower plasma concentrations of glucose and most amino acids. To our surprise, the hormone did not lower plasma tryptophan, but increased its concentration by 30 to 40% (Fernstrom and Wurtman, 1972b; Table 1). This effect was independent of the route by which the insulin was administered; it was associated with a 55% fall in plasma glucose, and with major reductions in the plasma concentrations of most other amino acids, including the neutral amino acids generally believed to compete with tryptophan for uptake into the brain (Blasberg and Lajtha, 1965; Guroff and Udenfriend, 1962; Table 1). Two hours after rats had received the insulin, brain tryptophan levels were elevated by 36% ($p < 0.01$) and brain serotonin levels by 28% ($p < 0.01$; Fernstrom and Wurtman, 1971b).

The increase in brain serotonin observed in rats receiving insulin might have been artefactual, resulting not from increased availability of substrate but from central reflexes activated by hypoglycemia. To determine whether the physiological secretion of insulin, in *normo-glycemic* rats, also increased plasma and brain tryptophan and brain

TABLE 1. *The effect of insulin on the concentrations of tryptophan, other amino acids, and glucose in rat plasma*

| Amino Acid | Concentration | | Percent Change |
	Control (μg/ml)	Insulin (μg/ml)	
L-Tryptophan	11.1 ± 0.6	16.6 ± 0.9	+50[a]
L-Tyrosine	11.4 ± 0.8	9.4 ± 1.0	–18
L-Phenylalanine	11.8 ± 0.3	10.8 ± 0.9	–9
L-Serine	24.6 ± 1.2	17.0 ± 1.3	–31[a]
L-Glycine	29.6 ± 1.4	18.3 ± 2.0	–38[a]
L-Alanine	25.3 ± 0.9	11.7 ± 0.7	–54[a]
L-Valine	21.5 ± 1.3	17.3 ± 1.8	–20
L-Isoleucine	13.7 ± 0.8	7.1 ± 0.4	–48[a]
L-Leucine	21.3 ± 1.2	16.6 ± 2.1	–22
	(mg/100 ml)		
Glucose	96.0 ± 2.5	43.0 ± 4.8	–55[a]

Groups of five to ten fasting 150- to 200-g rats were killed 2 hr after receiving insulin (2 I.U./kg, i.p.).
[a]$p < 0.01$.
(Reproduced from Fernstrom and Wurtman, 1972*b*.)

serotonin, these indoles were measured in rats fasted for 15 hr and then given free access to a carbohydrate diet. In a typical experiment, the animals ate an average of 5 g/hr during the first hour, and 2 g/hr during the second and third hours (Fernstrom and Wurtman, 1971*b*, 1972*b*). Plasma tryptophan levels were significantly elevated 1, 2, and 3 hr after food presentation; plasma tyrosine concentrations were depressed at all three times studied. Brain tryptophan and serotonin were significantly elevated at 2 and 3 hr (Fernstrom and Wurtman, 1971*b*, 1972*b*; Table 2).

Since carbohydrate consumption, by eliciting insulin secretion, raised plasma tryptophan levels, and ultimately the concentrations of tryptophan and serotonin in the brain, we anticipated that the simultaneous consumption of carbohydrates and protein would cause an even greater rise in brain serotonin. In addition to elevating plasma tryptophan by causing insulin secretion, the tryptophan in the dietary protein would contribute directly to plasma tryptophan; brain tryptophan and serotonin would presumably increase accordingly. However, when we gave fasted rats access to diets containing natural protein, we found that the expected major increase (about 60%, $p < 0.001$) in plasma

TABLE 2. *The effect of carbohydrate ingestion on brain serotonin concentrations and on plasma and brain tryptophan*

	Time after presentation of food (hr)			
	0	1	2	3
Plasma tryptophan	10.86 ± 0.55	13.56 ± 0.81[a]	14.51 ± 0.70[b]	13.22 ± 0.65[a]
Brain tryptophan	6.78 ± 0.40	8.32 ± 0.63[c]	11.24 ± 0.52[b]	9.81 ± 0.50[b]
Brain serotonin	0.549 ± 0.015	0.652 ± 0.046	0.652 ± 0.012[b]	0.645 ± 0.017[b]
Plasma tyrosine	13.03 ± 0.29	9.55 ± 0.34[b]	8.67 ± 0.26[b]	9.03 ± 0.21[b]

Plasma amino acid concentrations are in micrograms per milliliter. Brain tryptophan and serotonin concentrations are in micrograms per gram brain, wet weight. Average animal weight was 160 g.
[a] $p < 0.02$, differs from 0-hour group.
[b] $p < 0.001$, differs from 0-hour group.
[c] $p < 0.05$, differs from 0-hour group.
(Reproduced from Fernstrom and Wurtman, 1971b.)

tryptophan was not accompanied by increases in brain tryptophan or serotonin (Fernstrom and Wurtman, 1972a; Fernstrom, Larin, and Wurtman, 1973; Fig. 3).

It seemed possible that brain tryptophan failed to increase after protein ingestion because the plasma concentrations of other, competing, amino acids increased even more than that of tryptophan. To test this hypothesis, we allowed groups of animals to eat either a synthetic diet containing carbohydrates plus all of the amino acids in the same proportions as are present in an 18% casein diet, or this diet minus five of the amino acids thought to share a common transport system with tryptophan (tyrosine, phenylalanine, leucine, isoleucine, and valine) (Blasberg and Lajtha, 1965). Both diets significantly increased plasma tryptophan levels above those found in fasted controls. However, large increases in brain tryptophan, serotonin, or 5-HIAA occurred only when the competing neutral amino acids were deleted from the diet (Fernstrom and Wurtman, 1972a).

When this experiment was repeated omitting aspartate and glutamate (two acidic amino acids) from the diet instead of the five neutral amino acids, plasma tryptophan concentrations again increased 70 to 80% above those of fasted controls ($p < 0.001$); however, brain tryptophan, serotonin, and 5-HIAA remained unaffected.

We postulated that brain tryptophan and 5-hydroxyindole levels did not simply reflect plasma tryptophan, but depended also upon the plasma concentrations of other neutral amino acids. This relationship was confirmed by a correlation analysis comparing brain tryptophan and the ratio of plasma tryptophan to the five competing amino aicds in individual rats given diets containing various amounts of each amino acid (Fernstrom and Wurtman, 1972a). This analysis yielded a correlation coefficient of 0.95 ($p < 0.001$ that $r = 0$), whereas the correlation between brain tryptophan and plasma tryptophan alone was less striking ($r = 0.66$; $p < 0.001$ that $r = 0$). Similarly, the correlation coefficient for brain 5-hydroxyindoles (serotonin plus 5-HIAA) vs. the plasma amino acid ratio was 0.89 ($p < 0.001$), whereas that of 5-hydroxyindoles vs. tryptophan alone was only 0.58 ($p < 0.001$). The reason that brain tryptophan and serotonin appeared, in our earlier hypothesis, to depend upon plasma tryptophan alone was that all of the physiological manipulations tested at that time (tryptophan injections, insulin injections, carbohydrate consumption) raised the *numerator* in the plasma tryptophan:competitor ratio while either lowering the denominator or leaving it unaltered. Only when rats consumed protein were both the numerator and the denominator elevated. The effect of food consumption on 5-hydroxyindoles in rat brain may now be represented as in Fig. 4.

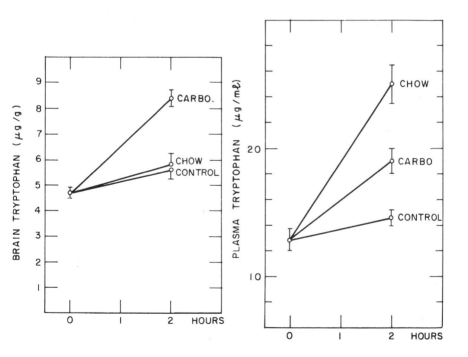

FIG. 3 Changes in brain and plasma tryptophan concentrations following the consumption of different foods. Groups of six rats were killed 1 or 2 hr after diet presentation. Vertical bars represent standard errors of the mean. Two-hour plasma tryptophan levels were significantly greater in rats consuming either diet than in fasting controls (chow: $p < 0.001$; carbohydrate: $p < 0.01$). Two-hour brain tryptophan levels were significantly elevated above controls only in rats consuming the carbohydrate-plus-fat diet ($p < 0.001$) (Fernstrom et al., 1973).

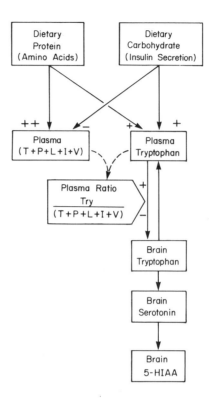

FIG. 4. Proposed sequence describing diet-induced changes in brain serotonin concentration in the rat (see text). The ratio of tryptophan to the combined levels of tyrosine, phenylalanine, leucine, isoleucine, and valine in the plasma is thought to control the tryptophan level in the brain.

In humans, although serum tryptophan does not increase after insulin injection as in rats, it does not substantially decrease (Lipsett, Madras, Wurtman, and Munro, 1973), as do the plasma concentrations of most other amino acids (including the competing neutral amino acids). Thus the plasma ratio of total tryptophan to the sum of the competing neutral amino acids probably increases, as it does in rats. We might then expect that brain tryptophan and serotonin levels would also rise in humans following insulin injection (or carbohydrate consumption). Furthermore, since the plasma concentrations of both tryptophan and the other neutral amino acids should rise in humans after the ingestion of protein, changes in the plasma amino acid ratio, and in brain tryptophan and serotonin levels, should again be similar in rats and man.

Tryptophan in plasma is distributed between two pools: about 10 to 20% circulates as the free amino acid, while the remainder is bound to serum albumin (McMenamy, Lund, and Oncley, 1957). None of the other amino acids binds appreciably (i.e., more than 1 to 2% of total concentration) to plasma proteins. Because binding in general implies storage, several investigators have suggested that the plasma free tryptophan pools determine the availability of circulating tryptophan to brain and other tissues (Knott and Curzon, 1972).

A variety of lipid-soluble compounds which bind to albumin in the blood [e.g., hormones, drugs, nonesterified fatty acids (NEFA)] may displace each other. Thus, for example, increasing the serum concentration of NEFA *in vitro* causes the concentration of albumin-bound tryptophan to fall, and that of free tryptophan to rise (McMenamy and Oncley, 1958). In collaboration with Drs. Bertha Madras and Hamish Munro, we examined this relationship *in vivo* by feeding rats diets that were expected to alter serum NEFA levels and then measuring the changes in serum free and albumin-bound tryptophan. We then attempted to determine whether diet-induced changes in brain tryptophan and 5-hydroxyindoles were best correlated with serum free or bound tryptophan (or, as described above, with changes in the ratio of serum tryptophan to the sum of the competing neutral amino acids) (Madras, Cohen, Fernstrom, Larin, Munro, and Wurtman, 1973). These data, described in detail elsewhere (Madras, Cohen, Munro, and Wurtman, *This Volume*), indicated that diet-induced changes in brain tryptophan are not at all correlated with what happens to serum free tryptophan. For example, the consumption by rats of a carbohydrate diet *decreases* serum free tryptophan (presumably via the insulin-mediated decline in serum NEFA), but *increases* brain tryptophan and 5-hydroxyindoles (probably by raising the serum ratio of tryptophan to competing neutral amino acids). Thus, the binding of serum tryptophan to albumin does *not* appear to limit the availability of the amino acid to the brain. Indeed, the physiological significance of tryptophan binding appears to be just the opposite: it allows serum tryptophan levels to remain elevated after insulin is secreted, at a time when the serum concentrations of its competitors for brain uptake are declining. Thus, albumin-binding makes it possible for carbohydrate consumption to elevate brain tryptophan and, consequently, 5-hydroxyindoles.

ACKNOWLEDGMENTS

These studies were supported in part by grants from the John A. Hartford Foundation, the National Aeronautics and Space Administration, and the U.S. Public Health Service (NS-10459 and AM-14228).

REFERENCES

Albrecht, P., Visscher, M. B., Bittner, J. J., and Halberg, F. (1956): Daily changes in 5-hydroxytryptamine concentration in mouse brain. *Proceedings of the Society for Experimental Biology and Medicine*, 92:702–706.

Blasberg, R., and Lajtha, A. (1965): Substrate specificity of steady-state amino acid transport in mouse brain slices. *Archives of Biochemistry and Biophysics*, 112:361–377.

Fernstrom, J. D., and Wurtman, R. J. (1971a): Brain serotonin content: Physiological dependence on plasma tryptophan levels. *Science*, 173:149–152.

Fernstrom, J. D., and Wurtman, R. J. (1971b): Brain serotonin content: Increase following ingestion of carbohydrate diet. *Science*, 174:1023–1025.

Fernstrom, J. D., and Wurtman, R. J. (1972a): Brain serotonin content: Physiological regulation by plasma neutral amino acids. *Science*, 178:414–416.

Fernstrom, J. D., and Wurtman, R. J. (1972b): Elevation of plasma tryptophan by insulin in the rat. *Metabolism*, 21:337–342.

Fernstrom, J. D., and Wurtman, R. J. (1974): Nutrition and the brain. *Scientific American* (in press).

Fernstrom, J. D., Larin, F., and Wurtman, R. J. (1973): Correlations between brain tryptophan and plasma neutral amino levels following food consumption. *Life Sciences*, 13:517–524.

Fishman, B., Wurtman, R. J., and Munro, H. N. (1969): Daily rhythms in hepatic polysome profiles and tyrosine transaminase activity: Role of dietary protein. *Proceedings of the National Academy of Sciences*, 64:677–682.

Guroff, G., and Udenfriend, S. (1962): Studies on aromatic amino acid uptake by rat brain *in vivo*. *Journal of Biological Chemistry*, 237:803–806.

Knott, P. J., and Curzon, G. (1972): Free tryptophan in plasma and brain tryptophan metabolism. *Nature*, 239:452–453.

Lipsett, D., Madras, B. K., Wurtman, R. J., and Munro, H. N. (1973): Serum tryptophan level after carbohydrate ingestion: Selective decline in non-albumin-bound tryptophan coincident with reduction in serum free fatty acids. *Life Sciences*, 12(Part 2):57–64.

Lovenberg, W., Jéquier, E., and Sjoerdsma, A. (1968): Tryptophan hydroxylation in mammalian systems. *Advances in Pharmacology*, 6A:21–36.

Madras, B. K., Cohen, E. L., Fernstrom, J. D., Larin, F., Munro, H. N., and Wurtman, R. J. (1973): Dietary carbohydrate increases brain tryptophan and decreases serum-free tryptophan. *Nature*, 244:34–35.

Marliss, E. B., Aoki, T. T., Unger, R. H., Soeldner, J. S., and Cahill, G. F. (1970): Glucagon levels and metabolic effects in fasting man. *Journal of Clinical Investigation*, 49:2256–2270.

McMenamy, R. H., Lund, C. C., and Oncley, J. L. (1957): Unbound amino acid concentrations in human blood plasmas. *Journal of Clinical Investigation*, 36:1672–1679.

McMenamy, R. H., and Oncley, J. L. (1958): The specific binding of L-tryptophan to serum albumin. *Journal of Biological Chemistry*, 233:1436–1447.

Moir, A. T. B., and Eccleston, D. (1968): The effects of precursor loading on the cerebral metabolism of 5-hydroxyindoles. *Journal of Neurochemistry*, 15:1093–1108.

Wurtman, R. J. (1970): Daily rhythms in mammalian protein metabolism. In: *Mammalian Protein Metabolism*, vol. 4, edited by H. N. Munro, pp. 445–479. Academic Press, New York.

Wurtman, R. J., and Fernstrom, J. D. (1972): L-Tryptophan, L-tyrosine, and the control of brain monoamine biosynthesis. In: *Perspectives in Neuropharmacology*, edited by S. H. Snyder, pp. 145–193. Oxford University Press, New York.

Wurtman, R. J., Rose, C. M., Chou, C., and Larin, F. (1968): Daily rhythms in the concentrations of various amino acids in human plasma. *New England Journal of Medicine*, 279:171–175.

Advances in Biochemical Psychopharmacology, Vol. 11
Raven Press, New York © 1974

Elevation of Serum Free Tryptophan, but Not Brain Tryptophan, by Serum Nonesterified Fatty Acids

B. K Madras, E. L. Cohen, H. N. Munro,
and R. J. Wurtman

*Department of Nutrition and Food Science, Massachusetts Institute of
Technology, Cambridge, Massachusetts 02139*

I. INTRODUCTION

Soon after plasma tryptophan is elevated in rats by the injection of tryptophan (Fernstrom and Wurtman, 1971*a*) or insulin (Fernstrom and Wurtman, 1972*a*), or by the consumption of a carbohydrate diet (Fernstrom and Wurtman, 1971*b*), the concentration of tryptophan in brain increases markedly. This is followed by an elevation in brain 5-hydroxyindoles, presumably mediated by the increased saturation of the enzyme tryptophan hydroxylase with its amino acid substrate (Fernstrom and Wurtman, 1971*a*). The increase in plasma tryptophan following insulin administration in rats (Fernstrom and Wurtman, 1972*a*), and the failure of serum tryptophan concentration to decline significantly in humans (Lipsett, Madras, Wurtman, and Munro, 1973) contrasts with the major decreases that insulin causes in levels of other plasma neutral amino acids (Fernstrom and Wurtman, 1972*a*, 1972*b*). Since tryptophan apparently competes with these neutral amino acids for transport into the brain, insulin is thereby able to elevate brain tryptophan (Fernstrom and Wurtman, 1972*b*).

The paradoxical response of plasma or serum tryptophan to insulin appears to result from an exceptional property of the amino acid: It is the only dietary amino acid that circulates in plasma in two forms, bound to albumin and unbound (McMenamy, Lund, and Oncley, 1957; McMenamy and Oncley, 1958). As described below, insulin increases serum tryptophan in rats by elevating the concentration of the albumin-bound fraction. Many other biologically active compounds that bind to albumin are rendered unavailable to certain tissues as a consequence. Hence, it seemed possible, *a priori*, that the unbound form of plasma or serum tryptophan would be most accessible for transport into brain, and that changes in this fraction after food consumption would be best correlated with brain tryptophan levels. We therefore elected to study the physiological control of the binding of tryptophan to albumin, and the consequences of this binding for brain tryptophan levels. Our studies suggest that serum nonesterified fatty acids (NEFA), which are tightly bound to albumin, can significantly decrease the binding capacity of albumin for tryptophan. Since insulin decreases NEFA, it causes a corresponding fall in serum free tryptophan and a simultaneous and large (in the rat) increase in albumin-bound tryptophan. Brain tryptophan levels change in an *opposite* direction from serum free tryptophan.

II. METHODS

In studies on humans, seven healthy males, 21 to 27 years old, were fasted overnight, from 8 p.m. to 8 a.m. After a fasting blood sample was taken at 8 a.m., the subjects were given glucose (75 g) to drink ("Glucola," Ames Company, Elkhart, Indiana). Blood samples, 20 ml, were collected into "vacutainers" at 30, 60, 90, 120, and 180 min, and allowed to clot. Following centrifugation at 2000 rpm (I.E.C. International Centrifuge head #279, 15 cm radius) for 10 min, the serum was transferred to tubes containing 5% CO_2 -95% N_2 to maintain pH.

In studies on rats, animals kept at 20 to 22°C in a room lit between 9 a.m. and 9 p.m. by "Vita-Lite" (50 to 75 foot-candles; Duro-Test Corp., North Bergen, N.J.) were deprived of food but not water for 15 to 20 hr before the start of each experiment. The animals were presented with a synthetic meal at 10 to 11 a.m. and were killed anternately with controls by decapitation 2 hr later. Blood was collected from the cervical wound and transferred to tubes which were kept at 4°C and which contained the gas mixture described above. Serum was pooled from two animals. Since both temperature and pH can significantly affect the binding of tryptophan to albumin [binding increases with decreasing temperature and increases with increasing pH (McMenamy and Oncley, 1958)], we selected a method for determining free tryptophan (equilibrium dialysis) which best enables us to maintain temperature and pH at physiological values, that is, 37°C ± 0.5, pH 7.4_5 to 7.5_5 .

Visking dialysis tubing (1 cm flat width, 18 cm lengths) was boiled for 15 min each time, twice in 0.0002 M disodium EDTA, twice in distilled water, and stored in distilled water at 4°C. The dialysis buffer was Krebs-improved Ringer bicarbonate, pH 7.4_5 (McMenamy, Lund, Van Marcke, and Oncley, 1961). In the human studies, the buffer also contained small molecular weight constituents of plasma as described by McMenamy et al. (1961). However, we later tested the value of these additions for rat serum and found them to have little influence on the results. Before dialysis, a length of tubing was knotted at one end, and 0.5 ml of buffer was pipetted into the sac. The dialysis tubing was knotted at the other end, approximately 6.5 cm from the first knot, folded in half, and introduced into a tube of 1-cm internal diameter containing 2 ml of serum. The same concentrations of free tryptophan are obtained using serum:buffer ratios of 4:1 and 15:1, after appropriate dilution corrections are made. After flushing with the $CO_2 : N_2$ mixture, the tube was stoppered tightly and placed in a shaking water bath maintained at $37°C$. Three-and-a-half hours later, at which time equilibrium is achieved, the dialysis sac was removed and the pH of the serum was measured. The pH was maintained at 7.4_5 to 7.5_5.

Serum total tryptophan and free tryptophan were determined by the method of Denckla and Dewey (1967). The concentration of bound tryptophan was calculated from the concentrations of free and total tryptophan, after volume corrections were made. NEFA were determined by the method of Dole and Meinertz (1960). Glucose was measured by the standard clinical method, and serum tyrosine by the fluorimetric procedure of Waalkes and Udenfriend (1957). Brains were removed immediately after decapitation, frozen on solid CO_2 and stored at $-70°C$ until tryptophan was assayed (Denckla and Dewey, 1967). When serotonin and 5-hydroxyindoleacetic acid were measured at the same time (Curzon and Green, 1970), the brains were divided in half at the midline, one-half being used for the tryptophan assay, the other for serotonin and 5-hydroxyindoleacetic acid.

III. RESULTS

Following glucose consumption by human subjects (Fig. 1), blood glucose levels rose to their maximum at 30 min and started to decline within 60 min. The concentration of serum NEFA fell progressively, reaching its lowest level 90 min after glucose. The concentration of total tryptophan fell slightly and progressively until 180 min, at which time it was 83% of control values (13.9 μg/ml to 11.6 μg/ml, $p < 0.02$). However, free tryptophan fell to 65% of control values between 90 and 180 min after glucose, from 3.4 μg/ml to 2.2 μg/ml ($p < 0.01$ at all times). The concentration of bound tryptophan did not change significantly from the initial value of 10.5 μg/ml. As a result of the large decline in the free moiety, the proportion of bound to total tryptophan rose significantly from 75% to 81% at 90 to 180 min after glucose. The decline in the unbound fraction paralleled the decline in NEFA. This observation led us to design experiments to determine whether the fall

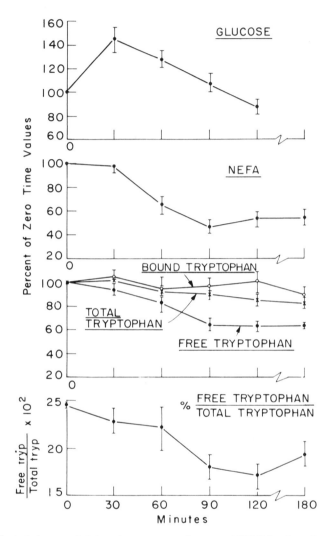

FIG. 1. Effect of glucose administration on serum glucose and NEFA levels, and on tryptophan levels and distribution between albumin-bound and free forms. The data are mean values ± standard errors for seven adult male subjects, and are expressed as deviations from the initial fasting level taken as 100 (initial glucose 84.8 ± 2.8 mg/100 ml, NEFA 0.73 ± 0.087 meq/liter, total tryptophan 13.9 ± 0.3 μg/ml, free tryptophan 3.4 ± 0.06 μg/ml, bound tryptophan 10.5 ± 0.3 μg/ml).

in NEFA caused the change in tryptophan binding. [Fatty acids are tightly bound to albumin and can discharge other molecules (Spector and Imig, 1971) from binding sites. Thus, a lowering of serum NEFA could result in an enhanced affinity of albumin for tryptophan.]

Five human subjects were given glucose to drink in order to produce an insulin-mediated decrease in serum NEFA. Blood samples were taken at 90 min, and the sera, harvested by centrifugation, were pooled.

Oleic acid in ethanol was added to the glass dialysis tubes in increments of 0.1 meq/liter (final serum concentration). Following evaporation of the ethanol, 2.0-ml aliquots of the pooled sera were added to each of the tubes, and equilibrium dialysis was performed. In the absence of added oleic acid, and in the presence of endogenous NEFA at a concentration of 0.290 meq/liter, free tryptophan levels were 1.7 μg/ml. In the samples to which oleic acid was added, bringing the total NEFA level to 0.826 meq/liter, the free tryptophan level rose to 2.3 μg/ml, an increase of 35% over the initial level. When the NEFA level was raised to 1.14 meq/liter by the addition of oleic acid (a concentration comparable with that found during fasting), freely dialyzable tryptophan levels rose to 2.8 μg/ml, an increase of 65% over initial values. Thus nonesterified fatty acids can profoundly influence the equilibrium between free and bound tryptophan. To assess the possibility that changes in serum free tryptophan altered the availability of the amino acid to tissues, we performed animal experiments. Fasted animals were killed after being allowed to consume one of two protein-free diets for 2 hr. One diet, used in previous studies in this laboratory (Fernstrom and Wurtman, 1971*b*), contained both carbohydrate and fat; fat was omitted in the other. Both diets *raised* serum total tryptophan, and brain tryptophan (Table 1), in confirmation of previous studies (Fernstrom and Wurtman, 1971*b*). Both diets *depressed* serum NEFA and the concentrations of serum free tryptophan and tyrosine, one of the neutral amino acids previously shown to compete physiologically with tryptophan for entry into brain (Fernstrom and Wurtman, 1972*b*). This observation supported the conclusion reached from the human studies: that insulin secretion increases the binding capacity of albumin for tryptophan probably by lowering albumin-bound NEFA. Animals fed the fat-containing diet showed much smaller decreases in NEFA and serum free tryptophan than the group fed the fat-free meal. Serum free tryptophan levels differed significantly between these two groups of rats (5.7 μg/ml *vs.* 3.4 μg/ml, $p < 0.001$). Even so, brain tryptophan levels were similar. To examine further the relationship between physiological changes in serum free tryptophan and brain tryptophan, groups of animals were presented with one of four diets, all containing protein, but differing in fat content. As the fat content of each synthetic diet increased, the concentrations of serum NEFA and of free tryptophan rose higher (Fig. 2); serum *total* tryptophan remained unchanged. Thus, a threefold increase in the level of free tryptophan was observed in the animals eating a meal containing 45% fat, compared with those consuming one containing no fat (i.e., from 4.1 ± 0.17 μg/ml to 12.0 ± 0.73 μg/ml;

TABLE 1. *Effects of carbohydrate and carbohydrate-fat diets on serum and brain tryptophan*

		Diets	
	Fasted controls	Carbohydrate + fat	Carbohydrate
Serum total tryptophan (μg/ml)	16.5 ± 0.3	18.4 ± 0.5[a]	19.1 ± 0.4[b]
Serum free tryptophan (μg/ml)	6.2 ± 0.1	5.7 ± 0.2[c]	3.4 ± 0.2[b]
Free (% of total)	37	33	18
Serum bound tryptophan (μg/ml)	10.3 ± 0.4	12.7 ± 0.7[c]	15.7 ± 0.5[b]
Serum tyrosine (μg/ml)	19.5 ± 0.7	11.7 ± 0.4	14.4 ± 0.5
NEFA (meg/liter)	0.831 ± 0.021	0.615 ± 0.029[b]	0.301 ± 0.024[b]
Brain tryptophan (μg/g)	2.24 ± 0.11	3.07 ± 0.18[b]	3.45 ± 0.19[b]

Two groups of 22 rats weighing 170 to 200 g were deprived of food but not water at 2 p.m. and presented with one of the experimental diets at 10:30 a.m the following day. Two hours later, the animals were decapitated and serum and brains taken for assay. Controls (44 rats) had free access to water and were killed during the duration of the experiment. Each serum value is obtained from two pooled samples. Each diet contained 270 g of dextrose, 221 g of sucrose, 270 g of dextrin, 40 g of Harper's salt mix, 10 g of vitamin mix (Wurtman et al., 1968), and 2 g of choline, to which was added 35 g of agar in 1 liter of water. The fat diet had an additional 150 g of Mazola oil. All values are given as mean ± SEM.

[a] $p < 0.01$, differs from controls.
[b] $p < 0.001$, differs from controls.
[c] $p < 0.05$, differs from controls.

$p < 0.001$). In spite of the profound rise in serum free tryptophan,* brain tryptophan, serotonin, and 5-hydroxyindoleacetic acid levels did not become elevated. In the group of animals eating the diet with the

---

*In a separate study with two groups of animals fed a diet containing 0% fat and 30% fat, free tryptophan was measured by three methods: equilibrium dialysis at 37°C, equilibrium dialysis at 0°C, and by ultrafiltration using diaflo cones (Amicon Corp.). Free tryptophan was 260%, 450%, and 200% greater in the rats fed the diet containing 30% fat than in those fed the diet containing no fat, respectively, with the three different methods. Nonesterified fatty acids were measured after equilibrium dialysis at both 37°C and 0°C and compared with the concentrations obtained before dialysis. At 37°C there was a small (10%) but insignificant increase in fatty acids; at 0°C there was no change.

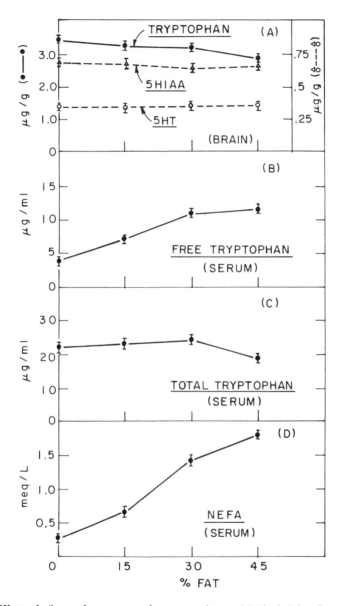

FIG. 2. Effect of dietary fat on serum free tryptophan and brain indoles. Four groups of animals (12 per group) were presented with four synthetic diets: *0% fat*: casein (200 g), sucrose (203 g), dextrose (254 g), dextrine (254 g); *15% fat*: Mazola oil (150 g), casein (200 g), sucrose (160 g), dextrine (200 g), dextrine (200 g); *30% fat*: Mazola oil (300 g), casein (200 g), sucrose (116 g), dextrose (147 g), dextrine (147 g); *45% fat*: Mazola oil (450 g), casein (200 g), sucrose (74 g), dextrose (93 g), dextrine (93 g). To each diet was added choline chloride (2 g/4 ml), vitamin mix (10 g) (Wurtman, Shoemaker, and Larin, 1968), and Harper's salt mix (40 g). Each diet was suspended in an agar base (35 g/liter H_2O). Rats were killed 2 hr after presentation of food, and assays were performed as described under Methods. All values are given as mean ± SEM.

highest fat content (45%), serum *total* tryptophan fell (from 22.0 ± 1.0 μg/ml to 18.7 ± 0.73 μg/ml; $p < 0.05$); brain tryptophan followed suit (3.43 ± 0.22 μg/g to 2.88 ± 0.12 μg/g). Thus, our observations indicate that the concentration of free tryptophan does *not* predict changes in brain tryptophan caused by physiological inputs such as eating. As described elsewhere (Fernstrom and Wurtman, *This Volume*), the best predictor of brain tryptophan appears to be the ratio of plasma or serum tryptophan to the sum of the neutral amino acids which compete with it for entry into the brain.

IV. DISCUSSION

These data suggest that serum nonesterified fatty acids, which circulate almost entirely bound to albumin, interfere with the binding of tryptophan to the protein molecule *in vivo*. This interaction permits an explanation of the paradoxical responses of serum tryptophan to insulin in rats and humans. Following carbohydrate consumption and insulin secretion, NEFA levels plunge, thus decreasing the amount of NEFA bound to circulating albumin. This allows more tryptophan to become bound, in rats, and prevents the dissociation of tryptophan from albumin, in humans. Hence, in rats, serum free tryptophan levels fall, albumin-bound tryptophan increases, and total tryptophan levels rise. In humans, serum free tryptophan levels also fall, albumin-bound tryptophan remains unchanged, and serum total tryptophan decreases slightly. Concurrently, in both species, the total serum concentrations of other neutral amino acids, which compete with tryptophan for entry into the brain, fall markedly. Perhaps the major significance of tryptophan binding is that it thus allows the ratio of serum total tryptophan to the sum of the neutral amino acids to increase with insulin secretion; this change occurs both in rats and humans. In rats it has been shown to cause elevations in brain tryptophan and serotonin (Fernstrom and Wurtman 1972*b*). That total serum tryptophan declines slightly in humans, yet increases in rats after carbohydrate consumption, may be accounted for by a species difference in the interaction of the tryptophan-binding mechanism with NEFA (Madras et al., *in preparation*). Under physiological conditions such as eating, there appears to be no direct correlation between changes in serum free tryptophan and in brain tryptophan.

ACKNOWLEDGMENTS

These studies were supported by grants from the National Aeronautics and Space Administration (NGR-22-009-627) and the National Institutes of Health (AM-14228).

REFERENCES

Curzon, G., and Green, A. R. (1970: Rapid method for the determination of serotonin and 5-hydroxyindoleacetic acid in small regions of rat brain. *British Journal of Pharmacology*, 39:653–655.

Denckla, W. D., and Dewey, H. K. (1967): The determination of tryptophan in plasma, liver, and urine. *Journal of Laboratory and Clinical Medicine*, 69:160–169.

Dole, V. P., and Meinertz, H. (1960): Microdetermination of long-chain fatty acids in plasma and tissues. *Journal of Biological Chemistry*, 235:2595–2599.

Fernstrom, J. D., and Wurtman, R. J. (1971a): Brain serotonin content: Physiological dependence on plasma tryptophan levels. *Science*, 173:149–152.

Fernstrom, J. D., and Wurtman, R. J. (1971b): Brain serotonin content: Increase following ingestion of carbohydrate diet. *Science*, 174:1023–1025.

Fernstrom, J. D., and Wurtman, R. J. (1972a): Elevation of plasma tryptophan by insulin in rat. *Metabolism*, 21:337–342.

Fernstrom, J. D., and Wurtman, R. J. (1972b): Brain serotonin content: Physiological regulation by plasma neutral amino acids. *Science*, 178:414–416.

Lipsett, D., Madras, B. K., Wurtman, R. J., and Munro, H. N. (1973): Serum tryptophan level after carbohydrate ingestion: Selective decline in nonalbumin-bound tryptophan coincident with reduction in serum free fatty acids. *Life Sciences*, 12 (Part II):57–64.

McMenamy, R. H., Lund, C. C., and Oncley, J. L. (1957): Unbound amino acid concentrations in human blood plasmas. *Journal of Clinical Investigation*, 36:1672–1679.

McMenamy, R. H., Lund, C. C., Van Marcke, J., and Oncley, J. L. (1961): The binding of L-tryptophan in human plasma at 37°C. *Archives of Biochemistry and Biophysics*, 93:135–139.

McMenamy, R. H., and Oncley, J. L. (1958): Specific binding of tryptophan to serum albumin. *Journal of Biological Chemistry*, 233:1436–1447.

Spector, A., and Imig, B. (1971): Effect of free fatty acid concentration on the transport and utilization of other albumin-bound compounds: Hydroxyphenylazobenzoic acid. *Molecular Pharmacology*, 7:511–518.

Waalkes, T. P., and Udenfriend, S. (1957): A fluorometric method for the estimation of tyrosine in plasma and tissues. *Journal of Laboratory and Clinical Medicine*, 50:733–736.

Wurtman, R. J., Shoemaker, W. J., and Larin, F. (1968): Mechanism of the daily rhythm in hepatic tyrosine transaminase activity: Role of dietary tryptophan. *Proceedings of the National Academy of Sciences*, 59:800–807.

Advances in Biochemical Psychopharmacology, Vol. 11
Raven Press, New York © 1974

Role of Active Transport of Tryptophan in the Control of 5-Hydroxytryptamine Biosynthesis

M. Hamon, S. Bourgoin, Y. Morot-Gaudry, F. Héry, and J. Glowinski

Groupe NB (INSERM U.114), Laboratoire de Biologie Moléculaire, Collège de France, Paris, France

I. INTRODUCTION

Since the study of Green, Greenberg, Erickson, Sawyer, and Ellizon (1962), many reports have revealed the importance of changes in the availability of tryptophan (TRY) in the modulation of serotonin (5-HT) synthesis in the central nervous system. Peripheral injection of this essential amino acid (Moir and Eccleston, 1968; Grahame-Smith, 1971a; Fernstrom and Wurtman, 1971; Carlsson, Bedard, Lindqvist, and Magnusson, 1972) or its direct introduction into the cerebral ventricle (Consolo, Garattini, Ghielmetti, Morselli, and Valzelli, 1965) increases the levels of 5-HT and of its metabolite 5-hydroxyindoleacetic acid (5-HIAA) in brain. Conversely, central 5-HT levels are reduced in rat fed with a TRY-free diet (Gál and Drewes, 1962). On the other hand, several treatments which stimulate 5-HT synthesis or turnover enhance brain TRY content (Tagliamonte, Tagliamonte, Perez-Cruet, Stern, and Gessa, 1971). In most cases, these physical or pharmacological states are associated with parallel changes in plasmatic free TRY concentrations (Knott and Curzon, 1972; Tagliamonte, Biggio, Vargiu, and Gessa, 1973). However, previously we have observed that changes in the active transport of the amino acid contribute also to the regulation of 5-HT synthesis (Glowinski, Hamon, and Héry, 1973). Thus, we would like to show: (1) that changes in 5-HT synthesis are not always associated with modifications of brain TRY content; (2) that changes in TRY concentrations may occur without parallel fluctuations in the transmitter formation; (3) that a good correlation exists between the alterations of the active transport of TRY in tissues and the variations of 5-HT synthesis. Finally, we would like to discuss the possible role in the control of 5-HT synthesis of the high-affinity transport process for TRY which is present in synaptosomes and in glial cells.

II. RELATIONS BETWEEN CHANGES IN TRY AVAILABILITY IN BRAIN AND 5-HT SYNTHESIS IN *IN VIVO* EXPERIMENTS

Various drugs which increase brain TRY content do not stimulate 5-HT synthesis. In our laboratory, this has been observed in the mouse brain after probenecid (200 mg/kg, i.p., 30 min), pargyline (75 mg/kg, i.p., 6 min) (Morot-Gaudry, Hamon, Bourgoin, and Glowinski, 1973*b*), and LSD (0.5 mg/kg, i.p., 1 hr) (Glowinski et al., 1973). A similar increase in TRY brain concentrations, with no change in 5-HT synthesis, has been seen by Korf, Van Praag, and Sebens (1972) after probenecid treatment. The LSD effect was particularly striking since, after ^{3}H-TRY intravenous injection, ^{3}H-5-HT formation was reduced although endogenous and labeled levels of TRY were increased in tissues (Morot-Gaudry, Bourgoin, Hamon, and Glowinski, 1973*a*). The inhibitory effect of LSD on 5-HT synthesis has also already been reported (Schubert, Nybäck, and Sedvall, 1970). Furthermore the increase of brain TRY content elicited by LSD was reported by Tonge and Leonard (1970) and by Freedman and Boggan (*preceding volume*).

However, as others, we have also found situations in which an increase in brain TRY content levels was associated with a parallel modification in 5-HT synthesis. Some stressful situations, such as electric foot shocks, stimulate central 5-HT turnover (Thierry, Fekete, and Glowinski, 1968). Therefore, changes in ^{3}H-TRY availability and 5-HT synthesis have been examined, 10 min after the intravenous injection of the labeled amino acid, in brain of mice submitted to two different stresses. As indicated by the increased conversion index of TRY into 5-HT, seen shortly after light diethyl ether anesthesia, the amine synthesis was accelerated. This effect was associated with a rise in endogenous and labeled levels of TRY (Bourgoin, Morot-Gaudry, Glowinski, and Hamon, 1973). Similarly, using the isotopic method but measuring the initial accumulation of ^{3}H-5-HT in tissues after MAO inhibition (pargyline 75 mg/kg, i.p., 1 min before ^{3}H-TRY injection), we found an increased conversion of TRY into 5-HT in brain of mice previously kept for 1 hr at 35°C. Again both tritiated and endogenous TRY levels were increased in tissues (Morot-Gaudry et al., 1973*a*). As already shown by Curzon, Friedel, and Knott (1973), a rise in the free form of TRY in plasma, induced by increased levels of free fatty acids, may be involved in this effect. However, the respective contribution in the stimulation of 5-HT synthesis of the changes in plasmatic free TRY concentrations or of those of TRY active transport at the membrane level cannot be distinguished.

One way to appreciate independently the role of TRY transport *in vivo* is to inject the labeled precursor into the cerebrospinal fluid. For instance, in a situation such as food deprivation, which increases both brain TRY levels and 5-HT synthesis (Curzon, Joseph, and Knott,

1972), ^{3}H-TRY as well as ^{3}H-5-HT levels were enhanced in the brainstem of rats 45 min after the intracisternal injection of the labeled amino acid (Glowinski et al., 1973). Similarly, 2 hr after progesterone (10 mg/animal, i.p.), the brain content of TRY, 5-HT, and 5-HIAA was increased.

Furthermore, after this latter treatment a marked rise in ^{3}H-TRY and in ^{3}H-5-HT (+68%) levels was observed in hypothalamus and whole brain of rats 15 min after injection of ^{3}H-TRY in the cisterna magna (Glowinski et al., 1973). Finally, although brain TRY levels are not affected after paradoxical sleep deprivation, 5-HT synthesis is stimulated (Héry, Pujol, Lopez, Macon, and Glowinski, 1970; Cramer, Tagliamonte, Tagliamonte, Perez-Cruet, and Gessa, 1973). In this situation, as in previous ones, marked increases in ^{3}H-TRY and ^{3}H-5-HT levels were found in the brainstem of rats shortly after the intracisternal ^{3}H-amino acid injection. In all these examples, the stimulation of 5-HT synthesis does not seem to be related only to eventual modifications of TRY metabolism in plasma. The increased initial accumulation of ^{3}H-TRY observed in tissues following its intracisternal injection strongly suggests the occurrence of changes in the TRY transport process at a cellular level. These results underline the importance of TRY newly taken up in tissues in the regulation of 5-HT synthesis, as already noticed by Shields and Eccleston (1972).

III. RELATIONS BETWEEN CHANGES IN TRY AVAILABILITY AND 5-HT SYNTHESIS IN BRAIN SLICES

In *in vivo* experiments, changes in the initial accumulation of ^{3}H-TRY in tissues may in some cases reflect a modification of the amino acid turnover rather than only a change of its active transport. Since TRY turnover is strongly reduced in brain slices, we have examined the initial accumulation of ^{3}H-TRY *in vitro* to appreciate as precisely as possible the changes occurring in the transport process of the amino acid. The role of TRY-active transport in the regulation of 5-HT synthesis is illustrated by two experiments (Table 1). Initial accumulation of ^{3}H-TRY and formation of ^{3}H-5-HT were increased in brainstem slices of rats previously deprived of paradoxical sleep for 96 hr, a situation which stimulates *in vivo* 5-HT synthesis (see Section II). Conversely, LSD pretreatment (1 mg/kg, i.p., 1 hr), which inhibits *in vivo* 5-HT synthesis, reduced the formation of ^{3}H-5-HT as well as the accumulation of ^{3}H-TRY in striatal slices of rats (Hamon, Bourgoin, Jagger, and Glowinski, 1973*b*). LSD added in the incubating medium affected neither ^{3}H-TRY transport nor ^{3}H-5-HT synthesis. The persistence *in vitro* of changes in TRY transport induced by *in vivo* treatments strongly suggests important modifications in the carrier

TABLE 1. *Effect of various treatments on ³H-TRY accumulation and ³H-5-HT synthesis in rat brain slices*

Treatments	³H-TRY (µC/g)		³H-5-HT (µC/g)		³H-5-HIAA (µC/g)	
	Control	Treated	Control	Treated	Control	Treated
Paradoxical sleep deprivation	4.009 ± 0.076	4.752[a] ± 0.146	0.224 ± 0.011	0.294[a] ± 0.011	—	—
LSD (1 mg/kg, i.p., 1 hr)	23.45 ± 0.91	20.20[a] ± 0.42	1.541 ± 0.137	1.151[a] ± 0.106	—	—
Dibutyryl cAMP (1 mM)	28.32 ± 0.46	34.76[a] ± 1.30	1.305 ± 0.093	1.632[a] ± 0.113	0.231 ± 0.009	0.289[a] ± 0.023
EGTA (0.5 mM)	24.98 ± 0.32	29.45[a] ± 0.58	1.026 ± 0.045	1.609[a] ± 0.044	0.194 ± 0.010	0.228 ± 0.018
La³⁺ (0.5 mM)	19.67 ± 0.29	23.10[a] ± 0.76	1.190 ± 0.059[a]	1.589[a] ± 0.088	0.203 ± 0.014	0.258[a] ± 0.013

In all cases, brain slices were incubated in physiological medium with ^{3}H-TRY (generally labeled) for 30 min at 37°C under a constant stream of O_2:CO_2 (95/5%). Brainstems of rats deprived of paradoxical sleep for 96 hr were incubated with 10 µC of ^{3}H-TRY, corresponding to a concentration of 0.46 µM. Effect of LSD treatment (1 mg/kg, i.p., 60 min before death) was studied on striatal slices (^{3}H-TRY:19 µC = 1.20 µM). Hippocampal slices were used when effects of *in vitro* addition of dibutyryl cyclic AMP (1 mM), EGTA (0.5 mM), or La^{3+} (0.5 mM) were analyzed. In the two latter cases, incubations were performed in a Ca^{2+}-free medium. Slices were incubated with 20 µC (1.60 µM), 19 µC (1.20 µM) and 18 µC (0.65 µM) of ^{3}H-TRY, respectively. ^{3}H-TRY was estimated in slices; ^{3}H-5-HT and ^{3}H-5-HIAA were estimated both in slices and in their incubating medium. Results are the mean ± SEM of six to eight determinations.

$^a p < 0.05$ when compared with respective control values.

activity. Similar observations have been recently made on synaptosomes (Belin, Chouvet, and Pujol, 1973).

A parallel stimulation of TRY transport and 5-HT synthesis can be directly demonstrated by incubating striatal or hippocampal slices with dibutyryl cyclic AMP (1 mM) (Table 1). This effect was mainly related to a decrease in the apparent K_m of TRY uptake in slices (1.25 mM and 0.65 mM in control and dibutyryl cyclic AMP hippocampal slices, respectively). A comparable activation of the TRY transport and of the 5-HT synthesis was seen in hippocampal slices depleted of Ca^{2+} (preincubation with 0.5 mM EDTA or EGTA) or incubated in the presence of lanthanum (La^{3+}) (Table 1), an inhibitor of Ca^{2+} transport (Miledi, 1971).

IV. CHARACTERISTICS OF TRY UPTAKE PROCESSES IN GLIAL CELLS AND SYNAPTOSOMES

The number of 5-HT terminals in isolated brain structures is very small (Kuhar and Aghajanian, 1973). Consequently, changes in TRY transport in slices do not occur only in 5-HT terminals. In fact, slices constitute a heterogeneous material, containing cell bodies and nerve terminals surrounded by glial cells. Therefore, in collaboration with A. Bauman and P. Benda, we have examined the characteristics of TRY transport in glial cells (C_6 clone; Benda, Lightbody, Sato, Levine, and Sweet, 1968) and in cortical synaptosomes. Two uptake processes for TRY could be found in these preparations: a high-affinity transport mechanism with a K_m of 60 μM and 7 μM in synaptosomes and glial cells, respectively (Fig. 1) and a low-affinity uptake process with a K_m of 0.8 mM in both tissues. Similarly, two TRY uptake processes, exhibiting different characteristics, were seen in cultured fibroblasts (3T3 clone; Bauman, Bourgoin, Benda, Glowinski, and Hamon, 1974). The occurrence of two different processes for TRY transport in clonal cells indicates that they are both confined in the same cell. These findings are in contrast with recent data (Wofsey, Kuhar, and Snyder, 1971), which demonstrate that the high-affinity uptake system for putative neurotransmitters, such as glutamic and aspartic acids and GABA, is solely present in a specific synaptosomal fraction.

The existence of two mechanisms for the uptake of TRY may reveal an intracellular distribution of the amino acid in two compartments or an allosteric negative cooperativity of the carrier (Fig. 2).

In diurnal variations of 5-HT metabolism (Héry, Rouer, and Glowinski, 1972) and after LAD treatment or paradoxical sleep deprivation (Table 1), changes in 5-HT synthesis depended on TRY newly taken up and not on the whole amino acid tissue content. This argues in favor of an intracellular compartmentation of TRY. A similar

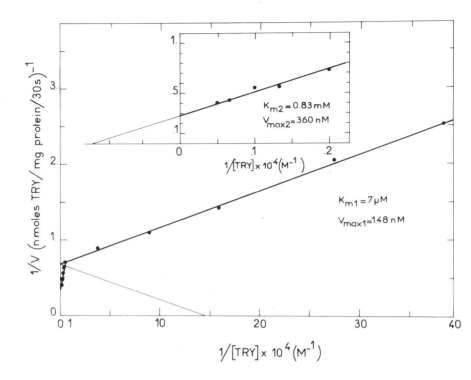

FIG. 1. *Double reciprocal plot of TRY accumulation in C_6 glial cells.* Cells were scraped from the culture flasks and washed three times with Hanks medium. Each determination of TRY uptake was made on 3.4 × 10⁶ cells corresponding to 0.3 to 0.4 mg of proteins. Incubations were performed in Hanks medium (1 ml) at 37°C for 30 sec under a constant stream of $O_2:CO_2$ (95/5%) in the presence of both ³H-TRY (2.5 µC of generally labeled amino acid, SA:3.2 C/mM) and L-TRY. Total TRY concentrations ranged from 2.6 µM to 2 mM. Accumulation of TRY was estimated in the cell pellet obtained by rapid centrifugation in cold (0°C). Values of TRY accumulated at 37°C have been corrected by substracting the corresponding amounts of TRY accumulated at 0°C (passive diffusion). Each point is the mean of four determinations.

hypothesis has been made by Grahame-Smith (1971b) and by Shields and Eccleston (1972).

Regulation of the carrier activity by TRY itself (Fig. 2) is supported by various observations. In Lineweaver-Burk representation, similar biphasic curves have already been described for various allosteric enzymes (Teipel and Koshland, 1969; Engel and Ferdinand, 1973). Furthermore, the K_m of the high-affinity TRY uptake process is 60 µM in synaptosomes and the concentrations of TRY in blood and in brain are, respectively, about 75 µM and 20 µM. On the other hand, the amino acid concentration in the culture medium of glial cells (Ham F 10 supplemented with 10% fetal calf serum) was also closely similar

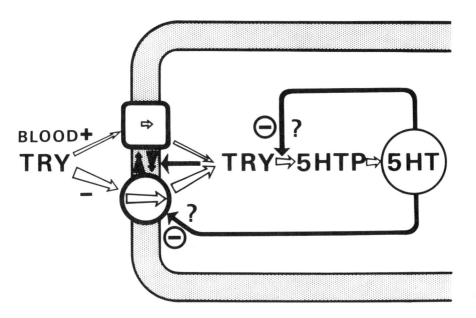

FIG. 2. *A hypothetical model for the regulation of 5-HT synthesis involving the TRY carrier.* As discussed in the text, the carrier molecule, localized in the membrane of surrounding glial cells and/or in that of the serotonergic neurons, may be an allosteric enzymatic molecule. According to the TRY concentration in the extracellular space (ex: blood) and/or in the cytoplasm, the affinity of the carrier for the amino acid could be increased (large arrow) or decreased (small arrow). Moreover, regulatory sites on the carrier molecule might explain the inhibition of the transmitter synthesis by high intraneuronal 5-HT levels.

(4μM) to the K_m of the high-affinity TRY uptake process in glial cells (7 μM) (Bauman et al., 1974). Moreover, regulatory sites are very likely localized on the carrier: whereas *para*-chlorophenylalanine is a potent competitive inhibitor of TRY uptake in glial cells as well as in synaptosomes, its ester derivative (methyl ester) stimulated the amino acid accumulation in glial cells (Bauman et al., 1974). This latter effect was related to a decreased K_m value of TRY uptake and not to a modification of its V_{max}. As already mentioned, the increased transport of TRY into cerebral slices induced by dibutyryl cyclic AMP is also related to an increased affinity of the carrier. On the other hand, Belin et al. (1973) have shown that the stimulating effect of reserpine pretreatment on TRY transport in brain synaptosomes resulted from a change in the K_m of the high-affinity uptake process.

V. CONCLUSION

There is little doubt that alterations of free TRY levels in plasma may affect central 5-HT synthesis. However, as illustrated by the few examples given in this report, regulation of 5-HT synthesis is also dependent on changes in the active transport of the amino acid. Since the affinity of the carrier molecule for TRY seems to be greater in glial cells than in synaptosomes, glia may regulate in some way the delivery of TRY into 5-HT neurons. However, regulatory events could as well occur at the level of the neuronal membrane. The limited number of 5-HT terminals in brain structures is a serious obstacle to the elucidation of this problem.

It should be remembered that other steps besides TRY availability or transport are involved in the regulation of the transmitter formation. This has been clearly indicated in various experiments. No changes in TRY transport were seen when 5-HT synthesis was inhibited *in vitro* by high 5-HT levels (Hamon, Bourgoin, and Glowinski, 1973a). Although TRY transport was stimulated by dibutyryl cyclic AMP, exogenous 5-HT taken up in 5-HT terminals was still able to inhibit the transmitter synthesis (Glowinski et al., 1973). The reduction in 5-HT synthesis seen 3 hr after MAO inhibition (Macon, Sokoloff, and Glowinski, 1971; Glowinski, Hamon, Javoy, and Morot-Gaudry, 1972) still persisted after peripheral injection of TRY (Carlsson et al., 1972). Finally, as suggested by the experiments of Carlsson, Lindqvist, Magnusson, and Atack (1973), the diminished rate of 5-HT synthesis seen in rat spinal cord after acute transection could not be reversed by TRY loading. Changes in the rate of TRY hydroxylation, possibly induced by modulations of the cofactor availability or by the alteration of the tryptophan hydroxylase conformation, may be involved in these experimental situations.

ACKNOWLEDGMENTS

This research was supported by grants from the Institut National de la Santé et de la Recherche Médicale (INSERM), the Direction de la Recherche et Moyens d'Essais (DRME), and the Société Chimique des Usines Rhône-Poulenc.

REFERENCES

Bauman, A., Bourgoin, S., Benda, P., Glowinski, J., and Hamon, M. (1974): Characteristics of tryptophan accumulation by glial cells. *Brain Research (in press)*.

Belin, M. F., Chouvet, G., and Pujol, J. F. (1973): Transport synaptosomal du tryptophane et de la tyrosine cérébrale; stimulation de la vitesse de capture après réserpine ou inhibition de la monoamine oxydase. Submitted to *Biochemical Pharmacology*.

Benda, P., Lightbody, J., Sato, G., Levine, L., and Sweet, W. H. (1968): Differentiated rat glial cell strain in tissue culture. *Science*, 161:370—371.

Bourgoin, S., Morot-Gaudry, Y., Glowinski, J., and Hamon, M. (1973): Stimulating effect of short term ether anaesthesia on central 5-HT synthesis and utilization in the mouse brain. *European Journal of Pharmacology*, 22:209—211.

Carlsson, A., Bedard, P., Lindqvist, M., and Magnusson, T. (1972): The influence of nerve-impulse flow on the synthesis and metabolism of 5-hydroxytryptamine in the central nervous system. *Biochemical Society Symposium*, 36:17–32.

Carlsson, A., Lindqvist, M., Magnusson, T., and Atack, C. (1973): Effect of acute transection on the synthesis and turnover of 5-HT in the rat spinal cord. *Naunyn Schmiedeberg's Archives of Pharmacology*, 277:1–12.

Consolo, S., Garattini, S., Ghielmetti, R., Morselli, P., and Valzelli, L. (1965): The hydroxylation of tryptophan *in vivo* by brain. *Life Sciences*, 4:625–630.

Cramer, H., Tagliamonte, A., Tagliamonte, P., Perez-Cruet, J., and Gessa, G. L. (1973): Stimulation of brain serotonin turnover by paradoxical sleep deprivation in intact and hypophysectomized rats. *Brain Research*, 54:372–375.

Curzon, G., Friedel, J., and Knott, P. J. (1973): The effect of fatty-acids on the binding of tryptophan to plasma protein. *Nature*, 242:198–200.

Curzon, G., Joseph, M. H. and Knott, P. J. (1972): Effects of immobilization and food deprivation on rat brain tryptophan metabolism. *Journal of Neurochemistry*, 19:1967–1974.

Engel, P. C., and Ferdinand, W. (1973): The significance of abrupt transitions in Lineweaver-Burk plots with particular reference to glutamate dehydrogenase negative and positive co-operativity in catalytic rate constants. *Biochemical Journal*, 131:97–105.

Fernstrom, J. D., and Wurtman, R. J. (1971): Brain serotonin content: Physiological dependence on plasma tryptophan levels. *Science*, 173:149–152.

Gál, E. M., and Drewes, P. A. (1962): Studies on the metabolism of 5-HT. II. Effect of tryptophan deficiency in rats (27520). *Proceedings of the Society for Experimental Biology and Medicine*, 110:368–371.

Glowinski, J., Hamon, M., and Héry, F. (1973): Regulation of 5-HT synthesis in central serotoninergic neurons. In: *New Concepts in Neurotransmitter Regulation*, edited by A. J. Mandell, pp. 239–257. Plenum Press, New York.

Glowinski, J., Hamon, M., Javoy, F., and Morot-Gaudry, Y. (1972): Rapid effects of monoamine oxidase inhibitors on synthesis and release of central monoamines. In: *Advances in Biochemical Psychopharmacology, Vol. 5: Monoamine Oxidases—New Vistas*, edited by E. Costa and M. Sandler, pp. 423–439. Raven Press, New York.

Grahame-Smith, D. G. (1971a): Studies "in vivo" on the relationship between brain tryptophan, brain 5-HT synthesis and hyperactivity in rats treated with a monoamine oxidase inhibitor and L-tryptophan. *Journal of Neurochemistry*, 18:1053–1066.

Grahame-Smith, D. G. (1971b): The transfer of certain small molecules across the synaptosome membrane. *Third International Meeting of the International Society for Neurochemistry*, Budapest. Abstract 425.

Green, H., Greenberg, S. M., Erickson, R. W., Sawyer, J. L., and Ellizon, T. (1962): Effects of dietary phenylalanine and tryptophan upon rat brain amine levels. *Journal of Pharmacology and Experimental Therapeutics*, 136:174–178.

Hamon, M., Bourgoin, S., and Glowinski, J. (1973a): Feed-back regulation of 5-HT synthesis in rat striatal slices. *Journal of Neurochemistry*, 20:1727–1745.

Hamon, M., Bourgoin, S., Jagger, J., and Glowinski, J. (1973b): Effects of LSD on synthesis, release and inactivation of 5-HT in rat brain slices. Submitted to *Brain Research*.

Héry, F., Pujol, J. F., Lopez, M., Macon, J., and Glowinski, J. (1970): Increased synthesis and utilization of serotonin in the central nervous system of the rat during paradoxical sleep deprivation. *Brain Research*, 21:391–403.

Héry, F., Rouer, E., and Glowinski, J. (1972): Daily variations of serotonin metabolism in the rat brain. *Brain Research*, 43:445–465.

Knott, P. J., and Curzon, G. (1972): Free tryptophan in plasma and brain tryptophan metabolism. *Nature*, 239:452–453.

Korf, J., Van Praag, H. M., and Sebens, J. B. (1972): Serum tryptophan decreased, brain tryptophan increased and brain serotonin synthesis unchanged after probenecid loading. *Brain Research*, 42:239–242.

Kuhar, M. J., and Aghajanian, G. K. (1973): Selective accumulation of ^{3}H-serotonin by nerve terminals of raphe neurons: An autoradiographic study. *Nature New Biology*, 241:187–189.

Macon, J. B., Sokoloff, L., and Glowinski, J. (1971): Feed-back control of rat brain 5-hydroxytryptamine synthesis. *Journal of Neurochemistry*, 18:323–331.

Miledi, R. (1971): Lanthanum ions abolish the "calcium response" of nerve terminals. *Nature*, 229:410–411.

Moir, A. T. B., and Eccleston, D. J. (1968): The effect of precursor loading in the cerebral metabolism of 5-hydroxyindoles. *Journal of Neurochemistry*, 15:1093–1108.

Morot-Gaudry, Y., Bourgoin, S., Hamon, M., and Glowinski, J. (1973a): Effects of various pharmacological and physical treatments on central 5-HT synthesis and utilization in the mouse brain (*in preparation*).

Morot-Gaudry, Y., Hamon, M., Bourgoin, S., and Glowinski, J. (1973b): Limitations of methods available for the estimation of the rate of 5-HT synthesis in the mouse brain (*in preparation*).

Schubert, J., Nybäck, H., and Sedvall, G. (1970): Accumulation and disappearance of ^{3}H-5-hydroxtryptamine formed from ^{3}H-tryptophan in mouse brain; effect of LSD 25. *European Journal of Pharmacology*, 10:215–224.

Shields, P. J., and Eccleston, D. (1972): Effects of electrical stimulation of rat midbrain on 5-hydroxytryptamine synthesis as determined by a sensitive radioisotope method. *Journal of Neurochemistry*, 19:265–272.

Tagliamonte, A., Biggio, G., Vargiu, L., and Gessa, G. L. (1973): Free tryptophan in serum controls brain tryptophan level and serotonin synthesis. *Life Sciences*, 12:277–287.

Tagliamonte, A., Tagliamonte, P., Perez-Cruet, J., Stern, S., and Gessa, G. L. (1971): Effect of psychotropic drugs on tryptophan concentration in the rat brain. *Journal of Pharmacology and Experimental Therapeutics*, 177:475–480.

Teipel, J., and Koshland, D. E., Jr. (1969): The significance of intermediary plateau regions in enzyme saturation curves. *Biochemistry*, 8:4656–4663.

Thierry, A. M., Fekete, M., and Glowinski, J. (1968): Effects of stress on the metabolism of noradrenaline, dopamine and serotonin (5-HT) in the central nervous system of the rat. II. Modifications of serotonin metabolism. *European Journal of Pharmacology*, 4:384–389.

Tonge, S. R., and Leonard, B. E. (1970): The effect of some hallucinogenic drugs on the amino acid precursors of brain monoamines. *Life Sciences*, 9:1327–1335.

Wofsey, A. R., Kuhar, M. J., and Snyder, S. H. (1971): A unique synaptosomal fraction, which accumulates glutamic and aspartic acids, in brain tissue. *Proceedings of the National Academy of Sciences, U.S.A*, 68:1102–1106.

Advances in Biochemical Psychopharmacology, Vol. 11
Raven Press, New York © 1974

The Major Role of the Tryptophan Active Transport in the Diurnal Variations of 5-Hydroxytryptamine Synthesis in the Rat Brain

R. Héry, E. Rouer, J. P. Kan, and J. Glowinski

*Groupe NB (INSERM U.114), Laboratoire de Biologie Moléculaire,
Collège de France, Paris 5e, France*

The existence of a circadian rhythm in the concentration of 5-hydroxytryptamine (5-HT) in the rat brain is well established (Dixit and Buckley, 1967; Montanaro and Graziani, 1967; Quay, 1965, 1968; Scheving, Harrison, Gordon, and Pauly, 1968; Bobillier and Mouret, 1971; Asano, 1971; Morgan and Yndo, 1973). A few years ago, this prompted us to study the diurnal variations of 5-HT metabolism (Héry, Rouer, and Glowinski, 1972). For this purpose, experiments were made in rats kept for at least 3 weeks in a constant environment and submitted to alternate periods of 12 hr of light (07.00 to 19.00 hr) and darkness (19.00 to 07.00 hr). ^{3}H-Tryptophan (TRY) was injected intracisternally at various times during the two successive periods of light and darkness. The total initial accumulation of ^{3}H-5-HT and of its metabolite ^{3}H-5-HIAA (5-hydroxyindoleacetic acid) in various parts of the brain was maximal during the light period. This effect was associated with a parallel increase in ^{3}H-TRY levels in tissues. Therefore a good correlation occurs between the diurnal changes in the initial accumulation of ^{3}H-TRY in tissues and the synthesis of ^{3}H-5-HT. Although no differences in the conversion index of TRY into 5-HT (^{3}H-5-HT + ^{3}H-5-HIAA/TRY specific activity) could be found in various structures in rats killed during the light or dark period, undoubtedly 5-HT synthesis was stimulated during light. This was demonstrated independently by measuring the rate of the initial accumulation of 5-HT after MAO inhibition (Héry et al., 1972). Consequently, the results obtained with the isotopic technic suggested that diurnal modulations of 5-HT synthesis were dependent on changes in TRY active transport occurring at a cellular level. In fact, the major contribution of TRY newly taken up in the regulation of 5-HT synthesis had already been noticed in previous studies on the effect of paradoxical sleep deprivation on 5-HT metabolism. In this state, the stimulation of 5-HT synthesis was also related to an increased accumulation in tissues of TRY newly taken up (Héry, Pujol, Lopez, Macon, and Glowinski, 1970).

In vitro studies made on brain slices of rats sacrificed at various times of the 24-hr cycle have confirmed the *in vivo* results. An increased synthesis of ^{3}H-5-HT from ^{3}H-TRY could be found in slices of brainstem and hypothalamus of rats killed at 15.00 hr during the light period when compared with those of rats killed during the night (21.00 hr). As in the *in vivo* studies, this effect was related to an enhanced accumulation of ^{3}H-TRY in tissues (Héry et al., 1972). More recently, we have further explored the diurnal physiological fluctuations in ^{3}H-TRY accumulation in slices. The time course of ^{3}H-TRY accumulation was examined in hypothalamic slices of rats killed during the light (15.00 hr) and dark periods (21.00 hr). Since high- and low-affinity transport systems for TRY were detectable in hypothalamic slices (K_m: 5×10^{-4} and 2×10^{-3}, respectively), tissues were incubated with low concentrations of ^{3}H-TRY (3.6×10^{-6} M). Figure 1 illustrates clearly that the diurnal changes in ^{3}H-TRY accumulation are related to modifications of the initial transport of the amino acid in tissues. It should be noticed that endogenous levels of TRY in tissues were similar at 15.00 and 21.00 hr. Although changes in plasmatic TRY concentrations were found between light (15.00 hr) and darkness (21.00 hr) (Héry et al., 1972), they were in the opposite direction to those observed for ^{3}H-TRY transport in slices. Indeed, plasmatic TRY levels were maximal at 21.00 hr, when ^{3}H-TRY uptake and 5-HT synthesis were reduced. As indicated by other studies (Knott and Curzon, 1972; Tagliamonte, Biggio, Vargiu, and Gessa, 1973), there is little doubt that in some physical or pharmacological states, changes in 5-HT synthesis regulation are linked to fluctuations in the concentrations of free TRY in plasma. However, our results demonstrate that 5-HT synthesis can be regulated by modifications of the active transport of the amino acid from plasma to brain. It remains to be demonstrated whether the diurnal changes observed are related to modifications of the affinity of the carrier molecule for its substrate or to the velocity of the transport process.

Complementary experiments were made to determine whether diurnal variations similarly occurred in the transport of ^{3}H-tyrosine. In experimental conditions similar to those used in the ^{3}H-TRY studies, differences in the transport of ^{3}H-tyrosine were also seen between the light and dark periods. Surprisingly, ^{3}H-tyrosine transport was enhanced in the dark period, when ^{3}H-TRY transport was reduced (Fig. 1). These data underline the specificity of the effects observed and strongly suggest the existence of different carrier molecules for the two

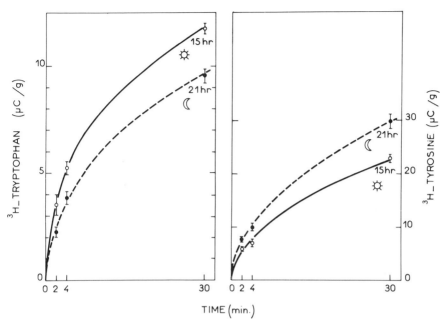

FIG. 1. *Diurnal changes in the active transport of tryptophan and tyrosine in hypothalamic slices.* Rats were submitted for 3 weeks to alternate 12-hr periods of light and darkness and were killed at 15.00 hr (light) and 21.00 hr (darkness). Hypothalamic slices were incubated for various times in the presence of ^{3}H-TRY (10 μC, 0.60 X 10^{-6}M) or ^{3}H-tyrosine (25 μC, 0.13 X 10^{-6} M) and the ^{3}H-amino acid content was estimated in tissues. Results are the mean ± SEM of values obtained with groups of eight rats.

amino acids. This statement is in agreement with recent results of Moir (1971).

Both the opposite diurnal changes in TRY and tyrosine transport and the complementary role of serotonergic and catecholaminergic systems in vigilance states (Jouvet, 1969) led us to investigate the diurnal changes in 5-HT synthesis after degeneration of catecholaminergic neurons. For this purpose, groups of rats received 6-hydroxydopamine (6-OH-DA) into the cerebrospinal fluid (250 μg intracisternally, 250 μg intraventricularly) and were sacrificed 3 weeks later. They, along with sham-injected animals, were submitted to alternate 12-hr periods of light and darkness during this time. 5-HT synthesis was estimated in hypothalamic slices at 15.00 hr (light) and 21.00 hr (darkness), as in previous experiments. After a 30-min incubation in the presence of ^{3}H-TRY, the total ^{3}H-5-HT formation (^{3}H-5-HT + ^{3}H-5-HIAA) was significantly reduced at 15.00 hr when compared with controls. The reverse effect was seen at 21.00 hr (Table 1). These changes were associated with parallel modifications in the accumulation of ^{3}H-TRY

TABLE 1. *Diurnal variations of ^{3}H-5-HT synthesis and ^{3}H-TRY accumulation in hypothalamic slices of 6-OH-DA-pretreated rats*

	15.00 hr (light)		21.00 hr (dark)	
	Sham-injected	6-OH-DA	Sham-injected	6-OH-DA
Hypothalamus				
^{3}H-5-HT + ^{3}H-5-HIAA (μC/g)	2.11 ± 0.13	1.67 ± 0.09[a]	1.58 ± 0.06[b]	2.09 ± 0.11[a,b]
^{3}H-TRY (μC/g)	16.46 ± 0.51	12.90 ± 0.46[a]	11.92 ± 0.75[b]	14.19 ± 0.52[a]
Brainstem				
NE (μg/g)	0.74 ± 0.03	0.31 ± 0.01[a]	0.65 ± 0.05	0.31 ± 0.02[a]
Forebrain				
NE (μg/g)	0.34 ± 0.02	0.10 ± 0.01[a]	0.37 ± 0.01	0.13 ± 0.01[a]
DA (μg/g)	1.91 ± 0.89	0.89 ± 0.11[a]	1.80 ± 0.11	1.16 ± 0.15[a]

Groups of rats received 6-OH-DA in the cerebrospinal fluid (250 μg intracisternally and 250 μg intraventricularly); sham-injected animals received a 6-OH-DA-free solution. All animals were then submitted to a regular cycle of alternate 12-hr light (07.00 hr to 19.00 hr) and dark (19.00 hr to 07.00 hr) periods for 3 weeks. Synthesis of ^{3}H-5-HT was estimated in hypothalamic slices of rats killed at 15.00 hr or 21.00 hr. Slices were incubated for 30 min in the presence of 18 μC of ^{3}H-TRY (3.2 C/mM). Total levels of ^{3}H-5-HT + ^{3}H-5-HIAA found both in tissue and incubating medium were estimated. ^{3}H-TRY accumulation was also measured in slices at the end of the incubation period. Catecholamines (norepinephrine, NE, and dopamine, DA) were estimated in the brainstem and the forebrain (without hypothalamus). Results are the mean ± SEM of values obtained with groups of eight rats.

[a] $p < 0.05$ when compared with respective sham-injected rats.
[b] $p < 0.05$ when compared with animals killed at 15.00 hr (light).

in tissues. Therefore, the important degeneration of catecholaminergic neurons reversed the normal diurnal cycle of 5-HT synthesis (Héry, Rouer, and Glowinski, 1973). These results reveal the role of central catecholaminergic neurons in the control of the diurnal activity of serotonergic neurons. They also demonstrate, once more, the major role of TRY active transport in the regulation of 5-HT synthesis (Glowinski, Hamon, and Héry, 1973).

REFERENCES

Asano, Y. (1971): The maturation of the circadian rhythm of brain norepinephrine and serotonin in the rat. *Life Sciences*, 10:883–894.

Bobillier, P., and Mouret, J. R. (1971): The alternations of the diurnal variations of brain tryptophan, biogenic amines and 5-hydroxyindole acetic acid in the rat under limited time feeding. *International Journal of Neuroscience*, 2:271–282.

Dixit, B. N., and Buckley, J. P. (1967): Circadian changes in brain 5-hydroxytryptamine and plasma corticosterone in the rat. *Life Sciences*, 6:755–758.

Glowinski, J., Hamon, M., and Héry, F. (1973): Regulation of 5-HT synthesis in central serotoninergic neurons. In: *New Concepts in Neurotransmitter Regulation*, edited by A. J. Mandell, pp. 239–257. Plenum Press, New York.

Héry, F., Pujol, J. F., Lopez, M., Macon, J., and Glowinski, J. (1970): Increased synthesis and utilization of serotonin in the central nervous system of the rat during paradoxical sleep deprivation. *Brain Research*, 21:391–403.

Héry, F., Rouer, E., and Glowinski, J. (1972): Daily variations of serotonin metabolism in the rat brain. *Brain Research*, 43:445–465.

Héry, F., Rouer, E., and Glowinski, J. (1973): Effect of 6-hydroxydopamine on daily variations of 5-HT synthesis in the hypothalamus of the rat. *Brain Research*, 56:135–146.

Jouvet, M. (1969): Biogenic amines and the states of sleep. *Science*, 163:32–41.

Knott, P. J., and Curzon, G. (1972): Free tryptophan in plasma and brain tryptophan metabolism. *Nature*, 239:452–453.

Moir, A. T. B. (1971): Interaction in the cerebral metabolism of the biogenic amines: Effect of intravenous infusion of L-tryptophan on tryptophan and tyrosine in brain and body fluids. *British Journal of Pharmacology*, 43:724–731.

Montanaro, N., and Graziani, G. (1967): Influenza della somministrazionne di fenelzina, di 5-idrossitriptofano e della loro associazione sulla oscillazione giornaliera di concentrazione encefalica di serotonina nel ratto albino. *Dal Bolletino della Societa Italiana Di Biologia Sperimentale*, 43:42–45.

Morgan, W. W., and Yndo, C. A. (1973): Daily rhythms in tryptophan and serotonin content in mouse brain: The apparent independence of these parameters from daily changes in food intake and from plasma tryptophan content. *Life Sciences*, 12:395–408.

Quay, W. B. (1965): Regional and circadian differences in cerebral cortical serotonin concentrations. *Life Sciences*, 3:379–384.

Quay, W. B. (1968): Differences in circadian rhythms in 5-hydroxytryptamine according to brain region. *American Journal of Physiology*, 215:1448–1453.

Scheving, L. E., Harrison, W. H., Bordon, N. P., and Pauly, J. E. (1968): Daily fluctuation (circadian and ultraradian) in biogenic amines of the rat brain. *American Journal of Physiology*, 214:166–173.

Tagliamonte, A., Biggio, G., Vargiu, L., and Gessa, G. L. (1973): Free tryptophan in serum controls brain tryptophan level and serotonin synthesis. *Life Sciences*, 12:277–287.

Advances in Biochemical Psychopharmacology, Vol. 11
Raven Press, New York © 1974

The Role of Serotonin in the Regulation of a Phasic Event of Rapid Eye Movement Sleep: The Ponto-geniculo-occipital Wave

Steven Henriksen, William Dement, and Jack Barchas

Department of Psychiatry, Stanford University School of Medicine, Stanford,
California 94305

I. THE "PROCESS" OF STATE

Much experimental evidence has been gathered over the last decade that implicates central monoaminergic neuronal systems in what is best described as "state" processes. By state processes we mean the central brain mechanisms, as yet obscure, whose activity underlies the periodic fluctuations of neuronal activity to which physiologists refer as wakefulness, sleep, and rapid-eye-movement (REM) sleep. The term "state processes" also infers a rather unified or, better yet, holistic nature to these underlying mechanisms as opposed to the rather disparate and even opposing subprocesses that have come to be implied in the terms wakefulness, sleep, and REM sleep. In fact, it is becoming increasingly evident, particularly with reference to neurochemical processes, that the putative neurohumors previously related to specific states, that is, serotonin, 5-HT [slow-wave sleep (see Jouvet, 1972)], norepinephrine, NE [electrocortical waking (see Jouvet, 1972)], and dopamine, DA [behavioral waking (Jones, Bobillier, Pin, and Jouvet, 1974)], can no longer be thought of as the sole mediators of specific state behaviors, functioning only during the expression of these behaviors. Rather, they would best be considered as having a *multiplicity of functions across these behavioral states.*

Such a perspective can be drawn from the experimental realization that these behavioral states, far from being entirely unique physiologically, appear to have complicated subprocesses that extend across the rather arbitrary boundaries defined by the terms wakefulness, sleep, and REM sleep. This leads to the possibility that rather than having completely antagonistic processes subserving these state behaviors, fluctuations in the interaction of subprocesses may be the determining factor in the wide variety of state alterations we observe. In point of fact, recent evidence from neurochemical studies as well as the reinterpretation of previously existing data has led Jouvet and his colleagues in Lyon, the undisputed leaders in the field, to conclude that the subprocesses of wakefulness, as well as REM sleep, may utilize NE as a primary neurochemical mediator, and that slow-wave sleep (SWS), as well as REM sleep mechanisms, may encompass 5-HT processes (Jouvet, 1972).

II. TONIC AND PHASIC PHENOMENA

Perhaps the best example of the interaction of subprocesses ultimately resulting in a specific state can be seen in REM sleep. It was Guiseppi Morruzzi (1963) who first noted that processes of REM sleep in the cat could be functionally divided into two categories. Morruzzi suggested that REM sleep could be descriptively subdivided into *tonic events*, that is, those events that last throughout the duration of the episode, and *phasic events*, those events that could be characterized by temporal "outbursts" of activity above a background of relative quiescence. Morruzzi included under the heading of *tonic events* the desynchronization of the cortical electroencephalogram, cervical muscle atonia accompanied by profound depression of spinal reflexes, and the augmentation of pupillary myosis. Under *phasic events* he included the rapid eye movements, pontine reticular potentials, and pyramidal tract discharges accompanied by clonic activity of distal musculature.

In the decade since Morruzzi's report, physiologists have been able to localize to brainstem areas the subprocesses subserving both tonic and phasic activities (see Pompiano, 1970; Jouvet, 1972).

With respect to the phasic aspects of REM sleep, it appears that all of the phasic activity may be actuated by a single or series of generating mechanisms located in the pontine tegmentum (Jouvet, 1962; Pompiano and Morrison, 1965; Brooks and Gershon, 1971). The most precise measure of all phasic activity has been the large phasic slow potentials that can be recorded from widespread areas of the neuraxis, including the pontine reticular formation, the lateral geniculate bodies, and the occipital cortex of the cat, hence the discriptor ponto-geni-culo-occipital or PGO waves (Jeannerod, Mouret, and Jouvet, 1965). It appears that the phasic activity of REM sleep as measured by PGO discharge, far from being an inconsequential electrical phenomenon, is a *sine qua non* for the state, and may be subserved by, and perhaps facilitate the precipitation of, the tonic REM sleep processes (Ferguson, Henriksen, McGarr, Belenky, Mitchell, Gonda, Cohen and Dement, 1968; Dement, 1972).

Consonant with the previously outlined process concept of states,

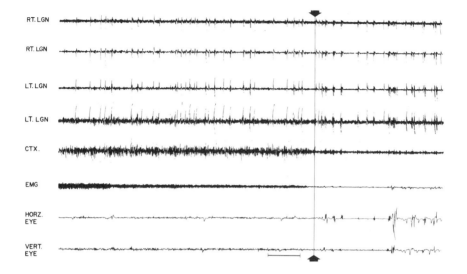

FIG. 1. The top four polygraphic tracings illustrate PGO activity during a SWS period just prior to the onset of a REM sleep episode (arrows). The tracing represents a period of approximately 1 min. Note the large amount of phasic activity seen prior to the REM episode itself. Abbreviations: RT.LGN, right lateral geniculate nucleus; LT.LGN, left lateral geniculate nucleus; CTX, cortical EEG; EMG, electromyogram; Horiz. and Vert. EYE, electrooculogram.

this PGO phenomenon, although predominantly occurring during REM sleep episodes, also occurs sporadically during slow-wave sleep (SWS). In addition, PGO waves are consistently seen in SWS just prior to REM sleep episodes, *preceding* the tonic activity of REM sleep by some 30 to 60 sec (Fig. 1). Thus, considerable temporal overlap exists between SWS processes per se and the primary phasic phenomenon, the PGO wave.

III. PGO WAVES AND 5-HT

When the serotonin depleting agent *p*-chlorophenylalanine (*p*-CPA) is administered to cats, a well-documented series of changes in waking as well as sleeping behavior occurs (Delorme, 1966; Koella, Feldstein, and Czicman, 1968; Dement, Mitler, and Henriksen, 1972). Following a single injection of the drug, undramatic changes in sleep behavior are seen during 1 to 2 days; this is somewhat dose-dependent, with higher *p*-CPA doses having an earlier effect. Thereafter, a precipitous drop in sleep is observed coincident with an increase in agitated waking behavior. This insomnia is somewhat paralleled by 5-HT depletion, and SWS returns to normal as 5-HT levels return to baseline values (Mouret,

Bobillier, and Jouvet, 1968; Koella et al., 1968). Although no causal relationship can be drawn from these results alone, these data do suggest that the behavior of sleep in some manner requires a 5-HT neurochemical mechanism for its normal occurrence.

In addition to *p*-CPA studies, a variety of experimental approaches, including lesion and other neuropharmacological approaches, suggest very convincingly that a 5-HT mechanism must be intact for normal sleep to occur (see Jouvet, 1972). However, not mutually exclusive to this evidence is the possibility (vida infra) that some subprocess of sleep may in some way utilize 5-HT, but as a consequence of the disruption of this mechanism brought about by the depletion of 5-HT, sleep processes in general may be *secondarily* disrupted.

Because of the observation that REM phasic phenomena are also altered following *p*-CPA administration, we have closely studied the effect of this 5-HT depletion on the distribution of REM PGO waves throughout the duration of *p*-CPA treatment (Henriksen and Dement, 1972). As we have previously reported (Dement et al., 1972), when cats are treated with *daily* injections of *p*-CPA (150 mg/kg, subcutaneous), in contrast to a single injection of the drug, a dissociation between the 5-HT depletion and insomnia occurs over time. That is, although daily injections of the drug result in the same initial insomnia seen with acute treatment, as daily injections continue, SWS and REM sleep partially return to normal values (up to 70% of baseline amounts) even though 5-HT levels remain maximally depleted (90%). Figure 2 illustrates the changes in SWS and REM sleep in a representative cat. It should be emphasized that with this dosage schedule, 5-HT levels in all brain areas assayed (cortex, midbrain, pons, medulla, hypothalamus) reach maximum levels of depletion by or during the fifth treatment day and *remain* at this level for as long as *p*-CPA is administered (in our experience up to 37 days).

As compared with elapsed time spent in SWS and REM sleep, however, distributional changes in PGO waves very closely parallel changes in 5-HT levels *throughout p*-CPA treatment and recovery. The earliest change in any PGO wave measure develops 16 to 24 hr following the first injection of the drug, literally *days before* the well-described insomnia is seen. This early change can be measured as a *progressive* decrease in the number of PGO waves observed in the last minute of SWS immediately prior to each REM episode. This decrease can be so dramatic as to result, by the second *p*-CPA treatment day, in REM sleep episodes beginning without a single foregoing PGO wave. Parallel to the change in SWS PGO waves is a frequent increase in the rate of discharge of REM PGO waves. Both of these PGO changes are illustrated in Fig. 3.

After the initial *facilitatory* effect on REM sleep PGO wave activity,

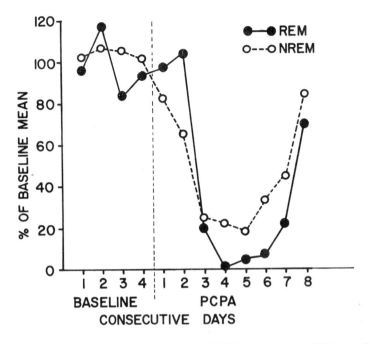

FIG. 2. Illustrated are the changes in amounts of REM sleep time and SWS time (NREM) across baseline days and chronic *p*-CPA treatment in a representative cat. Values are expressed as percentage of control. Note the increase in both REM sleep as well as SWS on *p*-CPA treatment days 6 to 8.

what might be described as the *second phase* of PGO wave changes develops. This second phase, lasting from about the middle of the second *p*-CPA treatment day to the sixth or seventh treatment day, is characterized in its earlier portion by a gradual and progressive breakdown in the REM state dependent distribution of the PGO waves. As illustrated in Fig. 3, by the third treatment day PGO waves again begin to occur in significant numbers during SWS. However, instead of being generally localized antecedent to REM sleep periods, the waves appear throughout SWS episodes. It is at this peak of a secondary increase in SWS PGO discharge that the first signs of sleep disruption are observed.

About the same time that we have observed the first sign of sleep disruption, PGO waves emerge for the first time into the waking state. This can be considered as the latter part of the second phase of drug effect. Within hours of the first waking waves, PGO waves become distributed throughout all states. At this time, insomnia begins to approach its peak (about *p*-CPA treatment day 4 to 5).

Concomitant with the emergence of the PGO waves into the waking state, periodic behavioral episodes best described as "pseudo hallucina-

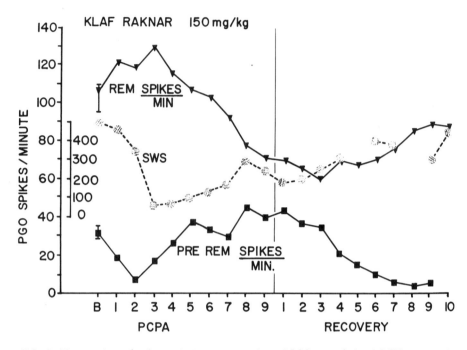

FIG. 3. Illustrated are the changes in the mean number of PGO waves/min of REM sleep, and the mean number of PGO waves in the last minute of SWS just prior to a REM episode in a representative cat. Note the early fluctuations in both the PGO measures as compared to the time course of changes in SWS time (stippled function).

tory" are observed. These episodes are *always* precipitated by bursts of at least two or more waking PGO waves. We believe that these episodes contribute considerably to the behavioral agitation and subsequent insomnia of p-CPA-treated animals. Fig. 4 illustrates a polygraphically recorded representative "hallucinatory"-like episode in one cat.

As can be seen in Fig. 2, the insomnia following chronic p-CPA treatment is rather transient. By the sixth or seventh p-CPA treatment day, SWS and REM sleep times begin slowly to return to near-normal values. However, even after this partial return of behavioral and polygraphic sleep, PGO waves continue to be distributed throughout all behavioral states. At about the same time as sleep begins to return, the "hallucinatory episodes" decrease in frequency and intensity. This suggests that cats habituate somewhat to the "disturbing" effect of these waking waves.

Therefore, the partial return of both SWS and REM sleep times, characterizing *the third phase* of drug action, is not paralleled by an appreciable regating of PGO waves into REM sleep episodes. Rates of discharge of PGO waves in REM sleep continue to decrease, and the

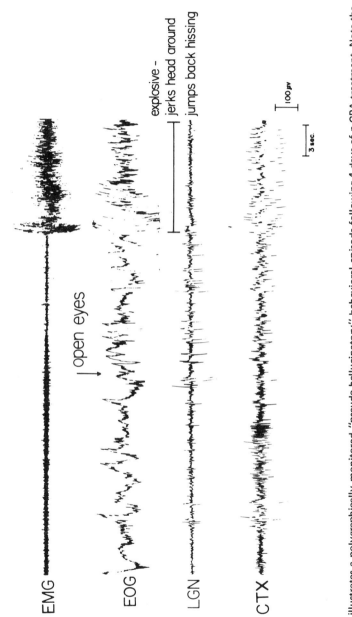

FIG. 4. This figure illustrates a polygraphically monitored "pseudo-hallucinatory" behavioral episode following 4 days of *p*-CPA treatment. Note the abrupt arousal and orienting reaction as noted in the tracing. Abbreviations: See Fig. 1.

frequency of PGO waves across all states can approach identity. However, slightly higher discharge rates are consistently observed in SWS and REM sleep as compared to wakefulness. The lower rate of PGO waves occurring in wakefulness appears in part to be due to the fact that highly activated behavior and/or drive-oriented behaviors (i.e., eating, drinking, etc.) tend strongly to suppress waking PGO discharge both during the behavior itself and for some minutes thereafter.

This rather stabilized condition between the distribution of SWS, REM sleep, and PGO waves obtains for as long as the drug is administered. However, if p-CPA administration is stopped, for example after nine p-CPA treatment days (Fig. 2), both sleep and REM rather abruptly make up the small differences between the stabilized p-CPA values and baseline amounts. On the other hand, the PGO waves—by all measures used including rates of discharge across states, amplitude histograms, and interval histogram profiles—are very slow in returning to baseline values and, in effect, very closely *parallel the recovery curves for 5-HT*. This slow recovery in PGO measures and 5-HT levels typically take from 15 to 20 days following nine days of p-CPA treatment.

In addition to the descriptive studies outlined above, we have employed another experimental approach to assess changes in PGO waves following p-CPA treatment. Single-pulse square-wave stimulation to localized areas of the pontine reticular formation is able to elicit waves that appear identical to REM PGO waves (Bizzi and Brooks, 1963; Henriksen and Dement, 1972). Furthermore, stimulation of these areas can only elicit stimulated-PGO waves during periods of spontaneous PGO activity, that is, REM sleep episodes. This has suggested the concept of a state-dependent gating mechanism for the occurrence of PGO waves (Bizzi and Brooks, 1963; Brooks and Bizzi, 1963). We have utilized this technique to study the changes in the ability to elicit stimulated PGO waves following chronic p-CPA treatment. Using this paradigm, results indicate that changes in the ability to elicit PGO waves (across various behavioral states) closely parallel the changes we have observed in the spontaneous alterations in PGO wave discharge following p-CPA.

IV. DISCUSSION

These data taken together, as well as pharmacological data not discussed here (Delorme and Jouvet, 1965; Jouvet, 1972; Dement, 1972; Jalfre, Monachon, and Haefely, 1972), strongly argue for 5-HT being an important, perhaps primary, inhibitory influence on the discharge of PGO waves. Furthermore, the fact that such a precise state-dependent gating of this phenomenon is observed suggests that this 5-HT PGO process may be linked to state processes with its primary inhibiting action occurring during wakefulness and to some extent SWS. Recent single-cell studies in certain raphe nuclei give neurophysiological substantiation to this suggestion. Raphe dorsalis cells are reported to have maximal rates of firing during wakefulness and SWS, while these same cells are either completely silent during REM sleep episodes or selectively silent during the PGO wave discharges of REM sleep (McGinty and Harper, 1972).

In summary, our data suggest that changes in PGO wave activity following *p*-CPA treatment more sensitively parallel 5-HT alterations than do any other behavioral measure. The effects on PGO wave discharge are the earliest measurable change after the initial *p*-CPA injection and the last measure to return to baseline following cessation of drug treatment. Furthermore, we feel that the behavioral result of the emergence of PGO waves into the waking state after daily or even single *p*-CPA treatment schedules may play an important role in the insomnia previously attributed to 5-HT depletion per se.

Does this mean that 5-HT mechanisms have little to do with SWS processes? Emphatically no! It is certainly possible, and even likely, that 5-HT mechanisms in the brain may be functioning in a multimodal way across a variety of states, as previously suggested. Anatomical projections of the 5-HT-containing cell bodies in the raphe nuclei do support this type of concept.

In addition, a great deal of very convincing evidence (see Jouvet, 1972) has linked 5-HT directly with sleep processes. It is possible that the partial return of REM sleep and particularly SWS following chronic *p*-CPA treatment is due, not only as we have suggested, to a gradual habituation to the loss or state-dependent gating of PGO discharge, but also to changes in postsynaptic 5-HT receptor sensitivity induced by prolonged 5-HT depletion. Even though our biochemical results indicate over 90% depletion of 5-HT with our dosage schedules, it is still possible that enough 5-HT remains in an active pool to explain the partial return of sleep, particularly when postsynaptic supersensitivity may be implicated. It is also intriguing to suggest that perhaps some predisposition for SWS following chronic *p*-CPA may result from negative feedback influences on arousal processes, presumably noradrenergic in nature (Jouvet, 1972), consequent on the "unopposed" insomnia following *p*-CPA.

Perhaps the best explanation for the *p*-CPA sleep effects is that all three of these alternatives, that is, habituation to state disruptions, postsynaptic 5-HT supersensitivity, and negative feedback on arousal mechanisms, play substantial roles in the effects that are observed.

Notwithstanding these complex SWS alterations following *p*-CPA, we have presented substantial evidence suggesting that 5-HT mechanisms have a primary influence on the state-dependent release of PGO waves. It is also suggested that this mechanism seems to be independent of the other processes mediating alterations in arousal that also utilize 5-HT neurons.

ACKNOWLEDGMENTS

We wish to thank George Mitchell, Bill Gonda, Pamela Angwin, Elizabeth Erdelyi, and John Lestian for their invaluable technical assistance in this research.

Research supported by Biological Sciences Training Grant MH8304-09 to Steven Henriksen, N.I.M.H. grant MH-13860, N.A.S.A. grant NGR 05-020-168 and Research Career Development Award MH5804 to William Dement, and Research Scientist Development Award MH24161 to Jack Barchas.

REFERENCES

Bizzi, E., and Brooks, D. (1963): Functional connections between pontine reticular formation and lateral geniculate nucleus during deep sleep. *Archives Italiennes de Biologie*, 101:666–680.

Brooks, D., and Bizzi, E. (1963): Brain stem electrical activity during deep sleep. *Archives Italiennes de Biologie*, 101:648–665.

Brooks, D., and Gershon, M. (1971): Eye movement potentials in the oculomotor and visual system: A comparison in reserpine induced waves with those present during wakefulness and rapid eye movement sleep. *Brain Research*, 27:223–239.

Delorme, F. (1966): *Monoamines et sommeil*. Étude polygraphique neuropharmacologique et histochimique des états de sommeil chez le chat. Thèse Université de Lyon, Imprimerie LMD.

Delorme, F., and Jouvet, M. (1965): Effets remarquables de la reserpine sur l'activité EEG phasique ponto-géniculo-occipitale. *Comptes Rendus des Seances de la Société de Biologie (Paris)*, 159:900–903.

Dement, W. (1972): Sleep deprivation and the organization of the behavioral states. In: *Sleep and the Maturing Nervous System*, edited by C. Clemente, D. Purpura, and F. Mayer. Academic Press, New York.

Dement, W., Mitler, M., and Henriksen, S. (1972): Sleep changes during chronic administration of parachlorophenylalanine. *Revue Canadienne de Biologie*, 31(suppl.):239–246.

Ferguson, J., Henriksen, S., McGarr, K., Belenky, G., Mitchell, G., Gonda, B., Cohen, H., and Dement, W. (1968): Phasic event deprivation in the cat. *Psychophysiology*, 4:238–239.

Henriksen, S., and Dement, W. (1972): Further studies in cats chronically treated with p-chlorophenylalanine (p-CPA). *Psychophysiology*, 9:126.

Jalfre, M., Monachon, M.-A., and Haefely, W. (1972): Drugs and PGO-waves in the cat. In: *Proceedings of the First Canadian International Symposium on Sleep*, edited by D. J. McClure. Hoffmann-La Roche Limited, Vaudreuil, Quebec.

Jeannerod, M., Mouret, J., and Jouvet, M. (1965): Effets secondaires de la déafferentation visuelle sur l'activité electrique phasique ponto-géniculo-occipitale du sommeil paradoxal. *Journal de Physiologie (Paris)*, 57:255–256.

Jouvet, M. (1962): Recherches sur les structures nerveuses et les mecanismes responsables des differentes phases du sommeil physiologique. *Archives Italiennes de Biologie*, 100:125–206.

Jouvet, M. (1972): The role of monoamines and acetylcholine-containing neurones in the regulation of the sleep-waking cycle. In: *Ergebnisse der Physiologie,* 64:116–307.

Jones, B., Bobillier, P., Pin, C., and Jouvet, M. (1974): The effect of lesions of catecholamine-containing neurons upon monoamine content of the brain in electro-encephalographic and behavioral waking in the cat. *Brain Research (in press).*

Koella, W., Feldstein, A., and Czicman, S. (1968): The effect of parachlorophenylalanine on the sleep of cats. *Electroencephalography and Clinical Neurophysiology* 25:481–490.

McGinty, D., and Harper, R. (1972): 5-HT-containing neurons: Unit activity during sleep. *Sleep Research,* 1:27.

Moruzzi, G. (1963): Active processes in the brain stem during sleep. In: *The Harvey Lectures, Series 58.* Academic Press, New York.

Mouret, J., Bobillier, P., and Jouvet, M. (1968): Insomnia following parachlorophenylalanine in the rat. *European Journal of Pharmacology,* 5:17–22.

Pompiano, O. (1970): Mechanism of sensorimotor integration during sleep. *Progress in Physiological Psychology,* 3:1–179.

Pompiano, O., and Morrison, A. (1965): Vestibular influences during sleep. I: Abolition of the rapid eye movements during desynchronized sleep following vestibular lesions. *Archives Italiennes de Biologie,* 103:569–595.

Advances in Biochemical Psychopharmacology, Vol. 11
Raven Press, New York © 1974

Serotonin — A Hypnogenic Transmitter and an Antiwaking Agent

Werner P. Koella

Research Department, Pharmaceuticals Division, CIBA-GEIGY Limited, Basel, Switzerland

For a long time sleep has been looked at as a mere passive phenomenon. The various functional changes that occur with the transition from waking to sleep were considered to be the manifestation of a simple drifting away from the active waking state (see, viz., Kleitman, 1939). However, increasing numbers of sleep researchers, following the lead of W. R. Hess (1924–1925, 1933, 1944), became aware that sleep is—at least in part—an active process. Moruzzi (1972), in his recent superb review of his own work and that of others, was led to conclude that there are indeed two aspects to sleep: a passive, yet indispensable, component which we may refer to as "de-waking" and a second, active one, comprising a variety of specific functional changes which clearly separate natural sleep from other "un-waking" states such as coma, general anesthesia, somnolence, and so on. Seen in this manner, sleep, in its organizational pattern, would not differ from most autonomic and somatic phenomena in that it would be the manifestation of a reciprocal behavior of two antagonistic systems, that is, an increase in tone in a "pro-sleep" system *and* a (reciprocal) decrease in tone in a "pro-waking" system.

Depletion of brain serotonin (5-HT) by pharmacological means (see, viz., Koella, Feldstein, and Czicman, 1968) or by destruction of the main source of cerebral 5-HT, that is, the raphe nuclei (Renault, 1967; Jouvet, 1967), is followed by a drastic lowering of sleep in the cat. Particularly in view of Jouvet's extended and beautiful work, part of which is presented in the preceding chapter, it is not unlikely that the raphe system with its serotonergic projections is involved in the induction of those functional changes in the various parts of the brain which we consider to be specific for sleep, and which constitute—at least in part—the active "pro-sleep" component. However, there seems to be an additional role of serotonin. During the last 12 years we have accumulated a number of data which indicate that serotonin also acts as an "anti-waking" agent, in that it promotes a decrease in the tone of the waking (i.e., arousal system) (Koella and Czicman, 1963, 1966). Figure 1 shows schematically this reciprocal influence upon the two antagonistic systems.

A first step in developing this idea was the observation that intracarotid injection of 5-HT in minute amounts (1 to 5 μg/kg) in the flaxedilized cat was followed, after a short initial phase of arousal, by a prolonged period of EEG hypersynchrony and mydriasis.

In an attempt to delineate the locus of action of 5-HT, we transected the brainstem of a number of cats and injected 5-HT in similar doses into the carotid artery. Invariably we found that after severing the posterior brainstem from the forebrain, serotonin induced signs of arousal only and there was no more indication of any hypersynchronizing effect. Evidently the synchronizing influence of 5-HT is due to an action of this substance somewhere in the posterior brainstem, from which the "hypnogenic" influence is transmitted to the forebrain via nervous pathways. This interpretation is supported by the additional observation that in intact, unanesthetized, immobilized cats injection of serotonin into the vertebral artery immediately induced a shift toward cortical synchrony without a preceding phase of arousal.

Since 5-HT does not penetrate the blood-brain barrier (BBB), we had to look for a 5-HT receptor area in the posterior brainstem situated

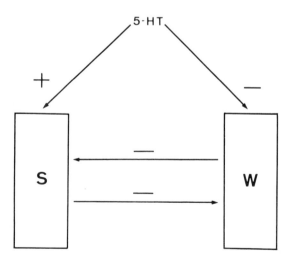

FIG. 1. Highly schematic representation of the proposed dual reciprocal involvement of 5-HT in the organization of sleep. S = "sleep system," including "effector areas." The + sign may indicate that 5-HT acts—within this system—as a promotor of sleep, that is, activities specific to days. W = "waking or arousal system." The − sign indicates within or without this system as an inhibitor. Note the reciprocal action of 5-HT an antagonistic S-W system, and the positive feedback arrangement between S and W.

outside of this barrier in order to explain adequately these central nervous system effects of intra-arterially applied serotonin. The area postrema, at the posterior angle of the fourth ventricle, had been shown to be invaded by substances otherwise blocked by the BBB (Wislocki and Putnam, 1920; Faber, 1955); it contains large quantities of MAO (Shimizu, Morikawa, and Okada, 1959). The area postrema was considered to be a likely candidate for this 5-HT-receptor function. The results of a number of additional experiments strongly indicated that this assumption was correct.

Following cauterization of the area postrema in a number of animals, the synchronizing effect of intracarotic serotonin was always reduced and, in a few cases, completely eliminated.

Application to the area postrema of cotton pellets soaked with LSD_{25} (50 to 100 $\mu g/ml$) or methysergide (UML, 50 to 100 $\mu g/ml$), two powerful 5-HT-blocking agents, led to partial or total suppression of the 5-HT-induced hypersynchronous phase; this blocking effect lasted as long as 4 hr.

Topical application of serotonin in concentrations of 5 to 10 $\mu g/ml$ to the posterior region of the fourth ventricle induced an increase in the recruiting responses, an increase (or appearance) of spindle bursts, and an increase in slow-wave output often lasting as long as 15 min. Application of solvent only (Tyrode® or saline solution) ordinarily did not produce any change in the EEG.

Since we published our initial experiments, Roth, Walton, and Yamamoto (1970) and Bronzino, Morgane, and Stern (1972) confirmed our findings by showing that injection of the amine into the arterial supply of this organelle or direct application of 5-HT to the area postrema induces EEG synchrony.

Furthermore, there are "information" channels capable of transmitting the 5-HT-induced signals from the area postrema to structures inside the BBB. Morest (1960) had found that nerve fibers leave the fiber plexus of the area postrema in the direction of the nucleus tractus solitarius. He also observed, in the area postrema, dendritic arborizations of neurons located in the medial edge of the tractus solitarius.

Finally, Magnes, Moruzzi, and Pompeiano (1961) demonstrated that low-frequency stimulation of the area of the solitary tract nucleus induced widespread synchronization of the EEG, which frequently outlasted the period of stimulation. Evidently this nuclear area is a hypnogenic structure. Corroborating evidence was offered by Bonvallet and collaborators (Bonvallet and Block, 1961; Bonvallet and Allen, 1962), who showed that arousal produced by reticular and nociceptive stimulation was more intense and more prolonged in cats in which discrete lesions had been placed in the cephalic part of the nucleus of the tractus solitarius. They postulated a negative feedback system operating between the reticular formation and the nucleus of the solitary tract whose function would be to check excessive arousal. Together with Bronzino (1968, 1971; see also Koella, 1970), we reinvestigated this idea and were able to produce further evidence for the existence of this negative feedback system. We were also able to show that in all likelihood it is this neuronal mechanism through which serotonin exerts its "anti-waking" effect. For this study we first made the following assumptions:

1. The first component of this feedback loop is a facilitating pathway leading from the mesencephalic reticular formation (RF) to the nucleus of the solitary tract (NTS); the amount of facilitatory information reaching NTS depends on the degree of arousal activity in the RF.

2. The NTS dispatches inhibitory messages which eventually feed back upon the elements constituting the mesencephalic activating system. The amount of this inhibitory feedback depends upon the degree of activity in NTS.

3. NTS projects as a whole to both sides of the reticular formation.

4. The "gain" in NTS is variable in the sense that the ratio between outgoing reticulopetal inhibitory information and incoming reticulofugal facilitating information depends on additional inputs to NTS. Signals reaching NTS through the sensory channels in the cranial nerves V, VII, IX, and X and also serotonin would constitute such "gain-modulating" factors.

5. The information, flowing from RF to NTS and back to RF, if made to consist of more or less synchronized volleys, could be visualized by means of evoked potential techniques.

In decerebellated cats we transected the brainstem at the mesencephalo-diencephalic border and placed a longitudinal midsagittal cut through the lower brainstem extending from the transection level down to the midpoint of the fourth ventricle. To avoid any possibility of "cross-talk," we separated the two halves of the brainstem by means of a plastic sheet of appropriate size and shape. A stimulation electrode was placed in the right mesencephalic reticular formation and recording electrodes were introduced into the NTS and the left RF, respectively. Single electrical shocks applied to the right RF were followed by evoked potentials in NTS as well as in the left RF, indicating information flow from the right RF to NTS and thence back to the left RF.

Placing small cotton pellets soaked with Xylocaine® (20 mg/ml) on the floor of the fourth ventricle just over NTS invariably led to marked reduction of the evoked potentials not only in NTS but in the left RF

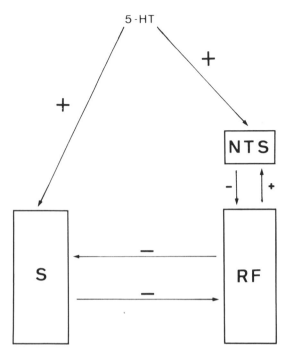

FIG. 2. Inclusion of reticulo-solitario-reticular feedback system (RF-NTS-RF) to realize "anti-waking" mechanism suggested by the experimental work. The + sign at the NTS side indicates "gain-enhancing" effect.

as well, an effect in all likelihood induced via gain reduction in NTS. In turn, placing 5-HT-soaked (3 µg/ml) pellets over the areae postremae was followed by a distinct increase of the evoked potentials in NTS and in the left RF by about 15 to 30%. Smaller doses had no effect. Evidently, serotonin led to an increase in gain in NTS but, probably in contrast to Xylocaine®, did so indirectly, that is, via action on 5-HT receptors in the area postrema and thence through neuronal pathways projecting into the NTS. Thus serotonin would increase the "anti-waking" activity of the RF-NTS-RF feedback system and in this way would promote the onset of, and would help to maintain, sleep.

Figure 2 depicts, again schematically, the extension of the "system" shown in Fig. 1 by inducing this gain-modulating mechanism of 5-HT.

Work is in progress in our laboratory to demonstrate the reticulo-solitario-reticular feedback system in a still better fashion and thus to furnish a still safer basis for the proposed dual (i.e., reciprocal) action of serotonin as a sleep-promoting and arousal-inhibiting agent.

REFERENCES

Bonvallet, M., and Allen, M. B., Jr. (1962): Localisation de formations bulbaries intervenent dans le contrôle de différentes manifestations de l'activation réticuliaire. *C.R. Soc. Biol.*, 156:597–601.

Bonvallet, M., and Bloch, V. (1961): Bulbar control of cortical arousal. *Science*, 133:1133–1134.

Bronzino, J. D. (1968): Verification of a neural feedback pathway in the brain stem of the cat between the midbrain reticular formation and the nucleus of the tractus solitarius. Dissertation, Worcester Polytechnic Institute.

Bronzino, J. D. (1971): Effect of serotonin and Xylocaine upon evoked responses established in neural feedback circuits associated with sleep-waking process. *Biological Psychiatry*, 3:217–226.

Bronzino, J. D., Morgane, P. J., and Stern, W. C. (1972): EEG synchronization following application of serotonin to area postrema. *American Journal of Physiology*, 223:376–383.

Faber, H. K. (1955): *Pathogenesis of Poliomyelitis*, pp. 1–157. Oxford, London.

Hess, W. R. (1924–1925): Über die Wechselbeziehungen zwischen psychischen und vegetativen Funktionen. *Schweiz. Arch. Neurol. Psychiat.*, 15:260–277, 1924, and 285–306, 1925.

Hess, W. R. (1933): Der Schlaf. *Klinische Wochenschrifte*, 12:129–134.

Hess, W. R. (1944): Das Schlafsyndrom als Folge dienzephaler Reizung. *Helvetica Physiologica Acta*, 2:305–344.

Jouvet, M. (1967): Mechanisms of the states of sleep: A neuropharmacological approach. In: *Sleep and Altered States of Consciousness*, edited by S. S. Kety, E. V. Evarts, and H. L. Williams. Williams & Wilkins, Baltimore.

Kleitman, N. (1939): *Sleep and Wakefulness*. University of Chicago Press, Chicago.

Koella, W. P. (1970): Serotonin oder Somnotonin? *Schweiz. Med. Wschr.*, 100:357–364.

Koella, W. P., and Czicman, J. S. (1963): Influence of serotonin upon optic evoked potentials, EEG, and blood pressure of cat. *American Journal of Physiology*, 204:873–880.

Koella, W. P., and Czicman, J. S. (1966): Mechanism of the EEG-synchronizing action of serotonin. *American Journal of Physiology*, 211:926–934.

Koella, W. P., Feldstein, A., and Czicman, J. S. (1968): The effect of para-chloro-phenyl-alanine on the sleep of cats. *Electroencephalography and Clinical Neurophysiology*, 25:481–490.

Magnes, J., Moruzzi, G., and Pompeiano, O. (1961): Synchronization of the EEG produced by low-frequency electrical stimulation of the region of the solitary tract. *Archives Italiennes de Biologie*, 99:33–67.

Morest, D. K. (1960). A study of the structure of the area postrema with Golgi methods. *American Journal of Anatomy*, 107:291–303.

Moruzzi, G. (1972): The sleep-waking cycle. In: *Ergebnisse der Physiologie*, Vol. 64. Springer-Verlag, Berlin.

Renault, J. (1967): Monoamines et sommeils. Rôle du système du raphé et de la sérotonine cérébrale dans l'endormissement. Thèse de Médicine, University of Lyon.

Roth, G. I., Walton, P. L., and Yamamoto, W. S. (1970): Area postrema: Abrupt EEG synchronization following close intraarterial perfusion with serotonin. *Brain Research*, 23:223–233.

Shimizu, N., Morikawa, N., and Okada, M. (1959): Histochemical studies of monoamine oxidase of the brain of rodents. *Zeitschrifte für Zellforschung*, 49:389–400.

Wislocki, G. B., and Putnam, T. J. (1920): Note on the anatomy of the areae postremae. *Anatomical Record*, 19:281–287.

Advances in Biochemical Psychopharmacology, Vol. 11
Raven Press, New York © 1974

On the Effects of Melatonin on Sleep
and Behavior in Man

H. Cramer, J. Rudolph, U. Consbruch, and K. Kendel

Department of Neurology, University of Freiburg, Freiburg, West Germany

Melatonin (5-methoxy-N-acetyltryptamine) has been identified as the principal hormone of the pineal gland (Lerner, Case, Takahashi, Lee, and Mori, 1958). Little is known about the functional role of this indoleamine in mammals, although remarkable contributions have been made to the control of its formation (Klein and Berg, 1970; Weiss and Crayton, 1970). In the rat, the synthesis of melatonin is increased several-fold during darkness, but few studies have been made on the synthesis and metabolism of melatonin in diurnal species: the melatonin content of the pineals of several birds was found to be greater during the daily dark period than during the light period (see Wurtman, Axelrod, and Kelly, 1968).

Although melatonin is thought to be manufactured almost exclusively in the pineal gland, recent findings point to the presence of melatonin in other parts of the nervous system, notably in the hypothalamus (Koslow, *This Volume*). Implants of melatonin in the hypothalamus of cats were followed by the appearance of behavioral and electroencephalographic signs of sleep (Marczynski, Yamaguchi, Ling, and Grodzinska, 1964). Systemic injection of the hormone into young chicks induced sleep and electrographic patterns of slow-wave sleep (Hishikawa, Cramer, and Kuhlo, 1969). In man the appearance of sleep following the administration of melatonin was observed by Antón-Tay, Diaz, and Fernandez-Guardiola (1971). In addition, these authors noted marked psychotropic effects of melatonin.

We studied the effects of melatonin in healthy young male volunteers. After an injection of melatonin (50 mg, i.v.) in daytime, sleep appeared within 15 to 40 min and lasted for 26 to 60 min, reaching stages 3 and 4. Further, the effect of melatonin (50 mg, i.v., at 21:30 hr) on night sleep was studied. All-night polygraphic recordings were performed on 15 subjects. After an adaptation night (injection of 0.9% NaCl), melatonin (50 mg) and its solvent alone were injected at 21:30 hr on night 2 and night 3 in random order. The EEG, EMG,

EOG, and respiration were recorded on a Siemens Universal. Evaluations were done under double-blind conditions independently by two experienced scorers. In addition, the EEG was recorded on tape and processed for frequency, amplitude, and power spectra on an IBM 1130 (Kendel, Cramer, and Wita, 1973). Psychic effects of melatonin were assessed by informal interviews and a battery of standardized questionnaires, including subjective rating scales for the refreshing effects of sleep, the Maudsley Medical Questionnary (MMQ) for neurotic tendency and extraversion (Eysenck, 1959, 1964), a Neurasthenia Supplementary Questionnary (DBKF) and a Mental State Questionnary (PZF) (Kendel, Beck, and Kruschke-Dubois, 1972). 5-Hydroxyindoleacetic acid (5-HIAA) was determined by a modification of the method of Udenfriend, Titus, and Weissbach (1955), adenosine $3', 5'$-monophosphate by the method of Gilman (1970). Melatonin significantly decreased the mean latency of sleep onset from 23.3 min in solvent control nights to 10.4 min in melatonin nights (Fig. 1). The mean latencies of deep sleep (stage 3) were 19.2 ($\pm$5.63 SEM) and 16.9 min ($\pm$ 1.30 SEM), and the latencies of REM sleep were 107.5 ($\pm$11.86 SEM) and 98.4 min ($\pm$12.88 SEM) in control and melatonin nights, respectively. No significant differences in total sleep and in the percentage of sleep stages 1 to 4 and REM were observed between the control and melatonin nights (Fig. 1). The frequency of total sleep stages was slightly increased, with more frequent waking periods and sleep stages 1 and 2 after melatonin. The average number of recorded body movements per night was 41 in control and 44 in melatonin nights.

The subjective experience of sleep and sleep quality was not significantly changed by melatonin. About half of the total nights for each group were reported as dreamless, 2 melatonin nights and 1 control night were characterized by frequent dreaming, and the rest were reported as having moderate dream activity. On the day following melatonin injections, psychomotor activity appeared to be slightly reduced, showing nonsignificant differences in the subjective rating scales. The standardized questionnaires comparing evening and morning values did not show significant changes following melatonin in the total group of subjects with respect to the main variables (sociability, emotional stability, mood, psychomotor activity) and additional variables tested. However, when the subjects were distributed according to their neurasthenia ratings, a subgroup with scores of a moderate degree of neurasthenia (MMQ above 13 points) showed negative influence of melatonin on psychic and motor activity (i.e., sedation) and positive influence on emotional stability as a trend ($p < 0.10$) (Table 1). The average excretion of 5-HIAA determined in 12 subjects was slightly decreased, by about 20%, from 1.77 and 2.35 mg to 1.64

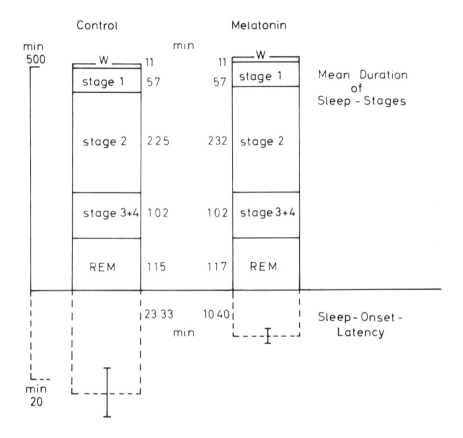

FIG. 1. Influence of melatonin (50 mg, i.v.) on average duration of sleep stages and on sleep latencies in 15 male subjects. The mean durations of the sleep stages remain unaltered by melatonin, while the latency of sleep onset is significantly reduced after injection of melatonin at 21.30 (p < 0.01 paired *t*-test; vertical bar ± SEM).

and 2.00 mg on the night and on the day following melatonin injections. The urinary excretion of adenosine $3',5'$-monophosphate determined in three subjects increased from 2.8 ± 1.83 to 25.0 ± 13.0 pmoles on the night after melatonin.

Melatonin, in the dosage used, appears to be a potent sleep inducer. Melatonin-induced sleep, behaviorally as well as by its polygraphic pattern, strikingly resembles natural sleep. In contrast to an earlier report (Antón-Tay et al., 1971), our polygraphic recordings and explorations did not reveal significant changes in the percentage and frequency of REM sleep and in dream activity. Computer analysis of EEG recordings during the first 10 min following injection failed to show marked changes in frequency spectra and amplitude histograms

TABLE 1. *Melatonin—Influences on affective variables*

	A		B		D		E		A - E	
	Co	Me	Co	Me	Co	Me	Co	Me	Co	Me
							(*)			
x̄ N	-1,4	-1,0	-1,2	+0,8	-0,8	-1,2	+3,6	-2,5	-9,2	-8,8
x̄ nN	-0,4	-1,3	-0,8	-1,3	-1,8	-0,7	-6,0	-2,8	-14,0	-12,9
x̄ N + nN	-0,8	-0,5	-0,1	+0,4	-1,3	-0,1	-1,7	-2,5	-11,3	-2,7

(*) $p < 0.10$ A = sociability
N = Neurasth. (MMQ > 13)   B = emotional stability
nN = nonneurasth. (MMQ < 13)   D = mood
Me = Melatonin   E = psychomotor activity
Co = Control group

except for a tendency toward desynchronization, which may be interpreted as an early sign of sleep not yet detectable by visual scoring. During deep sleep and REM sleep, however, an increase in degree of synchronization was found (Kendel et al., 1973), similar to experiments in chicks with very high doses which showed hypersynchronization of deep sleep after melatonin (Hishikawa et al., 1969). Our preliminary findings on psychotropic effects by use of a battery of standardized tests indicate that melatonin may have beneficial effects in emotional disorders in addition to its sleep-promoting action. Further studies exploring this possibility are in progress.

ACKNOWLEDGMENTS

This research was supported by the Deutsche Forschungsgemeinshaft (SFB 70). The technical assistance of Mrs. Mitta Weigel is gratefully appreciated.

REFERENCES

Antón-Tay, F., Diaz, J. L., and Fernandez-Guardiola, A. (1971): On the effects of melatonin upon human brain. Its possible therapeutic implications. *Life Sciences*, 10, 1:841.
Eysenck, H. J. (1959): Das Maudsley Personality Inventory als Bestimmer der neurotischen Tendenz und Extraversion. *Zeitschrift für Experimentalishe Angewandte Psychologie*, 6:167.
Eysenck, H. J. (1964): *Maudsley-Personlichkeitsfragebogen*, 2nd Ed. Hogrefe, Gottingen.
Gilman, A. (1970): A protein binding assay for adenosine 3', 5'-cyclic monophosphate. *Proceedings of the National Academy of Sciences (U.S.)*, 68:2165.

Hishikawa, Y., Cramer, H., and Kuhlo, W. (1969): Natural and melatonin-induced sleep in young chickens—A behavioral and electrographic study. *Experimental Brain Research*, 7:84.

Kendel, K., Beck, U., and Kruschke-Dubois, H. (1972): Die chronisch-neurasthenische Schlafstörung. *Arch. Psychiat. Nervenkr.*, 216:201.

Kendel, K., Cramer, H., and Wita, C. (1973): Influence of melatonin on EEG and human sleep: Automatic EEG-analysis and polygraphic night sleep investigations. *(In preparation.)*

Klein, D. C., and Berg, G. R. (1970): Stimulation of melatonin production by norepinephrine involves cyclic AMP-mediated stimulation of N-acetyltransferase. In: *Role of Cyclic AMP in Cell Function*, edited by P. Greengard and E. Costa. Advances in Biochemical Psychopharmacology, vol. 3. Raven Press, New York.

Lerner, A. B., Case, J. D., Takahashi, Y., Lee, T. H., and Mori, W. (1958): Isolation of melatonin, the pineal gland factor that lightens melanocytes. *Journal of the American Chemical Society*, 80:2587.

Marczynski, T. J., Yamaguchi, N., Ling, G. M., and Grodzinska, L. (1964): Sleep induced by the administration of melatonin (5-methoxy-N-acetyltryptamine) to hypothalamus in unrestrained cats. *Experientia*, 20:435.

Udenfriend, S., Titus, E., and Weissbach, H. (1955): The identification of 5-hydroxy-3-indoleacetic acid in normal urine and a method for its assay. *Journal of Biological Chemistry*, 216:499.

Weiss, B., and Crayton, J. W. (1970): Neural and hormonal regulation of pineal adenyl cyclase activity. In: *Role of Cyclic AMP in Cell Function*, edited by P. Greengard and E. Costa. Advances in Biochemical Psychopharmacology, vol. 3. Raven Press, New York.

Wurtman, R. J., Axelrod, J., and Kelly, D. E. (1968): *The Pineal*. Academic Press, New York.

Advances in Biochemical Psychopharmacology, Vol. 11
Raven Press, New York © 1974

Ventricular Fluid 5-Hydroxyindoleacetic Acid Concentrations During Human Sleep

Richard J. Wyatt*, Norton H. Neff**, Thomas Vaughan*,
John Franz***, and Ayub Ommaya***

*Laboratory of Clinical Psychopharmacology,
**Laboratory of Preclinical Pharmacology,
National Institute of Mental Health, and
***Surgical Neurology Branch, National Institute of Neurological Diseases
and Stroke, Bethesda, Maryland 20014

I. INTRODUCTION

The biogenic amine serotonin may be important for normal sleep (Jouvet, 1972; Wyatt, 1972). The major body of evidence suggests that serotonin either initiates or sustains non-rapid eye movement (NREM) sleep (Jouvet, 1972; Koella, Feldstein, and Czicman, 1968). If this hypothesis is valid, then serotonergic neuronal activity should be greater during NREM sleep than it is during waking and sleep associated with rapid eye movements (REM). Another hypothesis suggests that serotonin functions by suppressing phasic events (rapid eye movements and pontine geniculate occipital monophasic sharp waves) occurring during REM sleep from breaking through to NREM sleep and waking (Dement, Zarcone, Ferguson, Cohen, Pivik, and Barchas, 1969; Wyatt, 1972). If this hypothesis is valid, then serotonergic neuronal activity should be more active during waking and NREM sleep than it is during REM sleep. While recent studies in the cat measuring the electrical activity of neurons thought to contain serotonin have produced varied results, no measurement of brain serotonergic activity has been made in man (Bloom, Hoffer, Nelson, Sheu, and Siggins, 1973; McGinty, Harper, and Fairbanks, 1973; Jacobs, 1973).

It is thought that the production of 5-hydroxyindoleacetic acid (5-HIAA) in the brain is proportional to serotonergic neuronal activity (Aghajanian, Rosecrans, and Sheard, 1967). Thus, measuring the production of 5-HIAA during sleep and waking should give an index of serotonergic activity during these states. As an approximation of this, we have measured ventricular 5-HIAA concentrations during waking, NREM, and REM sleep as part of a study of presenile dementia (Slaby and Wyatt, 1974).

A few patients with presenile dementia are reported to have improved after a ventricular-peritoneal shunting procedure (Hakin and Adams, 1965; Slaby and Wyatt, 1974). We are studying the pressure dynamics and biochemical changes that take place with the shunting procedure.

II. METHODS

A silicone rubber ventricular fluid reservoir was installed in four patients with presenile dementia via right frontal burrholes, cannulating the frontal horn of the right lateral ventricle (Ratcheson and Ommaya, 1968). After the incision healed, access to ventricular fluid was gained by inserting a No. 21 gauge needle subcutaneously into the reservoir and connecting it to 3 m of polyethylene tubing. After withdrawing and discharging dead space fluid (tubing and reservoir volume was about 2 ml), 2 ml of ventricular fluid was removed, immediately frozen, and stored at $-20°C$ for up to 10 days. The total volume withdrawn was approximately 60 ml per night.

III. RESULTS

During the first 2 nights of the study, samples were withdrawn every 20 min and 5-HIAA concentrations determined(Korf and Valkenburgh-Sikkema, 1969). Since no clear relationship between the concentration of 5-HIAA in ventricular fluid and the time of night was found, in subsequent nights samples were withdrawn only when a patient had been in a clear sleep stage for at least 5 min. An attempt was made to distribute samples evenly between waking, NREM sleep, and REM sleep, but this was not always possible due to several nights of premature awakening.

Sleep was monitored electrophysiologically, and sleep stages were scored according to standard criteria (Rechtschaffen and Kales, 1968). The patients exhibited unusual EEG patterns; for example, spindles were sparse and although slow waves occurred they were not sufficient in number to be designated as stages III or IV. Because of the difficulty in scoring the stages of sleep, 5-HIAA values are reported only for waking, NREM sleep, and REM sleep. Mean 5-HIAA concentrations for each sleep classification and a mean concentration for the whole night were calculated. Whole-night means for the nine nights ranged from 50 to 115 ng per ml of ventricular fluid and were normalized to 100%. 5-HIAA concentrations for waking (50 samples), NREM sleep (76 samples), and REM sleep (34 samples) are expressed as a percentage of the mean value for the whole night.

Waking values (Fig. 1) were 98%, NREM sleep 104%, and REM sleep

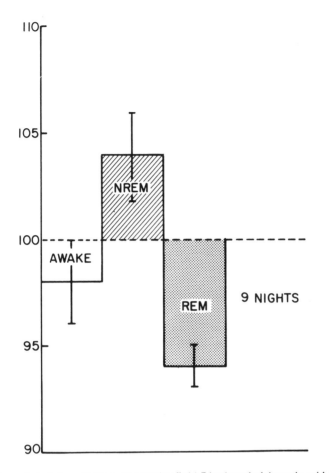

FIG. 1. Percent deviation of mean ventricular fluid 5-hydroxyindoleacetic acid levels in four patients over nine nights for awake, rapid eye movement (REM), and nonrapid eye movement (NREM) sleep. Mean whole-night values were normalized to 100%. NREM concentration is higher than either awake ($p < 0.05$) or REM ($p < 0.01$). Awake concentrations are higher than those for REM ($p < 0.05$).

94% of the mean for the total night. A Friedman's analysis of variance (Siegel, 1956) showed a significant ($p < 0.01$) difference between these means. A Wilcoxon signed-rank test (Siegel, 1956), comparing the means with each other was significant for each comparison: NREM higher than waking ($p < 0.05$); waking higher than REM ($p < 0.05$); and NREM higher than REM ($p < 0.01$). Five high values three standard deviations above the mean were not included in the calculations. They all occurred at the end of REM periods and may reflect either a physiological process or technical error.

IV. DISCUSSION

These patients were preselected because of a defect (dementia) during their waking state, and generalizations to the healthy brain may not be valid. These studies do, however, add to the increasing evidence that serotonin is involved in waking, NREM sleep, and REM sleep processes. By this method of estimation, it would appear that serotonergic neurons are most active during NREM, there being somewhat less activity during waking and REM sleep. Another possible explanation for these apparent changes is that they reflect alterations in cerebral blood flow or osmotic changes. If this were the case, other metabolites of biogenic amines might be expected to be changed in the same manner. Preliminary data, however, indicate that both homovanillic acid (dopamine metabolite), and 3-methoxy-4-hydroxy phenyl glycol (norepinephrine metabolite) concentrations alter in a very different manner than 5-HIAA during sleep.

Alterations of biogenic amine metabolism have been claimed for schizophrenic and depressive disorders in which the normal sleep-waking cycle is upset. Part of the argument in favor of these alterations is based on cerebral spinal fluid studies of metabolites of the parent amine. It seems possible that at least some of the confusion surrounding this field lies in our lack of understanding of the relationship of these amines to normal circadian and ultradian rhythms involved in NREM, REM sleep, and waking.

REFERENCES

Aghajanian, G. K., Rosecrans, J. A., and Sheard, M. H. (1967): Serotonin: Release in the forebrain by stimulation of midbrain raphe. *Science*, 156:402.

Bloom, F. E., Hoffer, B. J., Nelson, C., Sheu, Y-S, and Siggins, G. R. (1973): The physiology and pharmacology of serotonin mediated synapses. In: *Serotonin and Behavior*, edited by E. Usdin and J. Barchas, pp. 249–261. Academic Press, New York.

Dement, W., Zarcone, V., Ferguson, J., Cohen, H., Pivik, T., and Barchas, J. (1969): Some parallel findings in schizophrenic patients and serotonin-depleted cats. In: *Schizophrenia—Current Concepts and Research*, edited by D. V. Siva Sankar, PJD Publications. Hicksville, N.Y.

Hakin, S., and Adams, R. D. (1965): Special clinical problem of symptomatic hydrocephalus with normal cerebrospinal fluid pressure. *Journal of Neurological Science*, 2:307.

Jacobs, B. J. (1973): Amygdala unit activity as a reflection of functional changes in brain serotonergic neurons. In: *Serotonin and Behavior*, edited by E. Usdin and J. Barchas, pp. 281–289. Academic Press, New York.

Jouvet, M. (1972): The role of monoamines and acetylcholine-containing neurons in the regulation of the sleep-waking cycle. *Ergebnisse der Physiologie*, 64:166.

Koella, W. P., Feldstein, A., and Czicman, J. S. (1968): The effect of *para*-chloro-phenylalanine on the sleep of cats. *Electroencephalography and Clinical Neurophysiology*, 25:481.

Korf, J., and Valkenburgh-Sikkema, T. (1969): Fluorimetric determination of 5-hydroxyindoleacetic acid in human urine and cerebrospinal fluid. *Clinica Chemica Acta*, 26:301.

McGinty, D. J., Harper, R. M., and Fairbanks, M. K. (1973): 5-HT-Containing neurons: Unit activity in behaving cats. *Serotonin and Behavior*, edited by E. Usdin and J. Barchas, pp. 267–279. Academic Press, New York.

Ratcheson, R. A., and Ommaya, A. (1968): Experience with subcutaneous cerebrospinal fluid reservoir. *New England Journal of Medicine*, 279:1025.

Rechtschaffen, A., and Kales, A. (1968): *A Manual of Standardized Terminology, Techniques and Scoring System for Sleep Stages of Human Subjects*. U.S. Department of Health, Education and Welfare, Public Health Service. National Institutes of Health, National Institute of Neurological Disease and Blindness, Neurological Information Network, Bethesda, Md.

Siegel, S. (1956): *Nonparametric Statistics for the Behavioral Sciences*. McGraw-Hill Book Company, New York.

Slaby, A. E., and Wyatt, R. J. (1974): *Dementia in the Presenium*. Charles Thomas, Springfield, Ill.

Wyatt, R. J. (1972): The serotonin-catecholamine-dream bicycle: A clinical study. *Biological Psychiatry*, 5:33.

Advances in Biochemical Psychopharmacology, Vol. 11
Raven Press, New York © 1974

Effects of Central Alterations of Serotoninergic Neurons upon the Sleep-Waking Cycle

Michel Jouvet and Jean-Francois Pujol

Department of Experimental Medicine, University Claude Bernard School of Medicine, Lyon, France

The main experimental results which support the serotoninergic hypothesis of sleep have been the subject of recent reviews and symposia (Jouvet, 1969, 1972, 1973). They may be summarized shortly as follows:

Inhibition of synthesis of serotonin (5-HT) with *p*-chlorophenylalanine induces a secondary insomnia which is proportional to the decrease of cerebral 5-HT. This insomnia is immediately reversed and physiological sleep is induced by the injection of a small dose (2 to 5 mg/kg) of 5-hydroxytryptophan (5-HTP), which is rapidly converted into 5-HT as shown by the subsequent relative increase of 5-HT and 5-hydroxyindoleacetic acid (5-HIAA) (Bobillier, Froment, Seguin, and Jouvet, 1974).

The surgical inactivation of 5-HT-containing perikarya of the mesencephalic and pontine raphe system induces a very severe, long-lasting insomnia which is proportional to the decrease of 5-HT in the terminals. This insomnia is *not* reversible by secondary injection of a small dose of 5-HTP (probably because the 5-HTP cannot be metabolized into 5-HT in the degenerated 5-HT terminals).

Thus it is clear that both neuropharmacological and neurophysiological alterations of 5-HT synthesis lead to the same result, that is, a significant decrease of sleep. Evidently, the pharmacological approach cannot bring us much more precise information concerning the exact topography of the 5-HT neurons involved in sleep control. As a result, we have chosen some more sophisticated and specific techniques to alter the regional metabolism of 5-HT neurons, either by decreasing their activity with central injection of 5-6-hydroxytryptamine (5-6-HT) or by indirectly increasing their metabolism with localized lesions of the dorsal norepinephrine (NE) pathway.

I. CENTRAL INACTIVATION OF 5-HT NEURONS

The surgical destruction of raphe neurons by electrocoagulation is not selective enough, and it may be argued that such a lesion might also destroy non-5-HT-containing perikarya or ascending axons belonging to some putative synchronizing system (Mancia, 1969).

Fortunately, a much more selective destruction of 5-HT neurons can be achieved thanks to the toxic properties of 5-6-HT upon 5-HT neurons (Baumgarten, Bjorklund, Holstein, and Nobin, 1972; Daly, Fuxe and Jonsson, 1973; Bjorklund, Nobin, and Stenevi, 1973). Four different tactics are theoretically possible when using 5-6-HT to study the role of 5-HT neurons in sleep: direct injection into the perikarya of the raphe system, the ascending 5-HT pathways, some densely packed group of terminals, or the ventricles. Since the anatomy of the ascending 5-HT pathways is not yet mapped with precision in the cat, we have chosen the other three tactics, the results of which will be briefly summarized (Froment, Petitjean, Bertrand, Cointy, and Jouvet, 1974). In every case the experiments were performed in chronically implanted cats, recorded during a 1-week control period and for 2 weeks after the injection. Biochemical determination of brain monoamines was effected after the sacrifice of the animals.

A. Injection of 5-6-HT in the Rostral Raphe

The injection of either 5-6-HT (at a concentration of 0.5 μg in 1 μl) or solvent (Ringer solution containing 0.1 mg/ml acid ascorbic) in the rostral raphe (N. Raphe dorsalis or centralis superior) induced in both cases a short-duration increase of paradoxical sleep which lasted for 24 hr. Thereafter there was no significant alteration of either states of sleep or of PGO activity. The biochemical control did not show any alteration of monoamine endogenous level in the rostral part of the brain, but a significant decrease of 5-HT and 5-HIAA only in the cerebellum. Two conclusions may be drawn from this almost negative experiment:

1. Nonspecific chemical excitation of the raphe perikarya may induce some transitory increase of paradoxical sleep.

2. *In the dose we have used*, 5-6-HT does not seem to be toxic for the 5-HT perikarya since there was no alteration of brain monoamines.

B. Intraventricular Injection of 5-6-HT

In contrast to the preceding experiment, the injection of 1000 μg of 5-6-HT (diluted in Ringer and ascorbic acid) in the ventricles led to a very significant alteration of both states of sleep, whereas the control injection with Ringer did not change the sleep states at all. Immediately after the injection, the cats entered a state of prostration followed by some permanent discharges of PGO activity (as is always observed after injection of reserpine or after destruction of the raphe system.) During the days which followed, a permanent and significant decrease (−50%) of stage II (slow-wave) sleep and paradoxical sleep was observed while there was no qualitative alteration of the EEG pattern of sleep or of PGO activity during paradoxical sleep. In those cats the biochemical evaluation of brain monoamines did not reveal any change in catecholamine (NE or dopamine, DA) but a significant decrease of both 5-HT and 5-HIAA in most of the cerebral structures. Interestingly enough, there was a very significant correlation (R = 0.77, $p < 0.001$) between the decrease of both stage II and paradoxical sleep and the decrease of 5-HT and 5-HIAA; the correlation was not significant for stage I (spindles).

These results are in accord with the hypothesis that serotoninergic neurons play a major role in the determination of stage II and paradoxical sleep. They also confirm that in the cat, as in the rat, 5-6-HT has a very selective toxic effect upon the 5-HT terminals since we did not observe any change in catecholamine level.

Our results could be compared with those obtained after intraventricular injection of either 6-hydroxydopamine (6-OHDA) alone, which inactivates both 5-HT and catecholamine (CA) neurons, or 6-OHDA in chlorimipramine-pretreated cats, in which only CA neurons are inactivated while 5-HT neurons seem to be in a state of increased turnover as indirectly suggested by the increase of 5-HIAA (Laguzzi, Petitjean, Pujol, and Jouvet, 1972; Petitjean, Laguzzi, Sordet, Jouvet, and Pujol, 1972) (Fig. 1). A decrease of both stage II and paradoxical sleep follows the inactivation of 5-HT neurons, while stage II increases in the case of activation of 5-HT metabolism, which is reflected by the increase of 5-HIAA. On the other hand, stage I sleep appears to be the result of a complex interaction between CA (probably NE-activating mechanism) and 5-HT, since its proportion increases after inactivation of CA neurons. In this case stage I may represent mostly a passive decrease of waking rather than a more active sleep mechanism. Finally, paradoxical sleep seems to represent the result of the agonistic influence of some 5-HT and CA systems, since it decreases whenever one or both systems are inactivated.

Thus, the intraventricular injection of drugs which are selectively toxic for the monamine systems represents a useful tool for determining the monoaminergic mechanisms interacting during the sleep-waking cycle. Unfortunately, there is a great pitfall in the ventricular approach, since it lacks the topographical dimension which is of paramount importance. Indeed, any intraventricular injection of 5-6-HT

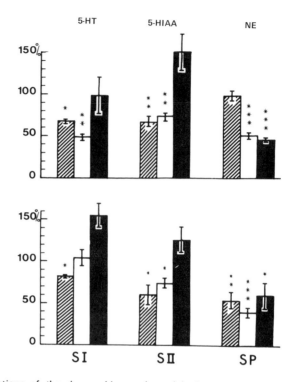

FIG. 1. Alterations of the sleep-waking cycle and brain monoamines after intraventricular injections of 5-6-HT and 6-OHDA. Above: Percentage of serotonin (5-HT), 5-HIAA, and norepinephrine (NE) in the mesencephalon 15 days after the intraventricular injection (as compared with control animals). Below: Percentage of stage I (SI), stage II (SII) and paradoxical sleep (SP) during the 15 days after the intraventricular injection as compared with the preinjection control for each group. Oblique hatchings: intraventricular injection of 1 mg of 5-6-HT. White rectangles: intraventricular injection of 2.5 mg of 6-OHDA after pretreatment with chlorimipramine (10 mg/kg). Black rectangles: intraventricular injection of 2.5 mg of 6-OHDA. Results are expressed in mean ± SEM.
  $*p < 0.05; **p < 0.01; ***p < 0.001.$

(or 6-OHDA) is followed by a diffuse alteration of monamine terminals belonging to different systems which may serve different functions, and this approach does not answer one of the most important questions regarding sleep mechanisms: Where are the 5-HT terminals located which play the most important role in the behavioral or EEG aspect of sleep? To solve this problem, we have chosen to inactivate the 5-HT terminals of the preoptic region directly.

C. Intracerebral Injection of 5-6-HT in the Preoptic Area

The preoptic area is certainly an interesting target. Indeed, numerous indirect data suggest that it plays an important role in sleep and that this role might be mediated through serotoninergic mechanisms: the destruction of the preoptic area is followed by a significant temporary decrease of sleep (McGinty and Sterman, 1968). The stimulation at high frequency (Sterman and Clemente, 1962), or the direct injection of 5-HT (Yamaguchi, Marczinski and Ling, 1963) in this area is followed by a significant increase of slow-wave sleep while the electrical stimulation is totally ineffective in inducing sleep in *p*-chlorophenylala-nine-pretreated cats (Wada and Terao, 1970). Our preliminary data strongly suggest that the 5-HT terminals located in this area are participating in the regulation of cortical synchronization during sleep, since the intracerebral injection of 5-6-HT in the preoptic area is followed by a significant decrease (−60%) of slow-wave sleep lasting at least 2 weeks while neither PS nor PGO activity is altered. Interestingly enough, after such selective destruction of the preoptic 5-HT terminals, the cat may present some period of slow-wave sleep-like behavior with an activated EEG. This suggests that some 5-HT terminals located in the brainstem may still act upon the postsynaptic mechanism responsible for the postural and ocular aspect of slow-wave sleep, while the preoptic 5-HT terminals for cortical synchronization are destroyed. If this hypothesis is valid, it is possible that some very localized lesion of the 5-HT-containing perikarya of the raphe system might alter either EEG or behavioral aspects of sleep, or both, as is the case in total destruction of the rostral raphe system (Jouvet, 1969).

Taken together, these results permit us to add the following precisions concerning the intervention of 5-HT neurons in sleep: Some group of 5-HT perikarya located mostly in the rostral part of the raphe system is responsible for the induction of sleep—by releasing 5-HT in some strategic midbrain and forebrain region; among them, the preoptic 5-HT terminals seem to play a major role in the induction of the postsynaptic events which lead to cortical synchronization. Finally, this concept may reconcile, on a histochemical and biochemical basis, the two main theories which are currently proposed in the explanation of sleep: the "ascending" hypothesis related to the role of the raphe (Jouvet, 1969) and the "descending" hypothesis related to the role of the preoptic area (see Moruzzi, 1972). Any selective inactivation of 5-HT neurons bearing upon the perikarya located in the raphe, the 5-HT ascending pathways, or the 5-HT terminals located in the preoptic area should lead to a significant decrease of sleep.

II. CENTRAL ACTIVATION OF 5-HT NEURONS

The direct activation of 5-HT neurons can theoretically be achieved through the stimulation of the 5-HT perikarya, as shown by the

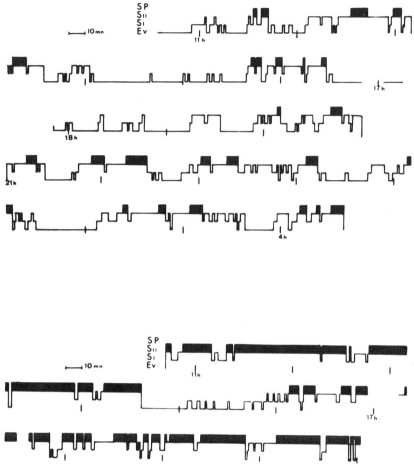

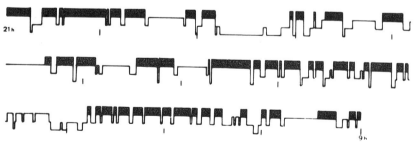

FIG. 2. Hypersomnia after bilateral lesion of the dorsal norepinephrine bundle. Above: Control recording from 11 a.m. to 4:30 a.m. Waking, stage I, stage II, and paradoxical sleep (in black) are in ordinates. Time scale 10 min. The percentage of paradoxical sleep is 13.5% of recording time. Below: The day following the coagulation, from 11 a.m. to 9 a.m., there is an enormous increase of PS which exceeds 60% of recording time.

subsequent increase of 5-HIAA in the terminals (Aghajanian, Rosecrans, and Sheard, 1967). However, this technique has given equivocal results upon the sleep states, either increase of sleep (Kostowski and Giacalone, 1969), or even increased waking (Polc and Monnier, 1970). Such contradictory results may be tentatively explained by the fact that electrical stimulation may activate other non-5-HT raphe neurons or may release 5-HT in a rather unphysiological way. Our method of activating 5-HT neurons is based upon the following neuropharmacological findings. It has been shown that the inactivation of CA neurons with α-methyl-p-tyrosine may induce an increased turnover of 5-HT neurons. In such a case, the decrease of endogenous CA is accompanied in the cat brainstem by an increase in the conversion of ^{3}H-tryptophan into ^{3}H-5-HT and by a significant increase of 5-HIAA (Jouvet and Pujol, 1972). If the hypothesis that CA neurons control 5-HT neurons is true, then the destruction of some well-defined CA pathway should also induce some activation of the 5-HT system. This problem was approached in the cat by a limited coagulation of the dorsal norepinephrine bundle (Maeda, Pin, Salvert, Ligier, and Jouvet, 1973). Indeed, previous investigations had shown that this bundle could be involved in the maintenance of tonic cortical arousal (Jones, Bobillier, Pin, and Jouvet, 1973). The bilateral coagulation of the dorsal norepinephrine bundle at the level of the isthmus, immediately adjacent to the periaqueducal gray matter and to the rostral raphe system, induces a most dramatic hypersomnia: As shown in Fig. 2, an increase of up to 400% of paradoxical sleep may be observed during the first day after the lesion. Usually this hypersomnia, first characterized by a selective increase of paradoxical sleep and secondarily by an increase of both states of sleep, lasts for 8 to 10 days, after which there is a slow return to baseline level. Control lesions effectuated either more dorsally (in the colliculi) (sham-operated cats) or more rostrally (in rostral mesencephalon) did not induce any significant alteration of sleep. Only the caudal extension of the lesion toward the group of perikarya belonging to the locus coeruleus and locus subcoeruleus totally impaired this phenomenon by suppressing paradoxical sleep.

The mechanism of this most characteristic hypersomnia (which is, by far, the most pronounced increase of paradoxical sleep we have ever observed) has been analyzed through neuropharmacological and biochemical means, and the results strongly suggest that the serotoninergic neurons of the raphe are involved.

1. The pretreatment of three cats with p-chlorophenylalanine, which induced an almost total insomnia, prevented the subsequent hypersomnia which normally follows the destruction of the dorsal norepinephrine bundle.

2. As shown in Fig. 3, the biosynthesis of 5-HT was measured during

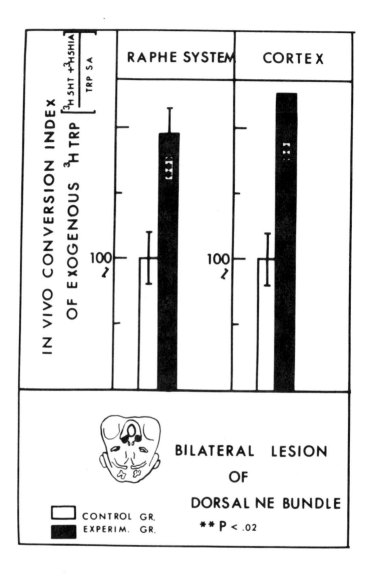

FIG. 3. Increase of activity of 5-HT neurons after lesion of the dorsal norepinephrine bundle. The lesion is schematized on a frontal section of the isthmus. The biochemical results are obtained 48 hr after the bilateral coagulation of the dorsal norepinephrine bundle. The rate of synthesis of 5-HT is estimated by following the initial accumulation of ^{3}H-5-HT and ^{3}H-5-HIAA endogenously synthesized from 1.5 mC of ^{3}H-tryptophan intravenously injected 20 min before sacrifice. The conversion index of tryptophan is given by the ratio (^{3}H-5-HT + ^{3}H-5-HIAA)/TRP. S.A. Results for the experimental group (black bars) are expressed as percent of the mean value for the control group (lesion of the inferior colliculi) ± SEM (five determinations). p values are calculated by a t test. Modified from Pujol, Stein, Blondaux, Petitjean, Froment, and Jouvet (1973).

the maximum period of hypersomnia, that is, 24 hr after the lesion. There was a very significant increase of the conversion of ^{3}H-trypto-phan into ^{3}H-5-HT in the cortex and the raphe, while no alteration of endogenous 5-HT or CA could be detected.

3. Specific alterations of endogenous level of monoamines could be detected in another series of experiments in which the cats were sacrificed 8 to 10 days after the lesion (Petitjean and Jouvet, 1970). There was a decrease of norepinephrine in the forebrain, mostly in the cortex, which can be explained by the destruction of the dorsal NE bundle. This decrease was accompanied by a significant increase of both tryptophan and 5-HIAA levels in the brain, while 5-HT did not change (Table 1). This result strongly suggests that the turnover of 5-HT neurons was still elevated 8 to 10 days after the lesion.

The intimate mechanisms by which the 5-HT perikarya of the raphe system are activated have not yet been discovered. A possible explanation is that some component of noradrenergic neurons, ascending from the rostral part of the locus coeruleus, would tonically control the activity of some serotonin-containing perikarya. There are, in fact, many catecholamine-containing terminals close to the 5-HT perikarya (Dahlstrom and Fuxe, 1964), but there is not yet any evidence of a direct noradrenergic-5-HT synapse. If such a control exists, it is possible that its suppression by a coagulation might release the activity of 5-HT neurons as is suggested by our two biochemical controls.

Such a mechanism would provide much plasticity and flexibility in

TABLE 1. *Neurophysiological and biochemical results obtained 10 days (A) after lesion of the dorsal NE bundle (M = 9) or (B) after sham operation in (n = 12)*

	SWS	PS	TRP	5-HT	5-HIAA	DA	NE
A	65 ± 3 [a]	15 ± 0.9 [a]	125 ± 14	91 ± 6	146 ± 14 [a]	98 ± 15	75 ± 6 [b]
B	45 ± 3	9 ± 0.8	−	102 ± 11	97 ± 10	80 ± 13	96 ± 4

SWS-PS: Percentage of slow-wave sleep and paradoxical sleep during the 10 days of recordings; mean ± SEM.

Percentage of tryptophan (TRP), 5-HT, 5-HIAA, dopamine (DA), and norepinephrine (NE) in the telediencephalon as compared with control cats. Absolute levels (in mg/g) are the following: TRP = 9700 ± 700; 5-HT = 599 ± 24; 5-HIAA = 305 ± 14; DA = 923 ± 43; NE = 415 ± 10.

[a] $p < 0.01$.
[b] $p < 0.05$.

Modified from Petitjean and Jouvet, 1970.

the control of norepinephrine neurons (which are strongly implicated in the control of cortical arousal) upon the 5-HT neurons involved in sleep.

III. SUMMARY

The biochemical demonstration of an increased activity of 5-HT neurons during hypersomnia and the converging data which demonstrate that any selective inactivation of 5-HT neurons leads to insomnia confirm the hypothesis that some group of 5-HT neurons plays a role of paramount importance in the regulation of sleep in the cat. We should no longer ask the question, do 5-HT neurons play a role in sleep mechanisms, but we should ask ourselves the question, how do 5-HT neurons regulate sleep?

ACKNOWLEDGMENTS

This work has been supported by Centre National de la Recherche Scientifique (LA 162), la Direction des Recherches et Moyens d 'Essais (72-108), et l'INSERM (U 52).

REFERENCES

Aghajanian, G. K., Rosecrans, J. A., and Sheard, M. (1967): Serotonin: Release in the forebrain by stimulation of midbrain raphe. *Science*, 156:402–403.

Baumgarten, H. G., Bjorklund, A., Holstein, A. F., and Nobin, A. (1972): Chemical degeneration of indolamine axons in rat brain by 5-6HT. An ultrastructural study. *Zeitung Zellforschtung*, 129:256–271.

Bjorklund, A., Nobin, A., and Stenevi, U. (1973): Effects of 5-6 hydroxytryptamine on nerve terminal serotonin and serotonin uptake in the rat brain. *Brain Research*, 53:117–127.

Bobillier, P., Froment, J. L., Seguin, S., and Jouvet, M. (1974): Effets de la P. Chlorophenylalanine et du 5 hydroxytryptophane sur le sommeil et le métabolisme central des monoamines et des protéines chez le chat. *Biochemical Pharmacology* (*in press*).

Dahlstrom, A., and Fuxe, K. (1964): Evidence for the existence of monoamine neurons in the central nervous system. I. Demonstration of monoamines in the cell bodies of brainstem neurons. *Acta Physiologica Scandinavica*, 62:suppl. 232.

Daly, J., Fuxe, K., and Jonsson, G. (1973): Effects of intracerebral injections of 5,6-dihydroxytryptamine on central monoamine neurons: Evidence for selective degeneration of central 5-hydroxytryptamine neurons. *Brain Research*, 49:476–482.

Froment, J. L., Petitjean, F., Bertrand, N., Cointy, C., and Jouvet, M. (1974): Effets de l'injection intracérébrale de 5-6 hydroxytryptamine sur les monoamines cérébrales et les états de sommeil du chat. *Brain Research* (*in press*).

Jones, B. E., Bobillier, P., Pin, C., and Jouvet, M. (1973): The effects of lesion of catecholamine containing neurons upon monamine content of the brain and EEG and behavioural waking in the cat. *Brain Research*, 58:157–177.

Jouvet, M. (1969): Biogenic amines and the states of sleep. *Science*, 163:32–41.

Jouvet, M. (1972): The role of monoamines and acetylcholine containing neurons in the regulation of the sleep-waking cycle. *Ergebnisse der Physiologie*, 64:166–307.

Jouvet, M. (1973): Serotonin and sleep in the cat. In: *Serotonin and Behaviour*, edited by J. Barchas and E. Usdin. Academic Press, New York.

Jouvet, M., and Pujol, J. F. (1972): Les médiateurs chimiques—Leur rôle dans la physiopathologie de la motricité, de la vigilance et du comportement. Étude neurophysiologique et biochimique. *Revue Neurologique*, 127:115–138.

Kostowski, W., and Giacalone, E. (1969): Stimulation of various forebrain structures and brain 5-HT, 5-HIAA and behaviour in rats. *European Journal of Pharmacology*, 7:176–180.

Laguzzi, R., Petitjean, F., Pujol, J. F., and Jouvet, M. (1972): Effets de l'injection intraventriculaire de 6-hydroxydopamine sur le cycle veille-sommeil du chat. *Brain Research*, 48:295–310.

McGinty, D. J., and Sterman, M. B. (1968): Sleep suppression after basal forebrain lesions in the cat. *Science*, 160:1253–1255.

Maeda, T., Pin, C., Salvert, D., Ligier, M., and Jouvet, M. (1973): Les neurones contenant des catécholamines du tegmentum pontique et leurs voies de projection chez le chat. *Brain Research* 57:119–152.

Mancia, M. (1969): EEG and behavioural changes owing to splitting of the brain stem in cats. *Electroencephalography and Clinical Neurophysiology*, 27:487–503.

Moruzzi, G. (1972): The sleep-waking cycle. *Ergebnisse der Physiologie*, 64:1–165.

Petitjean, F., and Jouvet, M. (1970): Hypersomnie et augmentation de l'acide 5-hydroxy-indolacétique cerebral par lesion isthmique chez le chat. *Comptes Rendus de la Société de Biologie* (Paris), 164:2288–2293.

Petitjean, F., Laguzzi, R., Sordet, F., Jouvet, M., and Pujol, J. F. (1972): Effets de l'injection intraventriculaire de 6-hydroxydopamine: I. Sur les monoamines cérébrales du chat. *Brain Research*, 48:281–293.

Polc, P., and Monnier, M. (1970): An activating mechanism in the ponto-bulbar raphe system of the rabbit. *Brain Research*, 22:47–63.

Pujol, J. F., Stein, D., Blondaux, C., Petitjean, F., Froment, J. L., and Jouvet, M. (1973): Biochemical evidences for interaction phenomena between noradrenergic and serotoninergic systems in the cat brain. In: *Frontiers in Catecholamines Research*, Edited by E. Usdin. Pergamon Press, Elmsford, N.Y.

Sterman, M. B., and Clemente, C. D. (1962): Forebrain inhibitory mechanisms: Sleep patterns induced by basal forebrain stimulation in the behaving cat. *Experimental Neurology*, 6:103–117.

Wada, J. A., and Terao, A. (1970): Effect of parachlorophenylalanine on basal forebrain stimulations. *Experimental Neurology*, 28:501–506.

Yamaguchi, N., Marczinski, R. J., and Ling, G. M. (1963): The effects of electrical and chemical stimulation of the preoptic region and some non-specific thalamic nuclei in unrestrained, waking animals. *Electroencephalography and Clinical Neurophysiology*, 15:145–166.

Advances in Biochemical Psychopharmacology, Vol. 11
Raven Press, New York © 1974

5-Hydroxytryptamine and Sleep in the Cat: A Brief Overview

Carl D. King

Department of Pharmacology, University of Tennessee Medical Units, Memphis, Tennessee 38103

Hobson (1969), in a brief review, asked the question: Is brain 5-hydroxytryptamine (5-HT) uniquely important for the genesis of sleep? On the basis of the then-available evidence, he concluded that the answer was at best doubtful. With a great deal of new data available, the same question may be asked again. The answer can now be more sanguine than was Hobson's. Despite this, doubts remain. The evidence, though strong, is incomplete. And there is no proof which establishes 5-HT as the *sole* central mediator of the states of sleep.

Most specifically, at least one aspect of sleep, the **pontine-geniculate-occipital** (PGO) wave, can be linked with some certainty to 5-HT (Jalfre, Ruch-Monachon, and Haefely, 1974). There is compelling evidence that brain 5-HT tonically suppresses the PGO wave, and that a lowering of 5-HT can unleash the PGO wave, so that it intrudes into all of the stages of sleep and even into wakefulness (Dement, Mitler, and Henriksen, 1972).

But what of the larger question? Is 5-HT, as the monoamine hypothesis of sleep (Jouvet, 1969, 1972) proposes, necessary for the production of sleep? The monoamine hypothesis has had a strong heuristic value, but how potent has been its predictive value?

I. THREE STAGES OF CAT SLEEP

To answer the questions posed above, it is essential to remember that somnolence does not necessarily equal physiological sleep. In the cat, sleep consists of a complex process which can nevertheless be divided into at least three distinct stages. One of these stages is marked by sleep spindles in the cortical electroencephalogram (EEG); these spindles are separated by low-voltage, fast EEG activity. This stage of cat sleep has been called by various names: "spindle sleep" (Jewett, 1968; King, 1971), for example, or "light slow-wave sleep" (Ursin, 1968, 1971), or "stage I" (Jouvet, 1972). The second stage is marked by a predominance in the cortical EEG of high voltage slow-waves, with or without spindles, and has been called "slow-wave sleep" (Jewett, 1968; King 1971), or "deep slow-wave sleep" (Ursin, 1968, 1971), or "stage II" (Jouvet, 1972). The third stage is, of course, rapid eye movement (REM) sleep. The first two stages together consistute "non-REM" sleep. Many investigators who use the cat as an experimental animal neglect to distinguish the two stages of non-REM sleep, and lump them together under the catch-all term "slow-wave sleep." This is a mistake, because the two stages of the cat's non-REM sleep can be selectively altered in different ways, both by physical and by pharmacological interventions (King and Jewett, 1971; Ursin, 1971, 1972). It is important, moreover, to distinguish the three stages of cat sleep, because many drugs which modify 5-HT metabolism—fenfluramine, for example (Johnson, Funderburk, Ruckart, and Ward, 1972)—can induce a state of somnolence which fails to show these stages. It is important that such a state not be confused with true, physiological sleep. It is an important point, too, theoretically, since 5-HT may be important for one stage of sleep, and not for another.

II. 5-HT AND SLEEP: POSITIVE CORRELATIONS

Amid the controversies of sleep research, one fact stands clear: 5-HT can induce EEG synchronization. The pioneering work of Koella and co-workers in this area (Koella, Trunca, and Czicman, 1965; Koella and Czicman, 1965, 1966) has been amply confirmed (for a recent example, see Bronzino, Morgane, and Stern, 1972). It is also clear that parenteral administration of 5-hydroxytryptophan (5-HTP) to cats can induce sedation and EEG synchronization. But is this the same thing as sleep? For the cat, the answer is probably no. Echols and Jewett (1972), for example, found that 5-HTP (30 mg/kg, i.p.) in cats caused miosis, sedation, but also a dissociation of the EEG and behavior; in other words, slow waves appeared in the EEG of awake animals. Because of this, exact scoring of the sleep records was precluded. Sleep spindles were abolished by the drug for 5 hr, and no REM sleep occurred for 6 hr. Such a state, by definition, cannot be called sleep.

The most suggestive evidence for a link between 5-HT and sleep in cats comes (1) from the insomnia produced by lesions of the 5-HT-rich raphe nuclei (for a summary, see Jouvet, 1972) and (2) from the demonstration of the insomnia which attends the depletion of brain 5-HT by p-chlorophenylalanine, first described by Jouvet and co-workers (Delorme, Froment, and Jouvet, 1966), and confirmed by Koella, Feldstein, and Czicman (1968). But for which stage or stages of cat sleep is 5-HT important? Is it important for sleep per se, or for one of the various stages? The data of Ursin (1972) may provide some clues. Ursin, using single doses of *p*-chlorophenylalanine in cats, found a rather specific inhibition only of "deep slow-wave sleep." In most of the animals, "light slow-wave sleep" and REM sleep were not changed. It is premature to draw final conclusions here, however, and Ursin's findings need corroboration from other laboratories. Moreover, the specificity of the effect of *p*-chlorophenylalanine needs to be checked by the use of other drugs which inhibit brain tryptophan hydroxylase or in some other manner diminish central 5-HT turnover.

III. POSSIBLE ROLE OF META-CHLOROTYROSINE IN THE INSOMNIA PRODUCED BY *p*-CHLOROPHENYLALANINE

It is possible that the insomnia produced in cats by *p*-chlorophenyl-alanine is not entirely a specific effect due to depressed 5-HT synthesis. In some behavioral tests, the effects of *p*-chlorophenylalanine correlate very poorly with the time course of brain 5-HT depletion. Stark and Fuller (1972) have described such an example. In animals with electrodes placed in the posterior hypothalamus, *p*-chlorophenylalanine inhibited self-stimulation, but this occurred well before the time of maximal 5-HT depletion; additionally, the behavioral parameter had returned almost to control values at the time brain 5-HT reached nadir. In an attempt to explain this discrepancy, the authors cited data (Guroff, Kondo, and Daly, 1966) which showed that *p*-chlorophenyl-alanine is a substrate for phenylalanine hydroxylase *in vitro*. The chief product of the reaction is meta-chlorotyrosine. Stark and Fuller found that meta-chlorotyrosine also inhibited self-stimulation, both in rats and dogs. This occurred at doses lower than those of *p*-chlorophenyl-alanine required to produce the effect. Further, meta-chlorotyrosine did not alter brain 5-HT content. The authors suggested that meta-chlorotyrosine may be converted to meta-chlorotyramine and meta-chlorooctopamine, which may act as false transmitters.

In our laboratory, cats implanted with electrodes for the measurement of sleep patterns have been treated with meta-chlorotyrosine (75 mg/kg, i.p., given at 8:30 a.m.). The drug produced a brief period of insomnia. The cats sat or reclined quietly in their experimental sleep chambers, awake and alert, but with no signs of agitation or odd behavior. This effect began about 30 min after the drug was given, and lasted about 6 hr. Then, when sleep first returned, it consisted chiefly of spindle sleep and REM sleep, with only brief periods of slow-wave sleep. After about 8 hr, slow-wave sleep was back to control levels. Thus meta-chlorotyrosine can mimic several of the effects of *p*-chloro-phenylalanine. It would appear possible, then, that at least some of the effects of *p*-chlorophenylalanine on sleep may be due to false transmitter formation. This possibility, of course, must await the demonstration that meta-chlorotyrosine can be produced *in vivo* from *p*-chlorophenylalanine. Also, it is possible that meta-chlorotyrosine may inhibit 5-HT synthesis or turnover in cat brain, while it does not do so in dog or rat brain, and this point begs investigation.

IV. 5-HT AND SLEEP: NEGATIVE CORRELATIONS

As noted above, excellent evidence indicates that the PGO wave is tonically inhibited by brain 5-HT. If this is so, then 5-HT cannot be the sole neurohumor involved in the genesis of REM sleep. Indeed, since normal REM sleep in cats is marked by numerous bursts of PGO waves, then there must occur during REM sleep phasic *wanings* of 5-HT neuronal activity; one or more transmitters other than 5-HT must now come into play.

Furthermore, in certain conditions it is possible to show an apparent complete dissociation between 5-HT and sleep. Dement et al. (1972) have thus shown that during chronic daily treatment of cats with *p*-chlorophenylalanine, a phase of insomnia develops, but then disappears, and sleep returns almost to predrug baseline despite a continued and persistant depletion of brain 5-HT. These were the data which led Hobson (1969) to question the unique importance of 5-HT for sleep. It is crucial to the 5-HT hypothesis of sleep that this discrepancy be resolved. Possible explanations might include a decreased activity of arousal mechanisms which might develop as a homeostatic response to *p*-chlorophenylalanine insomnia. Or it might be that a very active, small pool of 5-HT, untouched by the chronic *p*-chlorophenylalanine, mediates the return of sleep. Or perhaps one might invoke supersensitivity of functionally denervated 5-HT receptors. Such explanations seem strained. Logically they are unappealing, since they state that (1) insomnia occurs after *p*-chlorophenylalanine because brain 5-HT is depleted, but (2) sleep returns because brain 5-HT is *not* depleted. Furthermore, when sleep returns during chronic administration of *p*-chlorophenylalanine, PGO waves not only appear in REM sleep, but also invade periods of wakefulness and non-REM sleep (Dement et al., 1972). This argues for a continued depression of central 5-HT turnover during the time when sleep returns, and argues against a role of 5-HT in generating this return of sleep.

V. BEYOND 5-HT

The data currently available indicate that 5-HT almost certainly is involved in sleep mechanisms; the data, however, also strongly suggest that it is an oversimplification to suppose that only 5-HT is involved. Thus basic questions remain. What neurohumors, for example, come into play during the phasic inhibitions of 5-HT neuronal activity during REM sleep? And how does sleep return to cats treated chronically with *p*-chlorophenylalanine? To explain this phenomenon it might be necessary to hypothesize the existence of a back-up, redundant sleep system which is not dependent on now-known neurotransmitters. Perhaps the sedating factor isolated by Fencl, Koski, and Pappenheimer (1971) from the cerebrospinal fluid of sleep-deprived goats becomes operative in insomnious animals. If so, what role might this factor play in normal sleep? The same question may be asked of the sedating factor isolated by Monnier and co-workers from rabbit brain (see Monnier and Hatt, 1971; Monnier, Hatt, Cueni, and Schoenenberger, 1972). And how do cholinergic mechanisms (see Magherini, Pompeiano, and Thoden, 1971; Domino and Stawiski, 1971) take part in the production of sleep? And what other brain factors are there which play a role in the sleep-wakefulness cycle? And, finally, even if 5-HT becomes established as a prime regulator of sleep, what regulates the regulator? Why—in short—and how do we go to sleep? These very important questions await further study.

ACKNOWLEDGMENTS

This work was supported by U.S. Public Health Service grant RR-05423.

REFERENCES

Bronzino, J. D., Morgane, P. J., and Stern, W. C. (1972): EEG synchronization following application of serotonin to area postrema. *American Journal of Physiology*, 223:376–383.

Delorme, F., Froment, J. L., and Jouvet, M. (1966): Suppression du sommeil par la *p*-chlorometamphetamine et la *p*-chlorophenylalanine. *Comptes Rendus de la Société de Biologie*, 160:2347–2351.

Dement, W. C., Mitler, M. M., and Henriksen, S. J. (1972): Sleep changes during chronic administration of parachlorophenylalanine. *Revue Canadienne de Biologie*, 31 (supplement):239–246.

Domino, E. F., and Stawiski, M. (1971): Modification of the cat sleep cycle by hemicholinium-3, a cholinergic antisynthesis agent. *Research Communications in Chemical Pathology and Pharmacology*, 2:461–467.

Echols, S. D., and Jewett, R. E. (1972): Effects of morphine on sleep in the cat. *Psychopharmacologia*, 24:435–448.

Fencl, V., Koski, G., and Pappenheimer, J. R. (1971): Factors in cerebrospinal fluid from goats that affect sleep and activity in rats. *Journal of Physiology* (London), 216:565–589.

Guroff, G., Kondo, K., and Daly, J. (1966): The production of meta-chlorotyrosine from para-chlorophenylalanine by phenylalanine hydroxylase. *Biochemical and Biophysical Research Communications*, 25:622–628.

Hobson, J. A. (1969): Sleep: Biochemical aspects. *New England Journal of Medicine*, 281:1468–1470.

Jalfre, M., Ruch-Monachon, M. A., and Haefely, W. (1974): Methods for assessing the interaction of agents with 5-hydroxytryptamine neurons and receptors in the brain. In: *Advances in Biochemical Psychopharmacology*, Vol. 10, edited by E. Costa, G. L. Gessa, and M. Sandler. Raven Press, New York.

Jewett, R. E. (1968): Effects of promethazine on sleep stages in the cat. *Experimental Neurology*, 21:368–382.

Johnson, D. N., Funderburk, W. H., Ruckart, R. T., and Ward, J. W. (1972): Contrasting effects of two 5-hydroxytryptamine-depleting drugs on sleep patterns in cats. *European Journal of Pharmacology*, 20:80–84.

Jouvet, M. (1969): Biogenic amines and the stages of sleep. *Science*, 163:32–41.

Jouvet, M. (1972): The role of monoamines and acetylcholine-containing neurons in the regulation of the sleep-waking cycle. *Ergebnisse der Physiologie*, 64:166–307.

King, C. D. (1971): The pharmacology of rapid eye movement sleep. *Advances in Pharmacology and Chemotherapy*, 9:1–91.

King, C. D., and Jewett, R. E. (1971): The effects of alpha methyltyrosine on sleep and brain norepinephrine in the cat. *Journal of Pharmacology and Experimental Therapeutics*, 177:188–194.

Koella, W. P., and Czicman, J. S. (1965): The area postrema as a possible receptor site for EEG synchronization by 5-HT. *Federation Proceedings*, 24:646.

Koella, W. P., and Czicman, J. S. (1966): Mechanism of the EEG-synchronizing action of serotonin. *American Journal of Physiology*, 211:926–934.

Koella, W. P., Feldstein, A., and Czicman, J. S. (1968): The effect of para-chlorophenylalanine on the sleep of cats. *Electroencephalography and Clinical Neurophysiology*, 25:481–490.

Koella, W. P., Trunca, C. M., and Czicman, J. S. (1965): Serotonin: Effect on recruiting responses of the cat. *Life Sciences*, 4:173–181.

Magherini, P. C., Pompeiano, O., and Thoden, U. (1971): The neurochemical basis of REM sleep: A cholinergic mechanism responsible for rhythmic activation of the vestibulo-oculomotor system. *Brain Research*, 35:565–569.

Monnier, M., and Hatt, A. M. (1971): Humoral transmission of sleep. V. New evidence from production of pure sleep hemodialysate. *Pflugers Archiv*, 329:231–243.

Monnier, M., Hatt, A. M., Cueni, L. B., and Schoenenberger, G. A. (1972): Humoral transmission of sleep. VI. Purification and assessment of a hypnogenic fraction of "sleep dialysate" (factor delta). *Pflugers Archiv*, 331:257–265.

Stark, P., and Fuller, R. W. (1972): Behavioral and biochemical effects of p-chlorophenylalanine, 3-chlorotyrosine and 3-chlorotyramine. A proposed mechanism for inhibition of self-stimulation. *Neuropharmacology*, 11:261–272.

Ursin, R. (1968): The two stages of slow wave sleep in the cat and their relation to REM sleep. *Brain Research*, 11:347–356.

Ursin, R. (1971): Differential effect of sleep deprivation on the two slow wave sleep stages in the cat. *Acta Physiologica Scandinavica*, 83:352–361.

Ursin, R. (1972): Differential effect of para-chlorophenylalanine on the two slow wave sleep stages in the cat. *Acta Physiologica Scandinavica*, 86:278–285.

Advances in Biochemical Psychopharmacology, Vol. 11
Raven Press, New York ©1974

Possible Role of Brain Serotonin and Dopamine in Controlling Male Sexual Behavior

G. L. Gessa and A. Tagliamonte

Institute of Pharmacology, University of Cagliari, Cagliari, Italy

I. INTRODUCTION

The finding that *p*-chlorophenylalanine (PCPA), a rather selective inhibitor of serotonin synthesis (Koe and Weissman, 1966), arouses sexual activity in male animals (Tagliamonte, Tagliamonte, Gessa, and Brodie, 1969; Sheard, 1969; Shillito, 1970; Hoyland, Shillito, and Vogt, 1970; Ferguson, Henriksen, Cohen, Mitchell, Barchas, and Dement, 1970; Salis and Dewsbury, 1971; Perez-Cruet, Tagliamonte, Tagliamonte, and Gessa, 1971), and the clinical observation that L-dihydroxyphenylalanine (L-DOPA), the direct precursor of dopamine, has an aphrodisiac effect in humans (Mitler, Morden, Levine, and Dement, 1972; Mones, Elizan, and Siegel, 1970; Van Woert, Heninger, Rathey, and Bowers, 1970; Bowers, Van Woert, and Davis, 1971; Benkert, Crombach, and Kockott, 1972; Kuruma, Bartholini, Tissot, and Pletscher, 1972; Mars, Libman, Schwartz, Gillo-Joffroy, and Barbeau, 1972*a*), have stimulated a large number of studies on the possible role of brain serotonin and catecholamines in controlling sexual behavior in males.

In our first report on the aphrodisiac effect of PCPA in male rats, we observed that the sexual stimulant effect of this drug was greatly potentiated by pargyline, a potent monoamineoxidase inhibitor (Tagliamonte et al., 1969). Since the administration of pargyline to rats pretested with an inhibitor of serotonin synthesis produces a selective accumulation of brain catecholamines, we postulated the existence of a reciprocal control of male sexual behavior, with serotonin inhibiting, and catecholamines stimulating this behavior (Tagliamonte et al., 1969; Tagliamonte, Tagliamonte, and Gessa, 1971).

The purpose of this chapter is to evaluate this theory critically.

II. SEROTONIN AND SEXUAL BEHAVIOR

A. *Effect of PCPA on Sexual Behavior*

The stimulatory effect of PCPA on male sexual behavior has been shown with different experimental models:

1. PCPA produces homosexual mounting behavior in rats and rabbits: that is, male-to-male mounting behavior in conditions when this does not occur in untreated animals and/or in a percentage of animals significantly greater than in controls (Tagliamonte et al., 1969; Malmnäs and Meyerson, 1971), see Fig. 1.

2. PCPA restores to normal the copulatory behavior of sexually sluggish male rats (Tagliamonte et al., 1971; Malmnäs and Meyerson, 1971).

3. PCPA increases the sexual motivation: Male rats treated with PCPA will not only mount a female in heat, but will also fight trying to mount an unreceptive female (Sjoerdsma, Lovenberg, Engelman, Carpenter, Wyatt, and Gessa, 1970).

4. PCPA does not further increase the number of ejaculations in rats with a high basal sexual activity (Whalen and Luttge, 1970); however, it decreases their ejaculation latencies (Salis and Dewsbury, 1971; Ahlenius, Eriksson, Larsson, Modigh, and Södersten, 1971).

B. *Correlation Between PCPA-Induced Hypersexuality and Serotonin Depletion*

Several arguments support the concept that the effects of PCPA on male sexual behavior are the consequence of the depletion of brain serotonin:

1. In rats and rabbits, there is a temporal correlation between the maximal depletion of brain serotonin by PCPA and the occurrence of hypersexuality (Tagliamonte et al., 1969; Perez-Cruet et al., 1971).

2. Selective destruction of serotonin nerve endings produced by 5,6-dihydroxytryptamine (5,6-DHT), administered in the lateral ventricle, causes homosexual mounting behavior in male rats (Da Prada, Carruba, O'Brien, Saner, and Pletscher, 1972*a*). This effect, however, has been observed 8 hr after treatment; of greater interest would be to know whether this effect is only transient or persists long after treatment, when the serotonin nerve endings have degenerated.

3. Homosexual behavior has been observed with different serotonin antagonists, such as methysergide, mesorgydine, and WA-335-BS, in male rats treated with testosterone (Benkerd and Eversmann, 1972).

4. The homosexual behavior induced by PCPA or by 5,6-DHT is abolished by L-5-hydroxytryptophan (5-HTP), the direct precursor of

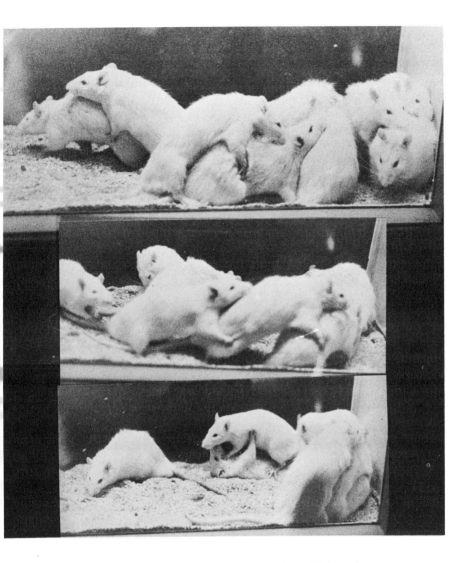

FIG. 1. Homosexual mounting behavior in male rats treated with PCPA and testosterone.

serotonin (Tagliamonte et al., 1969; Da Prada et al., 1972*a*).

5. The administration of 5-HTP also suppresses the copulatory behavior of normal rats with receptive females. The inhibition is potentiated by Ro 4-4602 (Tagliamonte, Fratta, Mercuro, Biggio, Camba, and Gessa, 1972), a peripherally acting decarboxylase inhibitor, suggesting that 5-HTP acts on the CNS (see also Table 1).

6. Different monoamineoxidase inhibitors (MAOI), such as pargyline, phenelzin, and iproniazide, suppress the copulatory behavior in

TABLE 1. *Inhibition by L-5-hydroxytryptophan of homo- and heterosexual behavior in normal and in PCPA-treated male rats*

	Percent of males mounting other males			Percent of males copulating with females		
	Controls	5-HTP	Ro 4-4602 + 5-HTP	Controls	5-HTP	Ro-4-4602 + 5-HTP
Normal rats	28	12	0	100	32	12
PCPA-treated rats	80	16	0	100	40	0

Each value was obtained from 25 sexually experienced male rats.

normal rats but not in rats pretreated with PCPA (Tagliamonte et al., 1971; Malmnäs and Meyerson, 1971; Dewsbury, Davis, and Jansen, 1972), indicating that the inhibition is secondary to the accumulation of brain serotonin.

C. Alternative Explanations

Although the arguments presented suggest that PCPA-induced hypersexuality is correlated with serotonin deficiency, we feel that the evidence is insufficient to resolve whether brain serotonin plays a role in controlling sexual behavior.

We are therefore further pursuing experiments in order to associate (or dissociate) serotonin changes induced by drugs with changes in sexual behavior. The results have thus far been contradictory. Recent observations from our laboratory (Fratta et al., *Journal of Pharmacy and Pharmacology, in press*), have shown that both dextro- and levo-PCPA are equally potent in inhibiting serotonin synthesis and in stimulating copulatory behavior in male sluggish rats. These results tend to *correlate* brain serotonin depletion with the occurrence of hypersexuality.

On the contrary, we found that the administration of L-tryptophan, unlike that of 5-HTP in a single or repeated doses which markedly increase brain serotonin synthesis, did not suppress the spontaneous copulatory behavior in male rats (Tagliamonte et al., 1972; see Table 2). Needless to say, these results do not support the theory of the inhibitory role of brain serotonin in sexual behavior!

Nevertheless, possible explanations for these contradictory results might be considered:

1. Serotonin formed after L-tryptophan loading may be metabolized intraneuronally and is functionally inactive, as suggested by Grahame-Smith (1971). However, this explanation does not clarify why 5-HTP is,

TABLE 2. *Failure of L-tryptophan (Try)*
to inhibit heterosexual behavior in male rats

Try (mg/kg)	Hours before test	Percent of males ejaculating
0	–	100
150	1	100
150 X 6	10, 8, 6, 4, 2, 1	100

Each value was obtained from 25 sexually experienced male rats exposed to female rats in estrus.

unlike tryptophan, active in suppressing copulatory behavior.

2. The concentration of brain serotonin required to inhibit the spontaneous copulatory behavior in rats may have to reach a certain critical level, which is obtained with 5-HTP but not with L-tryptophan administration. Table 3 shows that, indeed, much more serotonin and 5-HIAA accumulate in brain after 5-HTP than after tryptophan administration. This hypothesis may also explain the time course of the inhibitory effect of MAOI on mounting behavior. The inhibitory effect of pargyline and iproniazid is present from 8 to 24 hr after drug administration, when brain serotonin levels are increased to about 197% the normal values. On the other hand, the inhibition of spontaneous copulatory behavior is not present 72 hr after treatment, when brain serotonin levels are 72% above normal values.

TABLE 3. *Effect of L-tryptophan (Try), 5-hydroxytryptophan (5-HTP), and pargyline*
on brain serotonin metabolism in male rats

Treatment	mg/kg, i.p.	Hours before sacrifice	Try (μg/g)	5-HT (μg/g)	5-HIAA (μg/g)	Inhibition of sexual behavior
None	–	–	5.4 ± 0.07	0.61 ± 0.03	0.54 ± 0.04	
5-HTP	50	0.5	4.8 ± 0.12	1.6 ± 0.08	2.4 ± 0.11	Yes
Ro 4-4602 + 5-HTP	50 + 50	1, 0.5	4.4 ± 0.09	1.4 ± 0.07	2.1 ± 0.08	Yes
Try	150	1	20 ± 0.89	0.73 ± 0.03	1.2 ± 0.06	No
Try	150 X 6	10, 8, 6, 4, 2, 1	25 ± 0.65	0.78 ± 0.05	1.3 ± 0.06	No
Pargyline	80	18	5.2 ± 0.08	1.8 ± 0.06	0.09 ± 0.01	Yes
Pargyline	80	72	5.0 ± 0.05	1.1 ± 0.05	0.25 ± 0.02	No

Each value is the average ± SE of at least six determinations.

III. POSSIBLE STIMULATORY ROLE OF BRAIN DOPAMINE ON MALE COPULATORY BEHAVIOR

A stimulant role of catecholamines on male sexual behavior is suggested by the above-mentioned results on the potentiation by pargyline of the aphrodisiac effect of PCPA in rats; and by the recent reports indicating an aphrodisiac effect of L-DOPA in some parkinsonian patients.

However, the role of brain catecholamines in male sexual behavior is still controversial.

Thus, Bertolini and Vergoni found that p-chloro-N-methylamphetamine, a compound which inhibits serotonin synthesis and also releases brain catecholamines (Miller, Cox, Snodgrass, and Maickel, 1970), is more potent than PCPA in stimulating male-to-male mounting behavior (Bertolini and Vergoni, 1970). In addition, Benkert observed that the male-to-male mounting behavior produced by PCPA in rats is potentiated by L-DOPA, and reported that p-chloro-N-methylamphetamine has some beneficial effect in the treatment of impotence in male patients (Benkert, 1972b). Finally, Carruba et al. found that L-DOPA, when peripheral decarboxylase is inhibited by Ro 4-4602, produced male-to-male mounting behavior in rats (Da Prada, Carruba, O'Brien, Saner, and Pletscher, 1972b). On the contrary, Hyyppä et al. (Hyyppä, Lehtinen, and Rinne, 1971) failed to observe increased male-to-female copulatory behavior in male adult rats treated with L-DOPA.

With the experiments reported here we have attempted to clarify this problem.

From our experience with rats treated with PCPA, we learned that in studying a drug with a potential aphrodisiac effect, it is of critical importance to use animals with low levels of sexual activity, and vice versa, selected copulators should be used to ascertain whether a drug inhibits copulatory behavior. Therefore, in the present investigation, sexually sluggish rats were used to study the effect of drugs that stimulate brain catecholamine receptors, while rats with high levels of sexual activity were used to study the effect of those treatments which decrease central adrenergic activity. Experimental conditions are ascribed elsewhere in detail (Tagliamonte, Fratta, Del Fiacco, and Gessa, 1973).

A. Effect of Apomorphine and L-DOPA on the Heterosexual Copulatory Behavior of Male Rats with Receptive Females

The effect of apomorphine and the combination of L-DOPA with Ro 4-4602 on the copulatory behavior of sexually sluggish male rats with female rats in heat was studied. Ro 4-4602, a peripherally acting decarboxylase inhibitor, was given to prevent the decarboxylation of L-DOPA in peripheral tissues, and to increase the amount of L-DOPA reaching the brain (Kuruma et al., 1972). As Table 4 shows, a dose of 0.5 mg/kg of apomorphine significantly increased the number of rats showing mountings, intromissions, and ejaculations. On the other hand, 5 mg/kg of this compound produced a marked stereotyped behavior, which prevented the occurrence of other goal-directed behavior, including copulation.

TABLE 4. *Stimulation by L-DOPA (with Ro 4-4602) and apomorphine of the copulatory behavior of sexually sluggish male rats with receptive females*

Treatment	Percent of animals exhibiting at least one		
	Mounting[a]	Intromission[a]	Ejaculation[a]
Saline	58	54	14
Apomorphine, 0.5 mg/kg[b]	80	80	62
Ro 4-4602 + L-DOPA[c]	90	90	30
Ro 4-4602 + L-DOPA X 2	90	90	64

Each value was obtained from 50 rats. Each rat underwent different mating tests with and without treatment, at weekly intervals.

[a]Occurring within 30 min after male and female rats were paired.

[b]Apomorphine was given i.p. 15 min before the mating test.

[c]L-DOPA (100 mg/kg, i.p.) was given 20 min after Ro 4-4602 (50 mg/kg, i.p.). The experiment was performed half an hour after the last treatment.

Two doses of L-DOPA (100 mg/kg each) were injected i.p. 20 and 50 min after Ro 4-4602 respectively.

The experiment was performed half an hour after the last treatment.

A single injection of L-DOPA to rats pretreated with Ro 4-4602 markedly increased the number of male animals showing mounting and intromissions, while increasing only slightly the percentage of rats ejaculating. This effect was more pronounced when two doses of L-DOPA were given at a 30-min interval; this treatment also markedly increased the number of rats reaching ejaculation. On the other hand, no sexual stimulation was observed after the administration of L-DOPA alone at doses up to 150 mg/kg, i.p., given from 30 to 60 min before the test; this indicates that the aphrodisiac effect of L-DOPA has a central origin.

B. *Inhibitory Effect of Haloperidol on the Heterosexual Copulatory Behavior of Male Rats with Receptive Females*

Table 5 shows that the aphrodisiac effect elicited by apomorphine, or by the combination of L-DOPA with Ro 4-4602, was blocked by haloperidol. This drug also suppressed the spontaneous copulatory behavior of male rats with high basal levels of sexual activity.

TABLE 5. *Inhibition by haloperidol of the heterosexual copulatory behavior of normal male rats and of that induced by apomorphine or L-DOPA in sexually sluggish rats*

Rats	Treatment	Haloperidol	Percent of animals exhibition at least one		
			Mounting[a]	Intromission[a]	Ejaculation[a]
Selected copulators	None	–	100	100	100
Selected copulators	None	1	0	0	0
Sexually sluggish	Apomorphine	–	80	80	62
Sexually sluggish	Apomorphine	1	20	20	12
Sexually sluggish	Ro 4-4602 + L-DOPA	–	90	90	64
Sexually sluggish	Ro 4-4602 + L-DOPA	1	10	10	0

[a]Occurring within 30 min after male and female rats were paired.

Values for selected copulators were obtained from 40 rats. Values for sluggish rats were obtained from 50 rats. These treatments were given as described in Table 4.

C. *Effect of L-DOPA on the Copulatory Pattern of Male Rats with Receptive Females*

Experiments were carried out with selected copulators in order to study the effect of L-DOPA on the copulatory pattern.

As Table 6 shows, two doses of L-DOPA given at a 30-min interval to animals pretreated with Ro 4-4602 shortened the ejaculation latency, the postejaculatory interval, and also decreased the number of mounts and intromissions prior to ejaculations.

TABLE 6. *Effect of the combination of L-DOPA with Ro 4-4602 on the pattern of copulatory behavior of selected male copulators*

	Treatments		
	none	Ro 4-4602 + L-DOPA once	Ro 4-4602 + L-DOPA twice
Number of mounts before ejaculation	5.45	6.2	4.05[a]
Number of intromissions before ejaculation	9.35	8.85	6.73[a]
Ejaculation latency (min and sec)	9.60	8.76	6.20[a]
Postejaculation interval (min and sec)	6.32	6.12	4.12[a]

Each value is the mean of at least two experiments carried out with 28 rats. Every rat received one mating test under each of the three conditions, according to a latin square design. Treatments were given as described in Table 4.

[a] $p < 0.001$ with respect to control values.

D. Conclusion

The data reported show that the copulatory behavior of sexually sluggish male animals with receptive females is stimulated by a combination of L-DOPA with Ro 4-4602 and by apomorphine. Since the administration of L-DOPA to rats pretreated with Ro 4-4602 has been shown not only to increase the content of brain catecholamines (Bartholini, Costantinidis, Tissot, and Pletscher, 1971), but also to decrease that of serotonin (Butcher, 1972), the stimulatory effect on male copulatory behavior can be ascribed to either mechanism. However, since apomorphine is considered to act as a direct stimulant of the dopamine receptors in brain (Andén, Rubenson, Fuxe, and Hökfelt, 1967), the finding that this compound also stimulates the copulatory behavior in male rats supports the hypothesis that brain dopamine plays a stimulatory role in male sexual behavior.

Consistently, the effect of apomorphine and L-DOPA was prevented by haloperidol, a specific inhibitor of dopaminergic receptors in brain (Andén, Dahlström, Fuxe, and Hökfelt, 1966). Moreover, this drug also suppressed the spontaneous copulatory behavior of male rats with high basal levels of sexual activity.

IV. GENERAL CONCLUSION

Although the evidence presented here, that cerebral serotonin and dopamine exert a reciprocal control on male sexual behavior, is still circumstantial, this theory has fascinating implications, since it might offer a biochemical explanation for some disturbances in sexual behavior occurring in man. That this behavior in man and rats is controlled by not entirely different biochemical mechanisms is suggested by the finding that the aphrodisiac effect of L-DOPA, observed in a small percentage (2 to 4%) of parkinsonian patients, can be predictably obtained in rats, provided that they are chosen from among sexually sluggish animals.

One corollary of this study is that these rats may be a useful tool for screening drugs with potential aphrodisiac effects.

Finally, the shortening of the ejaculation latencies produced by reserpine (Dewsbury, 1971), PCPA (Perez-Cruet et al., 1971; Ahlenius et al., 1971), and L-DOPA, might be considered an animal model for experiments with drugs that may change the ejaculation precox in man.

REFERENCES

Ahlenius, S., Eriksson, H., Larsson, K., Modigh, K., and Södersten, P. (1971): Mating behavior in the male rat treated with p-chlorophenylalanine methyl ester alone and in combination with pargyline. Psychopharmacologia (Berlin), 6:383–388.

Andén, N.-E., Dahlström, A., Fuxe, K., and Hökfelt, T. (1966): The effect of haloperidol and chlorpromazine on the amine levels of central monoamine neurons. Acta Physiologica Scandinavica, 68:419–420.

Andén, N.-E., Rubenson, A., Fuxe, K., and Hökfelt, T. (1967): Evidence for dopamine receptor stimulation by apomorphine. Journal of Pharmacy and Pharmacology, 19:627–629.

Bartholini, G., Costantinidis, J., Tissot, R., and Pletscher, A. (1971): Formation of monoamines from various amino acids in the brain after inhibition of extracerebral decarboxylase. Biochemical Pharmacology, 20:1243–1247.

Benkert, O. (1972a): L-DOPA treatment of impotence: A clinical and experimental study. In: L-DOPA and Behavior, edited by Sidney Malitz, pp. 73–79. Raven Press, New York.

Benkert, O. (1972b): Pharmacological experiments to stimulate human sexual behaviour. Psychopharmacologia (Berlin), 26:133.

Benkert, O., Crombach, G., and Kockott, G. (1972): Effect of L-DOPA on sexually impotent patient. Psychopharmacologia (Berlin), 23:91–95.

Benkert, O., and Eversmann, T. (1972): Importance of the anti-serotonin effect for mounting behaviour in rats. Experientia, 28:532.

Bertolini, A., and Vergoni, W. (1970): Effetto eccito sessuale della p-cloro-N-metil-amfetamina associata a testosterone nel ratto maschio. Rivista di Farmacologia e Terapia, 1:423–426.

Bowers, B. M., Jr., Van Woert, M., and Davis, L. (1971): Sexual behavior during L-DOPA treatment for parkinsonism. American Journal of Psychiatry, 12:127–129.

Butcher, L. L. (1972): Behavioral, biochemical and histochemical analyses of the central effect of monoamine precursors after peripheral decarboxylase inhibition. Brain Research, 41:387–411.

Da Prada, M., Carruba, M., O'Brien, R. A., Saner, A., and Pletscher, A. (1972a): The effect of 5,6-dihydroxytryptamine on sexual behaviour of male rats. European Journal of Pharmacology, 19:288–290.

Da Prada, M., Carruba, M., O'Brien, R. A., Saner, A., and Pletscher, A.

(1972*b*): L-DOPA and sexual activity of male rats. *Psychopharmacologia* (Berlin), 26:135.

Dewsbury, A. D. (1971): Copulatory behavior of male rats following reserpine administration. *Psychonomic Science*, 22:177–179.

Dewsbury, A. D., Davis, H. N., Jr., and Jansen, E. P. (1972): Effect of monoamine oxidase inhibitors on the copulatory behavior of male rats. *Psychopharmacologia* (Berlin), 24:209–217.

Ferguson, J., Henriksen, S., Cohen, H., Mitchell, G., Barchas, J., and Dement, W. (1970): "Hypersexuality" and behavioral changes in cats caused by administration of *p*-chlorophenylalanine. *Science*, 168:499–501.

Grahame-Smith, D. G. (1971): Studies "in vivo" on the relationship between brain tryptophan, brain 5-HT synthesis and hyperactivity in rats treated with a monoamine oxidase inhibitor and L-tryptophan. *Journal of Neurochemistry*, 18:1053–1066.

Hyyppä, M., Lehtinen, P., and Rinne, U. K. (1971): Effect of L-DOPA on the hypothalamic, pineal and striatal monoamines and on the sexual behavior of the rat. *Brain Research*, 30:265–272.

Hoyland, V. J., Shillito, E. E., and Vogt, M. (1970): The effect of parachlorophenylalanine on the behaviour of cats. *British Journal of Pharmacology*, 40:659–667.

Koe, K. B., and Weissman, A. (1966): *p*-Chlorophenylalanine: A specific depletor of brain serotonin. *Journal of Pharmacology and Experimental Therapeutics*, 154:499–515.

Kuruma, I., Bartholini, G., Tissot, R., and Pletscher, A. (1972): Comparative investigation of inhibitors of extracerebral DOPA decarboxylase in man and rats. *Journal of Pharmacy and Pharmacology*, 24:289–294.

Malmnäs, C. O., and Meyerson, B. J. (1971): *p*-Chlorophenylalanine and copulatory behaviour in the male rat. *Nature*, 232:398–400.

Mars, H., Libman, I., Schwartz, A. M., Gillo-Joffroy, L., and Barbeau, A. (1972): L-DOPA in Parkinson's disease. *Canadian Psychiatry*, 17:123–131.

Miller, F. P., Cox, R. H., Jr., Snodgrass, W. R., and Maickel, R. P. (1970): Comparative effects of *p*-chlorophenylalanine, *p*-chloroamphetamine, and *p*-chloro-N-methyl-amphetamine on rat brain norepinephrine, serotonin and 5-hydroxyindoleacetic acid. *Biochemical Pharmacology*, 19:435–442.

Mitler, M. M., Morden, B., Levine, S., and Dement, W. (1972): *Physiology and Behavior*, 8:1147–1150.

Mones, R. J., Elizan, T. S., and Siegel, G. J. (1970): Evaluation of L-DOPA therapy in Parkinson's disease. *New York State Journal of Medicine*, 70:2309–2318.

Perez-Cruet, J., Tagliamonte, A., Tagliamonte, P., and Gessa, G. L. (1971): Differential effect of *p*-chlorophenylalanine (PCPA) on sexual behaviour and on sleep patterns of male rabbits. *Rivista di Farmacologia e Terapia*, II:27–34.

Salis, P. J., and Dewsbury, D. A. (1971): *p*-Chlorophenylalanine facilitates copulatory behaviour in male rats. *Reprinted from Nature*, 232:400–401.

Sheard, M. H. (1969): The effect of *p*-chlorophenylalanine on behaviour in rats: Relation to brain serotonin and 5-hydroxyindoleacetic acid. *Brain Research*, 15:524–528.

Shillito, E. (1970): The effect of parachlorophenylalanine on social interaction of male rat. *British Journal of Pharmacology*, 38:305–315.

Sjoerdsma, A., Lovenberg, W., Engelman, K., Carpenter, W. T., Wyatt, R. J., and Gessa, G. L. (1970): Serotonin now: Clinical implications of inhibiting its synthesis with para-chlorophenylalanine. *Annals of Internal Medicine*, 73:607–629.

Tagliamonte, A., Fratta, W., Del Fiacco, M., and Gessa, G. L. (1973): Evidence that brain dopamine stimulates copulatory behaviour in male rats. *Rivista di Farmacologia e Terapia*, IV:177–181.

Tagliamonte, A., Fratta, W., Mercuro, G., Biggio, G., Camba, R. C., and Gessa, G. L. (1972): 5-Hydroxytryptophan, but not tryptophan, inhibits copulatory behaviour in male rats. *Rivista di Farmacologia e Terapia*, III:405–409.

Tagliamonte, A., Tagliamonte, P., and Gessa, G. L. (1971): Reversal of pargyline-induced inhibition of sexual behaviour in male rats by *p*-chlorophenylalanine. *Nature*, 230:244–245.

Tagliamonte, A., Tagliamonte, P., Gessa, G. L., and Brodie, B. B. (1969): Compulsive sexual activity induced by *p*-chlorophenylalanine in normal and pinealectomized male rats. *Science*, 166:1433–1435.

Van Woert, M. H., Heninger, G., Rathey, U., and Bowers, M. B., Jr. (1970): L-DOPA in senile dementia. *Lancet*, 1:573–574.

Whalen, R. E., and Luttge, W. G. (1970): *p*-Chlorophenylalanine methyl ester: An aphrodisiac? *Science*, 169:1000–1001.

Advances in Biochemical Psychopharmacology, Vol. 11
Raven Press, New York © 1974

5-Hydroxytryptamine and Sexual Behavior in the Female Rat

Bengt J. Meyerson, Hugo Carrer, and Mona Eliasson

*Department of Medical Pharmacology, University of Uppsala,
Uppsala, Sweden*

I. INTRODUCTION

Neuropharmacological agents probably influence different components of sexual behavior in different ways. One neurotransmitter might play a predominant role in the activation or maintenance of a certain element, and another system might be more important to other constituents of sexual behavior. In the present neuropharmacological study we have been concerned with two different aspects of female sexual behavior: *copulatory behavior*, and the desire to seek sexual contact, *sexual motivation*. Copulatory behavior (the expression is here used synonymously with lordosis response) in the female rat is a fixed action pattern elicited by the mounting of the male. Once elicited, the lordosis response pattern is always the same. It is evident that this action pattern is not determined by practice; it is fully developed the first time it is elicited. Copulatory behavior is dependent on ovarian hormones. It disappears after ovariectomy but can be restored by injection of estrogen followed, after a certain time interval, by progesterone treatment. Thus the lordosis response in the female rat represents an easily recognizable hormone-dependent motor pattern. Far more complex is the study of sexual motivation. By sexual motivation we mean the condition which brings the animal to seek sexual contact with another animal. In the present study, this was measured in terms of the amount of an aversive stimulus the subject was willing to take to reach contact with a vigorous male and as the degree of preference for a sexually active male versus an estrous female in a choice situation. A recent study (Meyerson and Lindstrom, 1973) using these techniques has shown that in intact rats the urge to seek contact with a vigorous male fluctuates with the estrous cycle. This phenomenon disappeared after ovariectomy and was restored by estrogen treatment. Progesterone did not have the same significance in

229

the production of sexual motivation in the female rat as it had in the activation of the copulatory behavior.

II. BEHAVIOR TECHNIQUES USED

Details of the testing procedures and environmental conditions are described elsewhere (Meyerson, 1964a; Meyerson and Lindström, 1973). Ovariectomized rats (Sprague-Dawley rats purchased as specific pathogen free, 250 to 300 g) were kept under reversed day-night rhythm (12 hr of light, 12 hr of darkness) and tested during the dark period of the cycle. Drugs injected were as follows: Lysergic acid diethylamide (LSD, Sandoz), apomorphine (HCl, Sandoz), DL-p-chlorophenylalanine methylester HCl (PCPA, H 69/17, Hässle), DL-α-methyl-p-tyrosine (α-mTYR, H44/68, Hässle) were dissolved in saline. Pimozide (Janzen) was dissolved in glacial acetic acid, diluted with saline, and the pH adjusted to 5 by addition of 2 N NaOH.

A. Copulatory Behavior

The female was transferred to an observation cage which held a sexually active male and tested for display of copulatory behavior (lordosis response) on mounting by the male. The estimate of the percentage of lordosis response is based on the number of animals showing a clear-cut lordosis response after at least two out of five to six mounts. Copulatory behavior was induced by a single injection of estradiol benzoate followed 48 hr later by progesterone, 0.4 mg/rat. The hormones were dissolved in olive oil. Statistical significance was tested by the χ^2 test (Siegel, 1956).

B. Sexual Motivation

The *increasing-barrier technique* was used to record how much of an aversive stimulus the female was willing to take to reach contact with the sexually active male (Fig. 1). The subject was placed in the starting cage and had to pass an electrified grid in order to attain contact with the male in the goal cage. Only limited contact was possible through a wire mesh which separated the female from the incentive male; 15 sec were allowed in the goal cage, after which the female was brought back to the starting cage. Every second time the female passed the grid, the grid current was increased stepwise. When the subject had spent more than 5 min in the starting cage, the test session was ended. To test whether a treatment specifically affected the desire to seek contact with a male or whether it could be extended to another incentive, water was used as a reward after water deprivation. The same apparatus was

COPULATORY BEHAVIOR

SEXUAL MOTIVATION
"Urge to seek sexual contact"

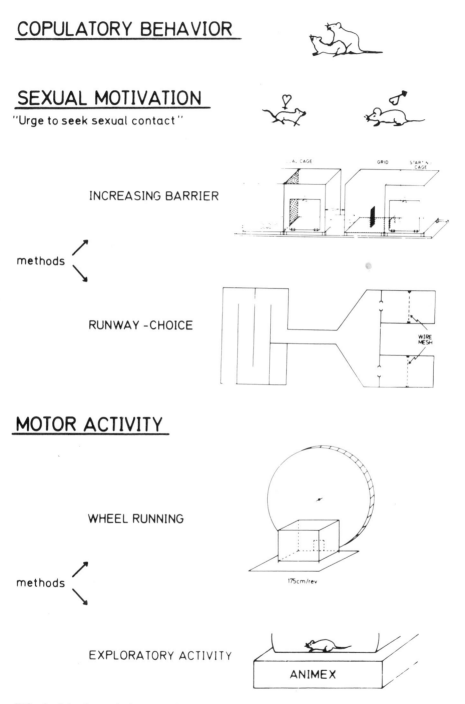

INCREASING BARRIER

methods

RUNWAY -CHOICE

MOTOR ACTIVITY

WHEEL RUNNING

methods

EXPLORATORY ACTIVITY

FIG. 1. Behavior techniques used. A detailed description of the procedures is given in Meyerson (1964*a*) and Meyerson and Lindström (1973).

used but the wire mesh in the goal cage was replaced by a water dipper. The subjects were deprived of water for 3 consecutive days. The water bottles were removed on day 0 after the test session had been performed that day. On the following days, a free supply of tap water was given for 45 min after the test session.

The *runway-choice method* (Fig. 1) was designed to investigate the choice between a sexually active male and an estrous female in a run-and-choice situation. The experimental animal was placed in a runway, the far end of which led the animal to two separate goal cages, one of which held a vigorous male and the other an estrous female. Each experimental female was subjected to 20 consecutive trials during each test session. The number of trials on which the female entered one of the goal cages and the percentage of these trials on which the female entered the goal cage with the male was recorded. Statistical significance was tested by the Wilcoxon matched-pairs signed ranks test (Siegel, 1956).

C. Motor Activity

Locomotor activity was measured by an "Animex" activity meter (Farad, Stockholm) and by a wheel-running apparatus (Fig. 1). In the activity meter, movements of the animal across a tuned oscillator coil system results in a change of tuning which is recorded as a count. One subject was placed in the apparatus at a time. Each recording session lasted 10 min. The activity measured during this time mainly reflects the animal's exploration of the cage. The running wheel was placed adjoining a $25 \times 25 \times 25$ cm cage in which the animal had free access to food and water. The subject was housed in the apparatus for at least 2 months before beginning the experiment. The number of revolutions (each 175 cm) per 12 hr (the dark period) was recorded. Estradiol benzoate, 5 μg/kg, s.c., produces an increase in wheel-running activity (see Table 3). Records were taken during 5 consecutive days before estrogen was given (preexperimental period). Experimental period wheel activity is expressed as percentage of the average preexperimental period activity. Statistical significance was tested by Student's t test.

III. COPULATORY BEHAVIOR

Previous neuropharmacological investigations have demonstrated that copulatory behavior is inhibited by drugs which increase mono-aminergic activity. This has been shown in several species, including rat, hamster, mouse, and rabbit (Meyerson, 1964a, b; Meyerson, 1966; Meyerson, 1970; Meyerson, Eliasson, Lindström, Michanek, and Söderlund, 1973).

A. 5-Hydroxytryptamine and Copulatory Behavior

The results obtained in the rat after selective increase of 5-hydroxy-tryptamine (5-HT) or catecholamines achieved by combined treatment with pargyline and amino acid precursors or synthesis inhibitors suggest that serotonergic pathways mediating inhibition of the lordosis response are of greatest importance (Meyerson, 1964a, b; Meyerson et al., 1973). This is further supported by the effect of lysergic acid

TABLE 1. *The effect of LSD and apomorphine on estradiol benzoate + progesterone activated copulatory behavior in ovariectomized rats*

Treatment (μg/kg)	Pre-inj.	Lordosis response (%)					N

Treatment (μg/kg)	Pre-inj.	Min after last injection 10	40	60	90	120	N
LSD, i.p.							
50	83	42^a	83		83		12
100	71	38^a	58^a		83		48
250	68	0	22^a		73		22
100/day × 7 Day 7	83	71	75		88		24
Apomorphine, s.c.							
50	92	92		96		96	24
200	93	43^a		90		91	58
500	86	16^a		50^a		93	44
Pimozide, s.c., 500 and 1 hr later apomorphine, 200	89	92		92		94	36
Saline, 0.2 ml and 1 hr later apomorphine, 200	80	46^a		77		83	36
Saline, i.p. or s.c., 0.2 ml (controls)	88	93		93		90	80

Estradiol benzoate (10 μg/kg, s.c.) was followed 48 hr later by progesterone (0.4 mg/rat). The pre-injection test was performed 4 hr and the 10-min test 6 hr after the progesterone injection. Controls had saline instead of LSD or apomorphine and were tested at the same time as test animals. Since no difference was obtained between the control groups, pooled data appear in the table. Significant difference between experimentals and controls tested by χ^2 test.
[a]$p < 0.001$.

diethylamide (LSD) (Table 1). Data are available to indicate that LSD can act as a 5-HT receptor agonist (Freedman, 1961; Andén, Corrodi, Fuxe, and Hökfelt, 1968; Aghajanian, 1972). The inhibitory effect of LSD on the lordosis response was already evident at a dose level of 50 µg/kg, i.p. This response almost disappeared after administration of 250 µg/kg, i.p. Clear-cut tolerance to the effect of LSD was noted when LSD was given once daily, 100 µg/kg, i.p., for 6 days, and tested with the same dose on the seventh day.

B. The Effect of Muscarinic Compounds

Muscarinic compounds such as pilocarpine decreased hormone-activated female copulatory response in rat and hamster (Lindström and Meyerson, 1967; Lindström, 1972). The monoamine oxidase inhibitor pargyline prolonged the effect of pilocarpine treatment (Lindström, 1970) (Table 2). p-Chlorophenylalanine (PCPA), 400 mg/kg, given 8 hr before pilocarpine treatment, prevented the inhib-

TABLE 2. The effect of pilocarpine alone and in combination with pargyline, PCPA, or α-mTYR on estradiol benzoate + progesterone activated copulatory behavior in ovariectomized rats

Treatment	Lordosis Response (%) Hours after last treatment		N
	½	1½	
Pilocarpine, 25 mg/kg	33[a]	67	43
Saline, 0.2 ml	85	90	48
Pargyline, 25 mg/kg, and 4 hr later pilocarpine, 25 mg/kg	20[a]	30[a]	46
Pargyline, 25 mg/kg, and 4 hr later saline, 0.2 ml	59	61	46
PCPA, 400 mg/kg, and 8 hr later pilocarpine, 25 mg/kg	71	71	34
PCPA, 400 mg/kg, and 8 hr later saline 0.2 ml	63	66	35
α-mTYR, 200 mg/kg, 22 and 8 hr before pilocarpine, 25 mg/kg	38[a]	86	21
α-mTYR, 200 mg/kg, 22 and 8 hr before saline, 0.2 ml	86	90	21

The data are taken from Lindström and Meyerson (1967) and Lindström (1970, 1971). The copulatory behavior was activated by estradiol benzoate (10 µg/kg) followed 48 hr later by progesterone (0.4 mg/rat). All injections were made subcutaneously.

[a]Significant difference: $p < 0.01$.

itory effect of pilocarpine whereas an analogous effect was not obtained after α-methyltyrosine (α-mTYR) (Lindström, 1971). It seems as if the effect of pilocarpine depends on the presence of 5-HT and that the inhibitory effect of pilocarpine on hormone-activated copulatory behavior in the female rat is mediated by serotonergic mechanisms.

C. Catecholamines and Copulatory Behavior

A brief interruption of copulatory behavior was also seen after pargyline in combination with dihydroxyphenylalanine (DOPA) in rat and hamster (Meyerson, 1964a; Meyerson, 1970). While this effect may be due to serotonin displacement, it could also be a direct effect of increased catecholamine levels. There is evidence that apomorphine increases central nervous dopaminergic activity (Ernst, 1967; Andén, Rubenson, Fuxe, and Hökfelt, 1967). Apomorphine, given at a dose level of 200 to 500 μg/kg, s.c., decreased hormone-activated copulatory behavior (Table 1). At this dose level, obvious stereotyped motor patterns were seen, especially movements of the head from side to side with aimless sniffing. The dopamine receptor blocking compound (Andén, Butcher, Corrodi, Fuxe, and Ungerstedt, 1970), pimozide (500 μg/kg), countered the effect of apomorphine (200 μg/kg).

D. Facilitation of Lordotic Behavior

To facilitate copulatory behavior, drugs might augment the hormone-activated response by (1) an effect on neuronal functions directly involved in the production of the behavior and/or (2) exerting an indirect influence by an effect on neuroendocrine function, thereby activating endogenous hormone production. The inhibitory effect of drugs which increase monoaminergic tone suggests the possibility of activating copulatory behavior by compounds which suppress mono-aminergic activity. In ovariectomized estrogen-treated rats, reserpine, tetrabenazine, PCPA (Meyerson, 1964c; Meyerson and Lewander, 1970; Ahlenius, Engel, Eriksson, and Södersten, 1972a), and α-mTYR (Ahlenius, Engel, Eriksson, Modigh, and Södersten, 1972b) could substitute for progesterone treatment which is otherwise necessary to obtain a response. Paris, Resko, and Goy (1971) found that reserpine treatment increased the concentration of plasma progesterone in ovariectomized rats. The possibility that amine depletion could activate the lordosis response by releasing adrenal progesterone was suggested by these authors. This effect appears to be more closely related to suppression of catecholamines than of 5-HT (Ahlenius et al., 1972a, b). However, there is also evidence that compounds which suppress 5-HT activity facilitate the lordosis response with no endocrine mediation.

Recently, Zemlan, Ward, Crowley, and Margules (1973) showed that PCA, methysergide, and cinanserin activated lordosis response in estrogen-primed ovariectomized-plus-adrenalectomized rats. These authors also included in their conclusions the important fact that although it was possible to activate lordosis response by suppression of serotonergic activity, soliciting patterns such as darting movements and ear wigglings were not evoked, suggesting that activation of the total "estrous behavior" pattern must also involve transmitters other than 5-HT.

IV. SEXUAL MOTIVATION

The techniques used to study sexual motivation require that the locomotor ability of the tested subject is not impaired. The treatments used in this investigation were chosen on the basis of a dose regimen which affected locomotor ability slightly or not at all, as established by measurements on the Animex activity device or wheel-running apparatus (Table 3).

A. Effects of LSD and Apomorphine

In the increasing-barrier apparatus, willingness to cross the grid was significantly decreased by LSD, 50 μg/kg, i.p. ($p < 0.01$) (Fig. 2). To test the specificity of this effect, another group of animals was deprived of water and run in the same apparatus with water instead of a male as a reward. In this test situation also, LSD (50 μg/kg, i.p.) decreased the number of crossings. This means that, measured by the increasing-barrier technique, sexual motivation was similarly decreased by a dose of LSD which decreased motor ability and running for water.

Given a choice between a vigorous male and an estrous female in the runway-choice apparatus, LSD (50 μg/kg, i.p.) did not significantly affect the increased preference for the male seen after estrogen treatment (preference day of estrogen treatment day 0, test: 54% controls: 52%; day 3, 10 min after LSD, 50 μg/kg, i.p., test: 75%, saline 0.2 ml, i.p., treated controls: 73%).

Apomorphine (25 μg/kg, s.c.) did not significantly affect locomotor activity as measured by the Animex, but clearly ($p < 0.01$, Fig. 2) decreased the number of grid crossings performed to reach contact with a vigorous male. When water was used as reward, apomorphine (50 μg/kg, s.c.) was not effective in decreasing the response. However, this dose significantly decreased locomotor activity (Table 3). In the runway-choice situation, preference for the male seen after estrogen treatment was not changed after apomorphine, 200 μg/kg, s.c. (preference day of estrogen treatment day 0, test: 59%, controls: 62%;

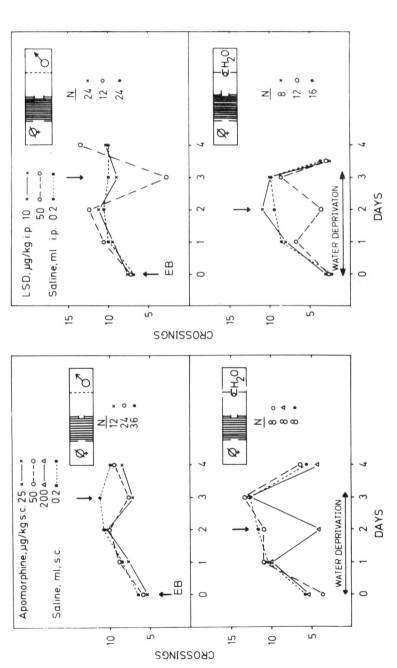

FIG. 2. The increasing barrier technique. The effect of apomorphine and LSD on the urge to seek contact with a sexually active male in estrogen-treated ovariectomized rats and on the urge to reach water in ovariectomized water-deprived rats. EB: estradiol benzoate (10 μg/kg, s.c.) was given on day 0, 4 to 6 hr before the test session started. Apomorphine and LSD were given 10 min before test, day 2 or 3.

TABLE 3. *Motor activity after LSD, apomorphine, α-mTYR, and PCPA in estradiol benzoate treated ovariectomized rats*

Treatment Estradiol benzoate, day 0 and:	Motor activity					
	Animex Counts/10 min, mean ± SEM	p	N	Wheel running Revolutions, percent of pre-experimental activity, mean ± SEM	p	N
LSD, μg/kg, i. p., day 2						
10	898 ± 82	NS	9			
50	532 ± 66	<0.01	9			
100	547 ± 59	<0.01	8			
Apomorphine, μg/kg, s. c., day 2						
25	890 ± 35	NS	10			
50	649 ± 29	<0.01	11			
200	756 ± 49	<0.01	9			
α-mTYR, mg/kg, i.p., day 2						
100	835 ± 71	NS	7			
200	501 ± 36	<0.01	5			
day 3				Controls / Exptls		
100				day 0: 150 ± 17 / 116 ± 24	NS	6
				day 3: 390 ± 49 / 144 ± 34	<0.01	6
PCPA, mg/kg, i. p.,				Controls / Exptls		
day 0: 100	865 ± 26	NS	12	85 ± 24 / 41 ± 12	NS	6
day 1: 50	810 ± 25	NS	12	143 ± 37 / 35 ± 15	<0.02	6
day 2: 50	788 ± 32	NS	12	245 ± 60 / 98 ± 32	<0.05	6
day 3: 50	866 ± 58	NS	12	415 ± 202 / 110 ± 31	NS	6
Saline, 0.2 ml i.p., day 2	951 ± 32		31			

Animex: Estradiol benzoate, 10 μg/kg, was given on day 0. LSD, apomorphine, and saline were injected 10 min, α-mTYR and PCPA, 5 hr before test. The effects of the drug treatments were tested against saline-treated controls run at the same test session.

Wheel running: Estradiol benzoate, 5 μg/kg, was given at day 0, just before the dark period started. Records were taken only during the 12-hr dark period of the light cycle. Preexperimental activity = the average of five consecutive 12-hr periods of recording. Controls preexperimental activity: 469 ± 203, Exptls: 462 ± 150 rev/12 hr. Controls were given hormone and saline treatment.

day 3, 10 min after apomorphine, 200 μg/kg, s.c., test: 77%, controls, saline, 0.2 ml, s.c.: 75%).

B. The Effect of PCPA and α-mTYR

The effect of PCPA was tested in the increasing-barrier apparatus. In the first experiment, no hormone was given (Fig. 3, oil blank). Some decrease ($p < 0.05$) in response was seen on the first two days of PCPA treatment. In contrast, a slight but insignificant increase of crossings was seen in the estradiol benzoate (2.5 μg/kg) experiment and a clearly significant ($p < 0.01$) increase in estradiol benzoate (10 μg/kg) experi-

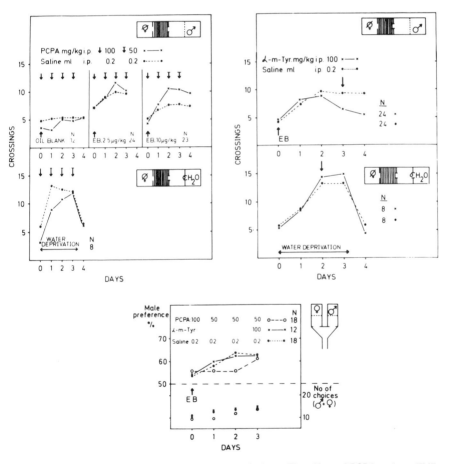

FIG. 3. The increasing-barrier and runway-choice technique. The effect of PCPA and α-mTYR on the urge to seek contact with a sexually active male in estrogen-treated ovariectomized rats and on the urge to reach water in water-deprived ovariectomized rats. EB: estradiol benzoate was given 4 to 6 hr before the test session on day 0. Preference % = choice of $\frac{\male}{\female + \male}$ × 100.

ment. It is interesting that the same dose regimen of PCPA decreased the number of crossings when water was used as reward and depressed the estrogen-activated wheel-running motor activity. A single dose of α-mTYR (100 mg/kg, i.p.), given 3 days after hormone treatment and 5 hr before the animals were tested, significantly decreased ($p.< 0.01$) the number of crossings in the increasing-barrier apparatus. The response was also significantly different from the response seen after hormone treatment only 1 day after the α-mTYR injection. In contrast, when water was used as a reward α-mTYR, if anything, increased the number of crossings on the day of treatment and the day after injection, although this effect was not statistically significant. Estrogen-induced increase in wheel-running activity was significantly decreased by α-mTYR, 100 mg/kg, i.p. (Table 3).

In the runway-choice situation the same treatment regimen was used as in the increasing-barrier experiments. There was a slight decrease in the number of trials the subject ran into the goal cages at days 2 and 3 in the PCPA experiment (Fig. 3). At day 3 there was an evident decrease in the preference for the male ($p < 0.001$). α-mTYR (100 mg/kg, i.p.) treatment did not affect behavior in the runway-choice experiment.

V. CONCLUSIONS

There is a mass of evidence to support the view that serotonergic pathways exist that mediate inhibition of the lordosis response in the female rat (for review, see Meyerson et al., 1973). Although 5-HT seems to play a predominant role in this response, the possibility cannot be ruled out that other transmitters also might be inhibitory. The inhibitory action of muscarinic compounds demands intact 5-HT mechanisms. In this context, the relationship between 5-HT and dopamine must be further elucidated.

At present there seems to be some controversy as to whether suppression of monoaminergic activity can substitute for progesterone treatment by a central effect directly involved in the production of copulatory behavior, or by an indirect effect mediated only by adrenal secretion of progesterone or progesterone-like steroids. If progesterone treatment can be substituted for by suppression of 5-HT activity and it can also be clearly established that endogenous hormone production is not involved, it would still be an open question whether progesterone acts by decreasing activity in certain 5-HT-containing neurons that normally suppress the lordosis response. Progesterone might also act by an alternative mechanism to trigger the system which activates the lordotic behavior synergistically with estrogen. Drugs which suppress 5-HT and progesterone activity might act on the lordotic behavior in a different manner, progesterone overcoming inhibition and 5-HT-suppressive compounds removing it. There is only scant experimental support as yet for such a hypothesis, for example, the fact that estrogen plus reserpine is more effective than estrogen plus a maximal dose of progesterone in producing the lordosis response in ovariectomized rats (Meyerson, 1964c). Further experiments are obviously necessary to clarify this point.

Experiments on the neuropharmacology of sexual motivation are at a very early stage. By running different methods in parallel, we hope to find a neuropharmacological profile which might provide evidence for the role of monoamines in the production of sexual motivation. The desire to reach contact with a vigorous male as measured by the increasing barrier apparatus was more sensitive to apomorphine than was copulatory behavior or the urge to cross the grid to reach water. This inhibitory effect of dopaminergic stimulation is difficult to interpret, considering the results of α-mTYR treatment. This drug, which decreases catecholaminergic activity, also had an inhibitory effect on grid crossings when a male was used as incentive. It was noteworthy that the number of crossings performed in the increasing-barrier apparatus increased after PCPA, provided that estrogen treatment was given. The significance of this effect is evident, considering that estrogen-activated wheel-running activity was significantly decreased as well as crossings performed to reach water. In spite of the increase of crossings to reach contact with a male in the increasing-barrier apparatus, preference for the male decreased after PCPA in the runway-choice situation. This raises the question of whether the PCPA increases the sexual drive, although heterosexual preference in the female is decreased. Further experiments are required before conclusive evidence can be reached on this point.

ACKNOWLEDGMENTS

The investigation was supported by grants from the National Institutes of Health, grant RO1-HD4108-03, and the Swedish Medical Research Council, grant 14X-64-08. For generous supplies of hormones we thank Organon, the Netherlands, through Erco, Sweden; Hässle, Sweden, for H44/68 and H69/17 and Janssen, Belgium, for pimozide.

REFERENCES

Aghajanian, G. K. (1972): LSD and CNS transmission. *Annual Review of Pharmacology*, 12:157–168.
Ahlenius, S., Engel, J., Eriksson, H., and Södersten, P. (1972a): Effects of tetrabenazine on lordosis behaviour and on brain monoamines in the female rat. *Journal of Neural Transmission*, 33:155–162.
Ahlenius, S., Engel, J., Eriksson, H., Modigh, K., and Södersten, P. (1972b):

Importance of central catecholamines in the mediation of lordosis behaviour in ovariectomized rats treated with estrogen and inhibitors of monoamine synthesis. *Journal of Neural Transmission*, 33:247–255.

Andén, N.-E., Butcher, S. G., Corrodi, H., Fuxe, K., and Ungerstedt, U. (1970): Receptor activity and turnover of dopamine and noradrenaline after neuroleptics. *European Journal of Pharmacology*, 11:303–314.

Andén, N.-E., Corrodi, H., Fuxe, K., and Hökfelt, T. (1968): Evidence for a central 5-hydroxytryptamine receptor stimulation by lysergic acid diethylamide. *British Journal of Pharmacology*, 34:1–7.

Andén, N.-E., Rubenson, A., Fuxe, K., and Hökfelt, T. (1967): Evidence for dopaminergic receptor stimulation by apomorphine. *Journal of Pharmacy and Pharmacology*, 19:627–629.

Ernst, A. M. (1967): Mode of action of apomorphine and dexamphetamine on gnawing compulsion in rats. *Psychopharmacologia* (Berlin), 10:316–323.

Freedman, D. X. (1961): Effects of LSD-25 on brain serotonin. *Journal of Pharmacology and Experimental Therapeutics*, 134:160–161.

Lindström, L. H. (1970): The effect of pilocarpine in combination with monoamine oxidase inhibitors, imipramine or desmethylimipramine on oestrous behaviour in female rats. *Psychopharmacologia* (Berlin), 17:160–168.

Lindström, L. H. (1971): The effect of pilocarpine and oxotremorine on oestrous behaviour in female rats after treatment with monoamine depletors or monoamine synthesis inhibitors. *European Journal of Pharmacology*, 15:60–65.

Lindström, L. H. (1972): The effect of pilocarpine and oxotremorine on hormone-activated copulatory behavior in the ovariectomized hamster. *Naunyn-Schmiedeberg's Archives of Pharmacology*, 275:233–241.

Lindström, L. H., and Meyerson, B. J. (1967): The effect of pilocarpine, oxotremorine and arecoline in combination with methyl-atropine or atropine on hormone-activated oestrous behaviour in ovariectomized rats. *Psychopharmacologia* (Berlin), 11:405–413.

Meyerson, B. J. (1964a): The effect of neuropharmacological agents on hormone-activated estrous behavior in ovariectomized rats. *Archives Internationales de Pharmacodynamie et de Thérapie*, 150:4–33.

Meyerson, B. J. (1964b): Central nervous monoamines and hormone induced estrus behaviour in the spayed rat. *Acta Physiologica Scandinavica*, 63:Suppl. 241.

Meyerson, B. J. (1964c): Estrus behaviour in spayed rats after estrogen or progesterone treatment in combination with reserpine or tetrabenazine. *Psychopharmacologia* (Berlin), 6:210–218.

Meyerson, B. J. (1966): The effect of imipramine and related antidepressive drugs on estrus behaviour in ovariectomized rats activated by progesterone, reserpine or tetrabenazine in combination with estrogen. *Acta Physiologica Scandinavica*, 67:411–422.

Meyerson, B. J. (1970): Monoamines and hormone activated oestrous behaviour in the ovariectomized hamster. *Psychopharmacologia* (Berlin), 18:50–57.

Meyerson, B. J., Eliasson, M., Lindström, L., Michanek, A., and Söderlund, A. C. (1973): Monoamines and female sexual behaviour. In: *Psychopharmacology, Sexual Disorders and Drug Abuse*, edited by T. A. Ban et al., pp. 463–472. North-Holland, Amsterdam.

Meyerson, B. J., and Lewander, T. (1970): Serotonin synthesis inhibition and estrous behaviour in female rats. *Life Sciences*, 9:661–671.

Meyerson, B. J., and Lindström, L. H. (1973): Sexual motivation in the female rat. *Acta Physiological Scandinavica*, Suppl. 389.

Paris, C. A., Resko, J. A., and Goy, J. A. (1971): A possible mechanism for the induction of lordosis by reserpine in spayed rats. *Biology of Reproduction*, 4:23–30.

Siegel, I. S. (1956): *Nonparametric statistics for the Behavioral Sciences.* McGraw-Hill, London.

Zemlan, F. P., Ward, L. I., Crowley, W. R., and Margules, D. L. (1973): Activation of lordotic responding in female rats by suppression of serotonergic activity. *Science*, 179:1010–1011.

Advances in Biochemical Psychopharmacology, Vol. 11
Raven Press, New York © 1974

Oppposite Effects of Serotonin and Dopamine on Copulatory Activation in Castrated Male Rats

Carl Olof Malmnäs

*Department of Medical Pharmacology, University of Uppsala, Uppsala,
Sweden*

I. INTRODUCTION

Copulatory behavior in the male rat consists of a series of mounts with or without penile intromission into the female vagina, terminated by ejaculation. It is well known that this behavior is dependent on gonadal hormones: After castration there is a gradual decline in copulatory activity which can be restored to the precastration level by means of testosterone propionate (TP) treatment (Shapiro, 1937; Stone, 1939; Beach and Holz-Tucker, 1949; Malmnäs, 1973a).

In the present investigation the influence on copulatory behavior of drugs which in different ways affect monoaminergic function was studied. In order to avoid drug effects on gonadal hormone production and to be able to maintain a submaximal response level in the predrug state, TP-treated castrated male rats were used.

II. MATERIAL AND METHODS

Male Wistar rats were tested once weekly for copulatory activity with spayed female rats in estrogen + progesterone induced behavioral estrus. The main variable measured was the percentage of subjects displaying at least one mount during a test (*mount percentage*). After castration there is a progressive decline in the percentage of subjects which display mounting with estrous females. By means of small amounts of TP (0.10 to 0.20 mg/kg/week, 3 to 4 days before testing), mount percentage can be maintained at a stable, submaximal level for several months.

Blank treated controls were tested parallel to subjects given neuropharmacological agents. Drug and blank injections were given intraperitoneally.

As a check on the specificity of drug effects on copulatory behavior, two additional techniques were used:

1. Motor activity measurements were performed with the Animex activity meter.

2. The specific orientation during copulatory tests toward (a) the female, (b) the environment, and (c) the subject itself was recorded.

For further information concerning methods and results, see Malmnäs (1973a,b,c,d).

III. DECREASED MONOAMINERGIC "TONE"

A. p-*Chlorophenylalanine*

p-Chlorophenylalanine (PCPA), which inhibits serotonin synthesis (Koe and Weissman, 1966), was given at a daily dosage of 100 mg/kg for 4 consecutive days. This treatment induced an increase in mount percentage (Table 1A). In order to exclude the possibility that this effect was mainly due to increased adrenal steroid hormone secretion, the experiment was repeated in adrenalectomized castrates (supplied with 0.9% saline in their home cages and injected once weekly with free dexamethasone, 0.5 mg/animal). The dose of PCPA was 100 mg/kg followed by 50 mg/kg for the next 3 days. In these subjects also, PCPA increased mount percentage (Table 1B). Thus, the facilitatory effects of PCPA on mount percentage were not secondary to increased steroid hormone secretion.

In contrast to the facilitatory effects on copulatory behavior, motor activity was decreased by PCPA treatment.

B. α-*Methyl*-p-*tyrosine*

Although PCPA mainly decreases serotonin synthesis, this drug also lowers catecholamine levels to some extent. Before the conclusion can be drawn, therefore, that the PCPA effect is due to impaired serotoninergic neurotransmission, the effect of catecholamine synthesis inhibition must be studied. α-Methyl-p-tyrosine (αMT), which blocks catecholamine synthesis at the tyrosine hydroxylase level (Spector, Sjoerdsma, and Udenfriend, 1965), was given 6 hr before the copulatory test, in doses of either 75 or 150 mg/kg. Motor activity was not significantly changed at the lower dose, but the higher dose decreased motor activity. After both doses, however, mount percentage was decreased (Table 1C).

TABLE 1. *Effect of certain drug treatments on testosterone propionate activated heterosexual mounting behavior in castrated male rats*

		Hours prior to exp. test	Mount percentage		N
	Treatment		Pre-exp.	Exp.	
A1	Saline, 4 × 0.4 ml	78, 54, 30, 6	45	45	40
A2	PCPA, 4 × 100 mg/kg	78, 54, 30, 6	49	90[a]	39
B1	Saline, 4 × 0.4 ml (adrenalect.)	78, 54, 30, 6	35	35	35
B2	PCPA, 100 + 3 × 50 mg/kg	78, 54, 30, 6	31	80[a]	35
C1	Saline, 0.4 ml	6	59	56	34
C2	αMT, 75 mg/kg	6	60	37[b]	35
C3	αMT, 150 mg/kg	6	58	8[a]	24
D1	Saline, 0.4 ml	4	53	53	40
D2	FLA-63, 10 mg/kg	4	55	48	33
E1	Saline, 0.4 ml	2	53	56	36
E2	Phenoxybenzamine, 3 mg/kg	2	57	47	30
F1	Saline, 0.4 ml	3	54	54	41
F2	Pimozide, 0.10 mg/kg	3	53	50	40
F3	Pimozide, 0.25 mg/kg	3	55	21[a]	42
G1	Saline, 3 × 0.4 ml	1.5, 1.0, 0.5	58	63	38
	Pargyline, 20 mg/kg	1.5			
	+ MK486, 50 mg/kg	1.0			
G2	+ saline, 0.4 ml	0.5	66	63	38
G3	+ DL-5-HTP, 2.5 mg/kg	0.5	68	37[a]	38
G4	+ L-DOPA, 2.5 mg/kg	0.5	56	95[a]	39
G5	+ DL-DOPA, 30 mg/kg	0.5	53	57	30
H1	Saline, 0.4 ml	0.2	45	48	44
H2	Apomorphine, 100 μg/kg	0.2	48	79[a]	42
H3	Clonidine, 30 μg/kg	0.2	52	45	29
H4	LSD, 30 μg/kg	0.2	56	28[b]	36

Probability of difference in mount percentage between the last no drug test (pre-exp.) and the drug test (exp.) was tested with the sign test.

[a] = $p < 0.001$.
[b] = $p < 0.01$.

C. Selective Impairment of Catecholaminergic Neurotransmission

The inhibitory effect of αMT on mount percentage raises the question of whether dopamine and norepinephrine are of equal significance for male copulatory behavior. In an attempt to obtain some information about this matter, an experiment was performed with the dopamine-β-hydroxylase inhibitor FLA-63 (Florvall and Corrodi, 1970). FLA-63, 10 mg/kg, was given 4 hr before testing. This treatment did not change mount percentage, but it did decrease motor activity (Table 1D). Nor did the norepinephrine receptor blocking agent phenoxybenzamine, 3 mg/kg, 2 hr before testing, change mount percentage (Table 1E). If the doses of FLA-63 and phenoxybenzamine were raised to the level that motor activity (including approach to the female) was completely abolished, copulatory behavior was of course also abolished. In contrast to the effect of FLA-63 and phenoxybenzamine, the dopamine receptor blocking agent, pimozide (Andén, Butcher, Corrodi, Fuxe, and Ungerstedt, 1970a), decreased mount percentage at a dose (0.25 mg/kg, 3 hr before testing) low enough not to decrease motor activity or otherwise disturb overt behavior (Table 1F).

IV. INCREASED MONOAMINERGIC "TONE"

A. Monoamine Precursors

Monoamine precursors were given after combined pretreatment with the monoamine oxidase inhibitor pargyline (20 mg/kg, 1.5 hr before testing), and the extracerebral decarboxylase inhibitor MK486 (50 mg/kg, 1.0 hr before testing). Such pretreatment combined with saline did not affect mount percentage (Table 1G). DL-5-HTP, 2.5 mg/kg, 0.5 hr before testing, induced a clear-cut decrease in mount percentage. In contrast, L-DOPA, 2.5 mg/kg, 0.5 hr before testing, increased mount percentage. L-DOPA, 2.5 mg/kg, but not DL-5-HTP, decreased motor activity. No effect on mount percentage was found after DL-DOPA 3.0, 10, or 30 mg/kg. The facilitatory effect of L-DOPA on mount percentage was blocked by pimozide, 0.10 mg/kg, 3 hr before testing; the pimozide + pargyline + MK486 pretreatment did not change mount percentage.

B. Monoaminergic Agonists

In another series of experiments, the effects of monoaminergic agonists were studied. The dopamine receptor stimulating agent, apomorphine (Andén, Rubenson, Fuxe, and Hökfelt, 1967), 30, 100, and 300 µg/kg (12 min before testing), increased mount percentage (Table 1H). The effect of apomorphine, 100 µg/kg, was blocked by pimozide, 0.10 mg/kg. Clonidine, which stimulates norepinephrine receptors (Andén, Corrodi, Fuxe, Hökfelt, Hökfelt, Rydin, and

Svennson, 1970*b*), did not affect mount percentage in any of the doses tested (3, 10, and 30 μg/kg, 12 min before testing). LSD, a serotonin receptor stimulating agent (Andén, Corrodi, Fuxe, and Hökfelt, 1968), decreased mount percentage when given in doses of 30 and 100 μg/kg (12 min before testing). Motor activity was decreased by apomorphine, 100 μg/kg, and clonidine, 30 μg/kg, but not by LSD, 30 μg/kg.

V. CONCLUSIONS

When the present data are taken together, the following working model of monoaminergic influence on display of mounting behavior can be set up (Table 2):

1. *Serotonin*: An increased serotoninergic tone induces a decreased response, while a decreased tone induces an increased response.

2. *Dopamine*: An increased tone induces an increased response, while a decreased tone induces a decreased response.

3. *Norepinephrine*: There is no support for any relationship similar to the ones for serotonin and dopamine.

TABLE 2. *Schematic representation of monoaminergic influences on testosterone propionate activated heterosexual mounting behavior in castrated male rats*

Amine	"Tone" at receptor sites	Mount percentage
Serotonin	Increased Decreased	Decreased Increased
Dopamine	Increased Decreased	Increased Decreased
Norepinephrine	Increased Decreased	Unaffected (?) Unaffected (?)

It should be mentioned briefly that among other copulatory parameters recorded, the latency period from the introduction of the female to the occurrence of the first mount (mount latency) was generally decreased when mount percentage was increased and vice versa. Furthermore, a shortening of time to ejaculation was found after PCPA and L-DOPA.

Of particular interest is that among nonmounting subjects, female-oriented activity was increased by apormorphine and decreased by pimozide, DL-5-HTP, and LSD. These findings indicate that components of the male's interaction with the female other than copulatory may be under serotoninergic/dopaminergic control.

It must be emphasized that these changes in sexual behavior related to changes in monoaminergic tone have a significance only so long as the treatment regimen used does not seriously impair well-being and/or motor capability of the subjects. The range of doses used in this investigation fulfilled these requirements: in instances where sexual behavior was suppressed, this was not associated with a disturbance of overt behavior.

REFERENCES

Andén, N.-E., Butcher, S. G., Corrodi, H., Fuxe, K., and Ungerstedt, U. (1970a): Receptor activity and turnover of dopamine and noradrenaline after neuroleptics. *European Journal of Pharmacology*, 11:303–314.

Andén, N.-E., Corrodi, H., Fuxe, K., Hökfelt, T. (1968): Evidence for a central 5-hydroxytryptamine receptor stimulation by lysergic acid diethylamide. *British Journal of Pharmacology*, 34:1–7.

Andén N-E., Corrodi, H., Fuxe, K., Hökfelt, B., Hökfelt, T., Rydin, C., and Svensson, T. (1970b): Evidence for a central noradrenaline receptor stimulation by clonidine. *Life Sciences* I, 9:513–523.

Andén, N-E., Rubenson, A., Fuxe, K., and Hökfelt, T. (1967): Evidence for dopamine receptor stimulation by apomorphine. *Journal of Pharmacy and Pharmacology*, 19:627–629.

Beach, F. A., and Holz-Tucker, A. M. (1949): Effects of different concentrations of androgen upon sexual behavior in castrated male rats. *Journal of Comparative and Physiological Psychology*, 42:433–453.

Florvall, L., and Corrodi, H. (1970): Dopamine β-hydroxylase inhibitors. *Acta Pharmaceutica Suecia*, 7:7–22.

Koe, B. K., and Weissman, A. (1966): p-Chlorophenylalanine: A specific depletor of brain serotonin. *Journal of Pharmacology and Experimental Therapeutics*, 154:499–516.

Malmnäs, C. O. (1973a): Testosterone-activated copulatory behavior in the castrated male rat. *Acta Physiologica Scandinavica*, Suppl. 395:9–46.

Malmnäs, C. O. (1973b): Monoamine precursors and copulatory behavior in the male rat. *Acta Physiologica Scandinavica*, Suppl. 395:47–68.

Malmnäs, C. O. (1973c): Copulatory behavior in the male rat after impaired monoaminergic neurotransmission. *Acta Physiologica Scandinavica*, Suppl. 395:69–95.

Malmnäs, C. O. (1973d): Effects of LSD-25, clonidine and apomorphine on copulatory behavior in the male rat. *Acta Physiologica Scandinavica*, Suppl. 395:96–116.

Shapiro, H. A. (1937): Effect of testosterone propionate upon mating. *Nature* (London), 139:588–589.

Spector, S., Sjoerdsma, A., and Udenfriend, S. (1965): Blockade of endogenous norepinephrine synthesis by α-methyl-tyrosine, an inhibitor of tyrosine hydroxylase. *Journal of Pharmacology and Experimental Therapeutics*, 147:86–95.

Stone, C. P. (1939): Copulatory activity in adult male rats following castration and injections of testosterone propionate. *Endocrinology*, 24:165–174.

Advances in Biochemical Psychopharmacology, Vol. 11
Raven Press, New York © 1974

Depressive Illness and Oral Contraceptives. A Study of Urinary 5-Hydroxyindoleacetic Acid Excretion

Jens Ancher Madsen*

Department of Gynecology and Obstetrics, University Hospital, Odense, Denmark

I. INTRODUCTION

Subjects suffering from depression have been found to have a significantly lower content of 5-hydroxyindoleacetic acid (5-HIAA) in their cerebrospinal fluid, with slow reversion to normal values on recovery (Ashcroft, Crawford, Eccleston, Sharman, MacDougall, Stanton, and Binns, 1966; Dencker, Malm, Roos, and Werdinius, 1966).

Some researchers have found that when depression is treated with a monoamine oxidase inhibitor (MAOI) the beneficial effect can be heightened with a tryptophan supplement (Glassman and Platman, 1969). Clinically, it has been known for a considerable time that oral contraceptives can cause depression. A survey of the literature (*British Medical Journal*, 1969) showed the frequency of this side effect to vary between 2 and 30%. In one controlled clinical trial, 12% were found to complain of tiredness and/or depression (Nilsson and Sölvell, 1967).

It is now well established that the metabolism of tryptophan along the kynurenine pathway after tryptophan loading is increased in: (1) pregnant women (Rose, 1969), (2) women about the time of ovulation (Brown, Rose, Price, and Wolf, 1969), (3) women taking a combination of estrogen-progestagen (mestranol-norethynodrel) (Brown et al., 1969), and (4) women being treated with estrogen (Brown et al., 1969)—as determined by the urinary excretion of kynurenine (KYN), kynurenic acid (KA), hydroxykynurenine (HK), xanthureninic acid (XA), and hydroxyanthranilic acid (HAA). Furthermore, increased excretion of HAA without a tryptophan load has been found in women taking oral contraceptives (Toseland and Price, 1969). In all cases a

---

*Present address: Clinical Chemical Department, Odense University Hospital, Odense, Denmark.

supplement of vitamin B_6 could either bring the output back to normal or bring it down toward normal.

II. METHODS AND MATERIALS

5-HIAA urinary output was determined in women taking oral contraceptives. To determine 5-HIAA, the method described by Contractor (1966) was used and fluorescence was measured on a Perkin-Elmer spectrophotofluorimeter MPF-2A. For the analysis, a morning urine sample acidified with 1 N HCl was used. This sample

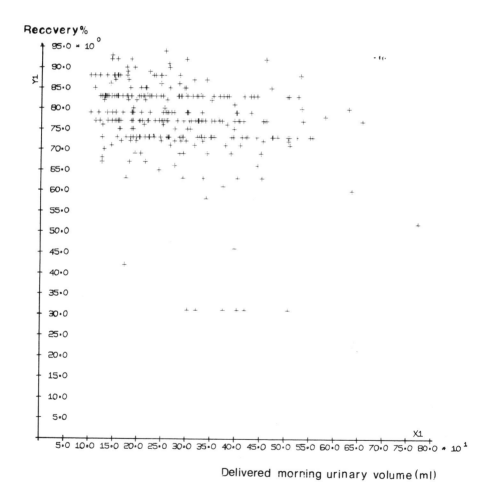

FIG. 1. The variation in recovery rate, as estimated by addition of 4 and 6 µg of 5-HIAA to the sample, independent of the delivered urinary volume.

was further diluted with 0.1 N HCl to a fixed volume corresponding to a diuresis in all subjects of 85 ml/hr (knowing the time over which the urine sample was collected). The subjects were not allowed to eat walnuts, bananas, old strong cheese, or drink red wine the day before collection. In addition, all subjects were between 15 and 35 years of age and had not received hormonal treatment nor been pregnant within the previous 6 months.

Three to 12 specimens of urine were obtained in each menstrual cycle (average 5), beginning with a cycle without oral contraceptives, followed by the first cycle with oral contraceptives and a varying number of subsequent cycles. The subjects took Neo-Delpregnin® (megestrol acetate, 4 mg, + ethinylestradiol 50 μg). A large individual variation in 5-HIAA excretion was found; recovery varied considerably from subject to subject, but was consistent in the same individual (Fig. 1). All values were therefore calculated in terms of 100% recovery from percentage recovery for each specimen. Using this procedure, there is a further factor which may have influenced the results adversely. Information about urine collection time supplied by the subject may have been incorrect. Excretion values of 5-HIAA are now being expressed relative to output of urinary creatinine.

III. RESULTS

Eleven subjects were followed through one cycle without oral contraceptives: 10 were followed in the first cycle with oral contraceptives, five in the second, four in the third, and four in the fourth. In eight of 11 subjects 5-HIAA excretion showed a tendency to increase during contraceptive treatment, but a significant increase was found in only one subject and then only after the third cycle (Table 1). It should be noted that only four of the subjects were followed beyond the third cycle.

In five further subjects, 5-HIAA was followed during the last cycle of contraceptives and again after stopping. Three subjects stopped because of depression and the other two because of tiredness. Four were followed only for three cycles after stopping the drug, and there was no significant variation in 5-HIAA excretion during this period. Two of them were unchanged, while symptoms disappeared in two more.

The fifth subject had become seriously depressed while taking oral contraceptives, and then had a 5-HIAA excretion value of 115 μg/hr ± 10. This subject was followed continuously through the six cycles after stopping the drug. A significant increase was found in the 5-HIAA excretion, output during the last cycle being 160 μg/hr ± 21 ($p < 0.01$). A significant increase, in fact, was already observed in the first cycle. Symptoms began to decrease by the end of the first cycle and had

TABLE 1. *Urinary excretion of 5-HIAA (in µg/hr) in different cycles*

	Cycle number				
	0	1	2	3	4
5-HIAA/hr	173 ± 44	197 ± 39			227 ± 28
by 100%	144 ± 30	129	155 ± 15		
recovery	181 ± 27	195 ± 6		214 ± 58	
(µg)	154 ± 25	132 ± 19		207 ± 81	165 ± 12
	132 ± 11	135 ± 23	123 ± 16		159 ± 31
	124 ± 20		120 ± 43		163 ± 26[a]
	95	92	136 ± 20	162 ± 20	
	117 ± 35	103 ± 37			
	158 ± 21	157 ± 29		130 ± 38	
	155 ± 28	154 ± 29			
	175 ± 21	186 ± 29	221 ± 31		

Cycle number 0 represents 5-HIAA analyses carried out without o.c. treatment. Cycle number 1 represents 5-HIAA excretion in the first o.c. period, and so on.

[a]Significant increase ($p < 0.05$) from cycle 0 to cycle 4.

totally disappeared by the middle of the second cycle.

IV. CONCLUSION

The investigations described above do not allow any final conclusion, although they are suggestive. Further studies are in progress in conditions associated with altered endocrine state such as depressive illness during pregnancy and puerperal psychosis.

V. SUMMARY

The urinary excretion of 5-HIAA was measured in 11 subjects under treatment with oral contraceptives, and the results were compared with output in the immediate pretreatment period. A significantly increased excretion of 5-HIAA was found in one of four subjects followed through four cycles using the drug. In five subjects who wished to withdraw from oral contraceptives, 5-HIAA was measured during one cycle. One of these was depressive and showed a significant increase of 5-HIAA output after discontinuance of the drug.

REFERENCES

Ashcroft, G. W., Crawford, T. B. B., Eccleston, D., Sharman, D. F., MacDougall, E.

J., Stanton, J. B., and Binns, J. K. (1966): 5-Hydroxyindole compounds in the cerebrospinal fluid of patients with psychiatric or neurological diseases. *Lancet*, 2:1049–1052.

Brown, R. R., Rose, D. P., Price, J. M., and Wolf, H. (1969): Tryptophan metabolism as affected by anovulatory agents. *Annals of the New York Academy of Sciences*, 166:44–56.

Contractor, S. F. (1966): A rapid quantitative method for the estimation of 5-hydroxyindoleacetic acid in human urine. *Biochemical Pharmacology*, 15:1701–1706.

Dencker, S. J., Malm, U., Roos, B.-E., and Werdinius, B. (1966): Acid monoamine metabolites of cerebrospinal fluid in mental depression and mania. *Journal of Neurochemistry*, 13:1545–1548.

Glassman, A. H., and Platman, S. R. (1969): Potentiation of a monoamine oxidase inhibitor by tryptophan. *Journal of Psychiatric Research*, 7:83–88.

Lead article (1969): Oral contraceptives and depression. *British Medical Journal*, 2:380–381.

Nilsson, L., and Sölvell, L. (1967): Clinical studies on oral contraceptives—a randomized doubleblind, crossover study of 4 different preparations. *Acta Obstetricia et Gynecologica Scandinavica*, 46:Suppl. 8.

Rose, D. P. (1969): The effects of gonadal hormones and contraceptive steroids on tryptophan metabolism. In: *Metabolic Effects of Gonadal Hormones and Contraceptive Steroids*, edited by H. A. Salhanick, D. M. Kipnis, and R. L. Vande Wiele, pp. 352–366. New York.

Toseland, P. A., and Price, S. (1969): Tryptophan and oral contraceptives. *British Medical Journal*, 1:777.

Advances in Biochemical Psychopharmacology, Vol. 11
Raven Press, New York © 1974

5-Hydroxytryptamine in Aggressiveness

Luigi Valzelli

Istituto di Ricerche Farmacologiche "Mario Negri," Milano, Italy

I. INTRODUCTION

In natural conditions of freedom, aggressiveness has to be considered as a normal aspect of animal behavior, being an instrument for survival of the living organism. Under these conditions, the stimulating effect of animals upon each other has been ascertained (Hamilton, 1916; Bayroff, 1936) and it has been noted that in a social group, aggressive behavior will be more or less evident according to the influence of the environment and the degree of socialization of the animal considered. The importance of such parameters has already been emphasized by Welch (1965). On the other hand, aggressiveness is a response depending, to a great extent, upon both genetically determined conditions and the influence of learning, as demonstrated by Seward (1954a,b) and Kahn (1951).

Among the most important causes driving the animal to aggressive behavior is competition for food (Fredericson and Birnbaum, 1954), for social rank (Kahn, 1954; King, 1957), and for reproduction (Kahn, 1961). Another very important factor is represented by the concept of "territoriality" by which the animal establishes its living space and, in addition, the level of interchanges with other animals so that all violations of the territory explode in fighting (Fredericson, 1950; King and Gurney, 1954; Scott, 1958, 1962; Clark, 1962a,b).

Laboratory animals lose their aggressiveness to a great extent when they are domesticated or bred in standard conditions. In such cases, the main motivation for aggressiveness, that is, competition for food, territory, and reproduction, is lost.

One of the methods available to induce aggressive behavior in laboratory animals is represented by the technique of prolonged isolation. This possesses the advantage of being a pure psychological reaction to an unusual situation. Following previous observations of Allee (1942a,b), Scott (1946, 1958, 1962), and Seward (1945a,b, 1946) concerning the behavior of mice in isolation, Yen, Stanger, and

Millman (1959) have established suitable experimental conditions for inducing aggressiveness in mice.

Not all strains of mice or rats are susceptible to becoming aggressive after isolation, probably because of a genetic component together with factors related to sex, since only male animals are susceptible to developing aggressive behavior patterns during isolation (Table 1). In other words, female mice do not show any active aggressiveness after isolation, and such a lack of response may be linked to a different basic predisposition to this type of behavior pattern. In fact, in many animal species, female animals living in a free and natural environment are never aggressive except during the period following delivery. In this case, female aggressiveness is only the expression of a motivated behavior pattern directed toward protection of the offspring.

TABLE 1. *Effect of prolonged isolation on male and female mice or rats of different strains*

| | | Percent aggressiveness | |
| | | 4 weeks of isolation | 7 weeks of isolation |
Strains	Sex		
Mice			
Albino Swiss	Male	100	100
Albino Swiss	Female	0	0
Albino CRl	Male		100
Albino CF1	Male	100	100
Albino CF1	Female		0
Albino CFW	Male	100	
Albino CFW/S	Female	0	
Albino BALB/C	Male	100	
Black DBA/2J	Female	0	
Black CBA	Male		0
Black CBA/J	Male	0[a]	
Black CBA/J	Female	0	
Black C57	Male		0
Rats			
Sprague-Dawley	Male	0	0
Sprague-Dawley	Female	0	0
Buffalo	Male	0	
Buffalo	Female	0	0
Wistar	Male		40
Wistar	Female		0

[a] Fighting among different strains.

II. BEHAVIORAL EFFECT OF ISOLATION

A. *In Mice*

Standard albino mice, living singly in cages 357 cm^2 in surface area for a period of about 4 weeks, develop a typical behavior pattern that passes from an initial stage of hyperreactivity to the usual environmental stimuli through a hyperactive phase, to aggressiveness. Aggressive behavior is manifested by actual fighting whenever other mice are introduced into the cage of the isolated mouse, and the fighting is fierce and vicious.

It is possible to score the degree of aggressiveness by evaluating the intensity of the behavioral pattern shown by the animals, so that the evaluation can be made as follows:

0   = The animals show no interest in their partners, except occasional nosing.
25 = Frequent vigorous nosing and tail rattling. The animals assume the position of readiness to fight and occasionally attack the partners (no more than three to four times in the period of observation).
50 = Tail rattling, squeaking, powerful attacks (no more than 10 to 11 times in the period of observation).
75 = The animals follow the partner, fierce wrestling and biting during most of the period.
100 = Fierce wrestling. The animals bite their partners hard enough to draw blood. The attacks cover practically the entire period of observation.

It should be underlined that this kind of score works only when three previously isolated mice are put in contact. A different method of evaluation must be adopted when an isolated aggressive mouse is put in contact with a normal inexperienced mouse.

B. *In Rats*

After a series of preliminary trials to find the rat strain most suitable for the isolation technique, Wistar male rats were isolated in single Makrolon cages for 6 weeks. At regular time intervals during the isolation period, the development of new behavioral patterns was followed by putting a mouse in contact with the isolated rat. In addition to the classical behavior pattern described as "muricide," which in these experimental conditions typically occurs about 10 min after contact with the mouse, it was possible to observe two other behavioral patterns, analyzed and described elsewhere (Valzelli and Garattini, 1972). Summarizing briefly, these two patterns have been defined as "friendly" and "indifferent." The "indifferent" rat is characterized by a complete lack of interest toward the mouse put into the cage. Such behavior is quite different from that typical of control animals: The rat remains motionless in a corner of the cage without displaying the normal ritual for identification of an intruder. The behavior of the rat defined as "friendly" is more complex. In fact, these animals have an active behavior pattern toward the mouse, resembling that of animals playing among themselves, with maternal components, such as preparation of a nest in which they repeatedly put the mouse. These animals also have episodes of tremor and uncoordinated hyperactivity. The typical "muricide" rat, on the other hand, kills the mouse in a few minutes, by breaking its neck. This behavior pattern is constant and repetitive and is not sustained or motivated by hunger; these animals, and all the others, have food and water *ad libitum*.

III. EFFECT OF ISOLATION ON BRAIN NEUROCHEMISTRY

A. In Mice

During recent years, a considerable effort has been made to determine whether behavioral modifications induced by isolation correspond to brain biochemical changes (Seward, 1946; Welch and Welch, 1966; Valzelli, 1967a; Welch, 1967; Valzelli and Garattini, 1968).

With respect to neurochemical transmitters, the first point to take into consideration is that levels of serotonin (5-HT), norepinephrine (NE), and dopamine (DA) do not change significantly either in whole brain or in different brain areas (Valzelli, 1967b; Giacalone, Tansella, Valzelli, and Garattini, 1968; Valzelli and Garattini, 1968). Nevertheless, even if the level of a brain amine does not change, the dynamic study of its turnover, which is essentially dependent on three factors—amine synthesis, release, and inactivation—may give interesting information. In such isolated aggressive mice, it is possible to observe clear modifications of amine turnover, particularly for brain serotonin whose turnover is decreased in comparison to normal mice (Garattini, Giacalone, and Valzelli, 1967; Valzelli, 1967a,b; Essman, 1969) (Table 2).

It should be emphasized that this biochemical modification occurs rapidly during the first or second day of isolation, while the aggressive behavior pattern develops gradually and is maximal only after 4 weeks of isolation (Giacalone et al., 1968; Garattini, Giacalone, and Valzelli,

TABLE 2. *Effect of prolonged isolation on brain concentration and turnover of 5-hydroxytryptamine (5-HT), norepinephrine (NE), and dopamine (DA) in aggressive mice and muricide rats*

	Brain amines					
	Concentration (μg/g)			Turnover rate (μg/g/hr)		
Type of Animal	5-HT	NE	DA	5-HT	NE	DA
Normally housed mice	0.65 ± 0.02	0.45 ± 0.02	1.07 ± 0.03	0.43	0.06	0.08
Isolated aggressive mice	0.64 ± 0.02	0.42 ± 0.03	1.02 ± 0.04	0.26^a	0.03^a	0.14^a
Normally housed rats	0.37 ± 0.01	0.38 ± 0.05	0.98 ± 0.04	0.33	0.08	0.50
Isolated muricide rats	0.36 ± 0.02	0.41 ± 0.03	0.94 ± 0.03	0.23^a	0.13^a	0.49

Each figure corresponds to at least eight determinations.
$^a p < 0.01$.

1969). Moreover, this difference in 5-HT turnover appears only in those animals which become aggressive and is not present, for example, in isolated female mice, which do not show aggressive behavior after isolation. Neither is there a change in those strains, such as $C_{57}B1/J6$, which are not susceptible to isolation-induced aggressiveness (Table 3).

TABLE 3. *Effect of prolonged isolation on brain 5-hydroxytryptamine turnover in female animals and in male animals belonging to strains not becoming aggressive by isolation*

Strains	Sex	Housing condition	Brain 5-HT turnover rate (μg/g/hr)	Aggressive behavior
Albino Swiss mice	Female	Grouped	0.42	0
Albino Swiss mice	Female	Isolated	0.42	0
$C_{57}B1/J6$ mice	Male	Grouped	0.58	0
$C_{57}B1/J6$ mice	Male	Isolated	0.62	0
Sprague-Dawley rats	Male	Grouped	0.27	0
Sprague-Dawley rats	Male	Isolated	0.29	0

Each figure corresponds to at least eight determinations.

The experimental analysis also shows a decrease of NE turnover and an increase of dopamine turnover in aggressive mice (Table 2).

Some relevant brain enzymes, including monoamine oxidases (MAO) and choline acetylase, are not affected by isolation. On the other hand, brain N-acetyl aspartic acid concentration is significantly decreased either in whole brain or in different brain areas of isolated aggressive mice (Marcucci, Mussini, Valzelli, and Garattini, 1968; Garattini et al., 1969). In normal animals, the level of this compound is affected by administration of 5-hydroxytryptophan, MAO inhibitors, reserpine, or LSD_{25} (McIntosh and Cooper, 1965), and seems to be correlated with brain serotonin metabolism. Such changes are not present either in isolated female animals or in those animals which do not become aggressive after isolation. Neither brain aspartic acid nor glutamic acid concentrations change in mice made aggressive by prolonged isolation.

It must also be stressed that aggressive mice seem to be less sensitive than their nonaggressive counterparts to blockade of brain 5-HT synthesis by p-chlorophenylalanine (Valzelli, 1973), a drug which has been claimed to decrease hyperirritability in septal rats (Dominguez and Longo, 1969) and to facilitate the muricide reaction in rats kept in isolation (Di Chiara, Camba, and Spano, 1971).

B. In Rats

As in mice, isolation does not modify the brain amine levels in rats with the three behavioral patterns described above (Table 2). However, it is possible to observe a decreased brain 5-HT turnover rate in muricide rats (Table 2), which is even greater in "indifferent" animals. In "friendly" animals, on the other hand, there is an increase in turnover rate of this amine (Valzelli and Garattini, 1972). NE turnover is increased in muricide and friendly rats, but is unchanged in indifferent animals. Dopamine turnover is significantly decreased only in friendly rats, which show tremors and uncoordinated hyperactivity (Valzelli, 1971).

The decrease in 5-HT turnover is particularly evident in the hemispheres of indifferent rats and in quadrigeminal bodies and in the posterior parts of hemispheres of muricide animals (Valzelli, 1971; Valzelli and Garattini, 1972). Other authors (Goldberg and Salama, 1969) have shown an increase of norepinephrine synthesis in the prosencephalon of muricide rats, but there are no published studies on 5-HT and DA turnover either in muricide or in friendly and indifferent animals.

It is interesting to observe that in muricide rats isolation does not induce any changes in brain levels of N-acetylaspartic acid as it does in aggressive mice (Valzelli, 1971).

IV. CONCLUSIONS

The isolation technique is one of the easiest and most reproducible methods of inducing behavioral alterations and aggressive responses in a number of laboratory animals. It seems possible that this has some relevance to altered behavioral patterns in human beings, as far as its psychosocial determinants are concerned.

Such a view is of particular interest when it is considered that brain neurochemical alterations present in isolated aggressive animals may be relevant to certain biochemical theories concerning alterations in the metabolism of brain monoamines in some mental diseases (Coppen, Eccleston and Peet, *This Volume*; Pare, Trenchard, and Turner, *This Volume*; Goodwin, *This Volume*; Lloyd, Farley, Deck, and Horny-kiewicz, *This Volume*; Wyatt, Gillin, Kaplan, Stillman, Mandel, Ahn, Vander Heuvel, and Walker, *This Volume*; Brune, 1967; Brune and Himwich, 1963; Himwich, Kety, and Smythies, 1967; Pollin, Cardon, and Kety, 1961; Schildkraut and Kety, 1967; Silverman, Cohen, Shmavonian, and Kirschen, 1961; Woolley and Shaw, 1954).

In this context, the alteration of brain 5-HT turnover present in both isolated aggressive mice and isolated muricide rats seems to assume a peculiar relevance, when it is considered that animals which do not become aggressive by isolation fail to manifest this change. However, one must consider that a functional interplay exists between the various brain neurotransmitters (Lloyd and Bartholini, 1974; Pepeu, Garau, and Mulas, 1974; Everett, 1974; Chase, *This Volume*) so that any modification of brain 5-HT turnover might affect brain NE, DA, and acetylcholine function. Thus it seems unlikely that a decrease in brain 5-HT turnover alone could be considered responsible for the aggressive behavior pattern resulting from prolonged isolation.

REFERENCES

Allee, W. C. (1942a): Group organization among vertebrates. *Science*, 95:289–293.

Allee, W. C. (1942b): Social dominance and subordination among vertebrates. *Biological Symposia*, 8:139–145.

Bayroff, A. G. (1936): Experimental social behavior of animals. I. Effect of early isolation of white rats on their later reactions to other white rats as measured by two periods of free choice. *Journal of Comparative Psychology*, 21:67–70.

Brune, G. G. (1967): Tryptophan metabolism in psychoses. In: *Amines and Schizophrenia*, edited by H. E. Himwich, S. S. Kety, and J. R. Smythies, pp. 87–96. Pergamon Press, Oxford.

Brune, G. G., and Himwich, H. E. (1963): Biogenic amines and behavior in schizophrenic patients. In: *Recent Advances in Biological Psychiatry*, vol. 5, edited by J. Wortis, pp. 144–160. Plenum Press, New York.

Clark, L. D. (1962a): Experimental studies of the behavior of an aggressive predatory mouse (*Onychomys leucogaster*). In: *The Roots of Behavior*, edited by E. L. Bliss, pp. 179–186. Harper & Row, New York.

Clark, L. D. (1962*b*): A comparative view of aggressive behavior. *American Journal of Psychiatry*, 119:336–341.

Di Chiara, G., Camba, R., and Spano, P. F. (1971): Evidence for inhibition by brain serotonin of mouse killing behavior in rats. *Nature* (London), 233:272.

Dominguez, M., and Longo, V. G. (1969): Taming effect of para-chlorophenyl-alanine on septal rats. *Physiology and Behavior*, 4:1031–1033.

Essman, W. B. (1969): "Free" and motivated behaviour and mice metabolism in isolated mice; In: *Aggressive Behaviour*, edited by S. Garattini and E. B. Sigg, pp. 203–208. Excerpta Medica Foundation, Amsterdam.

Everett, G. M. (1974): Effect of 5-hydroxytryptophan on brain levels of dopamine, norepinephrine, and serotonin in mice. In: *Advances in Biochemical Psychopharmacology*, Vol. 10, edited by E. Costa, G. L. Gessa, and M. Sandler. Raven Press, New York.

Fredericson, E. (1950): The effects of food deprivation upon competitive and spontaneous combat in C57 black mice. *Journal of Psychology*, 29:89–100.

Fredericson, E., and Birnbaum, E. A. (1954): Competitive fighting between mice with different hereditary background. *Journal of Genetic Psychology*, 85:271–280.

Garattini, S., Giacalone, E., and Valzelli, L. (1967): Isolation, aggressiveness and brain 5-hydroxytryptamine turnover. *Journal of Pharmacy and Pharmacology*, 19:338–339.

Garattini, S., Giacalone, E., and Valzelli, L. (1969): Biochemical changes during isolation-induced aggressiveness in mice. In: *Aggressive Behaviour*, edited by S. Garattini and E. B. Sigg, pp. 179–187. Excerpta Medica Foundation, Amster-dam.

Giacalone, E., Tansella, M., Valzelli, L., and Garattini, S. (1968): Brain serotonin metabolism in isolated aggressive mice. *Biochemical Pharmacology*, 17:1315–1327.

Goldberg, M. E., and Salama, A. I. (1969): Norepinephrine turnover and brain monoamine levels in aggressive mouse-killing rats. *Biochemical Pharmacology*, 18:532–534.

Hamilton, G. V. (1916): A study of perseverance reactions in primates and rodents. *Behavior Monograph*, 3:65–70.

Himwich, H. E., Kety, S. S., and Smythies, J. R. (1967): *Amines and Schizophre-nia*. Pergamon Press, Oxford.

Kahn, M. W. (1951): The effect of severe defeat at various age levels on the aggressive behavior of mice. *Journal of Genetic Psychology*, 79:117–130.

Kahn, M. W. (1954): Infantile experience and mature aggressive behavior of mice: Some maternal influences. *Journal of Genetic Psychology*, 84:65–75.

Kahn, W. H. (1961): The effect of socially learned aggression on submission mating behavior of C57 mice. *Journal of Genetic Psychology*, 98:211–217.

King, J. A. (1957): Relationship between early social experience and adult aggressive behavior in inbred mice. *Journal of Genetic Psychology*, 90:151–166.

King, J. A. and Gurney, N. L. (1954): Effect of early social experience on adult aggressive behavior in C57 BL/10 mice. *Journal of Comparative and Physiological Psychology*, 47:326–330.

Lloyd, K. G., and Bartholini, G. (1974): The effect of methiothepin on cerebral monoamine neurons. In: *Advances in Biochemical Psychopharmacology*, Vol. 10, edited by E. Costa, G. L. Gessa, and M. Sandler. Raven Press, New York.

McIntosh, J. C., and Cooper, J. R. (1965): Studies on the function of N-acetyl aspartic acid in brain. *Journal of Neurochemistry*, 12:825–835.

Marcucci, F., Mussini, E., Valzelli, L., and Garattini, S. (1968): Decrease in N-acetyl-L-aspartic acid in brain of aggressive mice. *Journal of Neurochemistry*, 15:53–54.

Pepeu, G., Garau, L., and Mulas, M. L. (1974): Does 5-hydroxytryptamine influence cholinergic mechanisms in the central nervous system? In: *Advances in*

Biochemical Psychopharmacology, Vol. 10, edited by E. Costa, G. L. Gessa, and M. Sandler. Raven Press, New York.

Pollin, W., Cardon, P. V., Jr., and Kety, S. S. (1961): Effects of amino acid feedings in schizophrenic patients treated with iproniazid. *Science*, 133:104–105.

Schildkraut, J. J., and Kety, S. S. (1967): Biogenic amines and emotion. Pharmacological studies suggest a relationship between brain biogenic amines and affective state. *Science*, 156:21–30.

Scott, J. P. (1946): Incomplete adjustment caused by frustration of untrained fighting mice. *Journal of Comparative Psychology*, 39:379–390.

Scott, J. P. (1958): *Aggression*. University of Chicago Press, Chicago.

Scott, J. P. (1962): Hostility and aggression in animals. In: *The Roots of Behavior*, edited by E. L. Bliss, pp. 167–178. Harper & Row, New York.

Seward, J. P. (1945a): Aggressive behavior in the rat. I. General characteristics, age and sex differences. *Journal of Comparative Psychology*, 38:175–197.

Seward, J. P. (1945b): Aggressive behavior in the rat. II. An attempt to establish a dominance hierarchy. *Journal of Comparative Psychology*, 38:213–238.

Seward, J. P. (1946): Aggressive behavior in the rat. IV. Submission as determined by conditioning extinction and disuse. *Journal of Comparative Psychology*, 39:51–75.

Silverman, A. J., Cohen, S. I., Shmavonian, B. M., and Kirshen, N. (1961): Neurochemical bases of mental illness. In: *Recent Advances in Biological Psychiatry*, Vol. 3, edited by J. Wortis, pp. 104–115. Plenum Press, New York.

Valzelli, L. (1967a): Drugs and aggressiveness. *Advances in Pharmacology*, 5:79–108.

Valzelli, L. (1967b): Biological and pharmacological aspects of aggressiveness in mice. In: *Neuropsychopharmacology*, Proceedings of the 5th C.I.N.P. Congress, edited by H. Brill, J. O. Cole, P. Deniker, H. Hippius, and P. B. Bradley, pp. 781–788. Excerpta Medica Foundation, Amsterdam.

Valzelli, L. (1971): Aggressivité chez le rat et la souris: Aspects comportementaux et biochimiques. *Actualités Pharmacologiques*, 24:133–152.

Valzelli, L. (1973): Environment influences upon neural and metabolic processes related to learning and memory. In: *Biochemical Bases of Learning and Memory*, edited by W. B. Essman and S. Nakajima. Spectrum Publications Inc., New York.

Valzelli, L., and Garattini, S. (1968): Behavioral changes and 5-hydroxytryptamine turnover in animals. *Advances in Pharmacology*, 6B:249–260.

Valzelli, L., and Garattini, S. (1972): Biochemical and behavioural changes induced by isolation in rats. *Neuropharmacology*, 11:17–22.

Welch, B. L. (1965): Psychophysiological response to the mean level of environmental stimulation: A theory of environmental integration. In: *Symposium of Medical Aspects of Stress in the Military Climate*, edited by D. Mck. Rioch, pp. 39–96. U.S. Government Printing Office, Washington, D.C.

Welch, B. L. (1967): Discussion of the paper by A. B. Rothballer, "Aggression defense and neurohumors." In: *Aggression and Defense. Neural Mechanism and Social Patterns (Brain Function, Vol. V)*, edited by C. D. Clemente and D. B. Lindsley, pp. 150–162. University of California Press, Berkeley.

Welch, B. L., and Welch, A. S. (1966): Graded effect of social stimulation upon D-amphetamine toxicity, aggressiveness and heart and adrenal weight. *Journal of Pharmacology and Experimental Therapeutics*, 151:331–338.

Woolley, D. W., and Shaw, E. (1954): A biochemical and pharmacological suggestion about certain mental disorders. *Proceedings of the National Academy of Sciences*, 40:228–231.

Yen, C. Y., Stanger, R. L., and Millman, N. (1959): Ataractic suppression of isolation-induced aggressive behavior. *Archives Internationales de Pharmacodynamie et de Thérapie*, 123:179–185.

Advances in Biochemical Psychopharmacology, Vol. 11
Raven Press, New York © 1974

Brain 5-Hydroxytryptamine and Memory Consolidation

Walter B. Essman

Queens College of the City University of New York, Flushing, New York
11367

I. INTRODUCTION

There has been some evidence to support a role for brain 5-hydroxytryptamine (5-HT) in both learning and the processing of acquired information. Such evidence derives largely from studies in which either impaired learning ability in animals or interference with the consolidation of a behavior has been associated with changes in brain 5-HT concentration and/or metabolism (Woolley, 1965; Essman, 1970*a*). More specifically, several agents and/or events capable of interfering with either task acquisition proactively or task retention retroactively have been associated with brain 5-HT alterations. Such anterograde or retrograde amnesias have been demonstrated both in animals and man with such conditions as hypothermia, diethyl ether anesthesia, hypoxia, or electroconvulsive shock (ECS), and it is of interest to note that a common denominator of such agents is their ability to elevate brain 5-HT level (Essman, 1971*a*). The relationship between the retrograde amnesic properties of specific physical agents such as ECS and brain 5-HT concentration and metabolism have been considered in a number of earlier studies (Essman, 1970*a*, 1971*a*, 1973*a*); a very compelling argument in favor of the hypothesis that disruption of the memory consolidation process eventuating in a retrograde amnesia for those stimulus events thereby disrupted may be related to 5-HT changes produced by the amnesic stimulus through the observation that the incidence of ECS-induced retrograde amnesia depends upon the ability of ECS to elevate forebrain 5-HT (Garattini, Valsecchi, and Valzelli, 1957; Essman, 1968*a*). This has been demonstrated in developmental studies wherein at a critical age in the CF-1S mouse (17 days), where endogenous forebrain 5-HT level is low and 5-HT turnover is high, post-training ECS failed significantly to elevate forebrain 5-HT level, modify its turnover, or produce any appreciable incidence of retrograde amnesia. This effect could not be observed in younger or older animals, where the same treatment resulted in both 5-HT changes as well as a retrograde amnesia for the passive avoidance response to which these animals were trained (Essman, 1970b).

We have previously indicated that a retrograde amnesia for an otherwise stable conditioned response could be produced in mice by either intracranial or intraventricular administration of small amounts of exogenous 5-HT leading to elevated tissue 5-HT levels consistent with those produced by some of the more standard amnesic treatments (Essman, 1970a, 1973b). It has also been observed that one site where such an effect appears reliably to elevate brain 5-HT as well as producing a retrograde amnesic effect is the medial portion of the hippocampus, a brain region consistently implicated by electrophysiological, ablation, and electrical stimulation studies as being relevant to memory processing.

One purpose of the present communication has been to consider several aspects of the relationship between intrahippocampal 5-HT, memory processing, and those neurochemical events with which these may be associated. More specifically, as we have previously considered for ECS, there is a considerable basis upon which the role of 5-HT in memory consolidation may be related to protein synthesis in the brain (Essman, 1971b, 1973c). One aspect of the relationship between memory processing and brain protein synthesis has been approached methodologically utilizing antimetabolites, which produce extensive inhibition of cerebral protein synthesis, to interrupt either the processing of memory, its storage, or possibly its retrieval (Agranoff, 1969; Barondes, 1970; Flexner and Flexner, 1966). One problem inherent in this approach has been that the effect upon proteins may be unrelated to the memory phenomenon and an apparently marked resistance of the protein synthesis process to such disruption at sites where it could indeed be relevant to memory consolidation. The approach considered herein has been concerned with an emphasis upon 5-HT and cerebral protein synthesis within the limits of those temporal factors by which 5-HT changes in brain relate to the course of postexperiential memory consolidation.

II. CHANGES IN BRAIN 5-HT AND AMNESIC TREATMENT

An early demonstration that one amnesic event, electroconvulsive shock, is capable of changing brain 5-HT was derived from studies in which a biphasic temporal effect of such stimulation was demonstrated (Garattini, Valsecchi, and Valzelli, 1957). Subsequently, it was shown that a *single* ECS treatment capable of effecting a retrograde amnesia for a conditioned avoidance response in mice was also capable of elevating brain 5-HT level (Essman, 1968a), and modifying brain metabolism (Essman, 1968b). It has also been shown that several pharmacological agents which block ECS-induced brain 5-HT elevation are also capable of significantly attenuating the amnesic effect of ECS (Essman, Steinberg, and Golod, 1968; Essman, 1971a). Changes in brain 5-HT have also been related to alterations in related cerebrospinal fluid constituents, at times beyond those at which any behavioral effect was in evidence. For example, in both monkey and man, a single ECS or multiple ECS treatment has been shown to elevate tryptophan and 5-hydroxyindoleacetic levels in the cerebrospinal fluid (Essman, 1973d). Such changes have, to some extent, been relatable to alterations in brain 5-HT content and metabolism by such amnesic treatment. It is of some interest to note that the extent of an ECS-induced retrograde amnesia seems to depend more intimately upon the degree to which such treatment does, indeed, modify brain 5-HT; aside from the developmental differences mentioned earlier in this paper, we have observed that the elevation of brain 5-HT in the mouse is highly dependent upon the site at which the electrodes are placed and the path of the current delivered; in this same regard, the incidence of ECS-induced retrograde amnesia also depends upon electrode placement to the extent that bitemporal current delivery results in the most highly consistent incidence of retrograde amnesia and also produces the most appreciable degree of forebrain 5-HT elevation (34%) within 20 min following stimulation. Frontal electrode placement and current delivery produced a lowered incidence of retrograde amnesia (55%) with a reduced magnitude of subsequent brain 5-HT elevation (11%). In the majority of our studies with ECS, transcorneal electrode placement and current delivery was chosen, inasmuch as this provided for a consistent degree of amnesic potency (approximately 90%) as well as a statistically significant elevation of forebrain 5-HT.

In Table 1, a summary of some of those changes in brain 5-HT content and metabolism following a single ECS have been presented. These measures were taken at a time (15 min) at which peak elevation of forebrain 5-HT content occurred following a single ECS (20 mA, 200 msec, 400 V) delivered to male CF-1S mice through transcorneally applied salinized electrodes. Forebrain 5-HT and 5-hydroxyindoleacetic acid (5-HIAA) concentrations were measured utilizing standard techniques (Welch and Welch, 1969), and 5-HT turnover was estimated under identical treatment conditions at several time intervals following tranylcypromine-induced monoamine oxidase inhibition. These results, consistent with several previous findings from our laboratory, have indicated that the initial peak response to a single ECS is represented by increased 5-HT ($p < 0.02$), decreased 5-HIAA ($p < 0.01$) and decreased 5-HT turnover ($p < 0.02$).

TABLE 1. *Mean (± 0) level of forebrain 5-HT and measures of its metabolism in sham ECS ($\overline{ECS}$) and electroconvulsive shock-treated (ECS) mice*

Determination	$\overline{ECS}$	ECS
5-HT (μg/g)	0.78 (0.10)	0.91 (0.16)
5-HIAA (μg/g)	0.46 (0.12)	0.22 (0.08)
5-HT turnover rate (μg/g/hr)	0.32 (0.14)	0.37 (0.11)
5-HT turnover time (min)	58.86 (19.22)	81.81 (15.44)

All values were obtained 15 min after treatment.

Some further consideration was given to the issue of the potential sites at which the effects of ECS upon brain 5-HT might possibly be related to the disruption of memory consolidation. One such site is the synaptic region, which may be methodologically examined through the preparation of the isolated presynaptic nerve ending (synaptosome). Mice were given a single transcorneal ECS, as described above, and were killed by cervical dislocation at 15 min following treatment. Brain tissue was dissected under cold conditions and cerebral cortex, medulla, and cerebellum isolated. These regions were prepared following homogenization in 0.32 M sucrose with differential and density gradient centrifugation to provide for a synaptosome fraction (Whittaker, 1969). The synaptosome preparation was incubated with ^{14}C-5-HT for 20 min; the synaptosomes were pelleted, washed several times, and then resuspended for subsequent measurement of 5-HT incorporation. The data obtained have been summarized in Table 2, where it is apparent that in all three regions of the brain considered, a significant reduction of 5-HT incorporation occurred *in vitro* for the tissue derived from those mice treated with ECS *in vivo* as compared with control animals receiving only sham ECS ($\overline{ECS}$). These data indicated quite convincingly that 5-HT uptake by the isolated presynaptic nerve ending has been altered by changes in that organ induced by a single electroconvulsive shock.

With some consideration given generally to the issue of reliable changes in brain 5-HT to at least one amnesic event, ECS, it might be appropriate to consider more specifically, on a behavioral level, what relationship between ECS-induced 5-HT changes and the direct effects of 5-HT per se may bear upon the issue of memory consolidation and/or retrograde amnesia.

TABLE 2. *Incorporation of* ^{14}C-*5-hydroxytryptamine (cpm/mg original tissue) into synaptosomes derived from several brain regions following a single electroconvulsive shock*

	^{14}C-5-hydroxytryptamine incorporation	
Region	$\overline{ECS}$	ECS
Cerebral cortex	98	10[a]
Midbrain	72	7[a]
Cerebellum	91	19[a]

[a] $p < 0.01$.

III. RETROGRADE AMNESIA: EFFECTS OF ECS AND 5-HT AND THEIR TEMPORAL CHARACTERISTICS

A considerable number of studies in which ECS has been used as an amnesic event have pointed toward a rather consistent observation that this treatment, like several other amnesic stimuli, provides for a temporal gradient for retrograde amnesia; that is, as a time interval between a learning situation or training event and the presentation of the amnesic event is increased, the incidence of subsequently demonstrated retrograde amnesia is decreased. Therefore, the most effective utilization of amnesic agents or events appears to be under conditions where they are presented as proximally as possible in time with the task for which retention is to be subsequently evaluated.

In several experiments, this classic relationship was again evaluated, wherein a single ECS treatment (400 V, 20 mA, 200 msec) was given at several time intervals after training to groups of mice conditioned to a simple passive avoidance response with a single training trial (Essman and Alpern, 1964). The retention of the conditioned avoidance response was tested 24 hr following ECS treatment; this was the time at which it had been established in previous studies that no behavioral or biochemical effects of a single treatment were in evidence. A parallel experiment was carried out in which at several times following training for acquisition of the avoidance behavior, mice were given an injection of 2 μg of 5-HT directly into the medial hippocampus, the amine was delivered in 5 μl of 0.9% NaCl through a 30-gauge stylette-mounted needle, perforating the skull over predesignated landmarks after localization of the region to be affected (Essman, 1973c). The dose of

5-HT utilized was based upon measures of the relationship between intracranially administered exogenous 5-HT and the elevation of forebrain 5-HT levels; at the dose utilized in the previous studies, there was an 89% increase in 5-HT content of the forebrain tissue over the first 10 min following injection, with a significant increase in tissue levels of the amine persisting as long as 30 min. An injection of 5-HT as described was given at 10 sec (0) or 1, 2, 4, 8, 16, or 32 min following a training trial. Control animals were given an equivalent volume, injected intrahippocampally with 0.9% NaCl. All animals were tested for retention of the conditioned avoidance response 24 hr following training. At this time, there was no evidence to suggest any overt behavioral difference between experimental and control animals. The results of postconditioning ECS treatment or 5-HT injection have been summarized in Fig. 1. These data indicate that unlike control animals for each experimental series (ECS, or intrahippocampal NaCl), both ECS or intrahippocampal 5-HT treatments following a single conditioning experience provided for a time-related retrograde amnesia for that conditioned response. It has not been the purpose of this study to evaluate the efficacy of ECS versus 5-HT, but to illustrate that a temporal gradient for retrograde amnesia can be effected by both treatments over comparable time intervals. Since such post-treatment times were not coincident for both treatment conditions, it further seems inappropriate to attempt any direct comparison of efficacy.

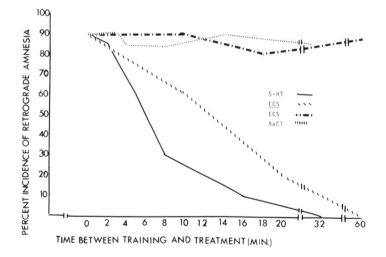

FIG. 1. Percent incidence of retrograde amnesia among mice given postconditioning electroconvulsive shock or intracranial 5-HT.

One issue relevant to the present study concerns the specificity of 5-HT for the amnesic effect observed. To evaluate this more extensively, a similar behavioral study was initiated wherein postconditioning treatment at 10 sec consisted of the intrahippocampal injection of 0.9% NaCl, 5-HT (2 μg) or equimolar concentrations of either norepinephrine, N-acetyl-5-HT, or 5-methoxy-1-methyltryptamine, given in a volume of 5 μl. When tested for retention of the conditioned avoidance response at 24 hr following training, only 5-HT-injected mice showed a significant incidence of retrograde amnesia ($p < 0.001$), as compared with groups of animals treated intracranially with related analogues or derivatives. Under the latter treatment conditions, from 80 to 100% of the animals in each group showed retention of the conditioned avoidance response, whereas 80% of the 5-HT-treated animals showed a retrograde amnesia for that response.

One related question concerning the potential amnesic effect of 5-HT given directly into hippocampal tissue is the extent to which this treatment might produce other changes which could also explain the amnesic effect. There are two phenomena which are capable of producing a retrograde amnesia which have suggested themselves as possibilities. One of these is the induction of seizure activity by hippocampal injection and the other is the possibility that such treatment might initiate spreading cortical depression, possibly through an increase in extracellular potassium. Both these possibilities were investigated in mice in which bilateral electroencephalographic records were monitored during intrahippocampal injection of 5-HT. Under these conditions, a basal record of approximately 13 cps at an amplitude of ~50 μV showed no alteration in frequency, but a 20% amplitude reduction persisting for approximately 240 sec following injection. Clearly, these electroencephalographic findings would not support either cortical seizure activity or cortical spreading depression as possible consequences of the 5-HT treatment, and therefore seem excludable as primary or secondary principles in the amnesic phenomenon.

IV. BRAIN 5-HYDROXYTRYPTAMINE AND THE INHIBITION OF CEREBRAL PROTEIN SYNTHESIS

Previous indications have suggested that the synthesis of cerebral proteins may be inhibited when 5-HT is injected intracranially, intraventricularly, or intrahippocampally in mice. This effect has also been demonstrated *in vitro* in that protein synthesis, as measured by incorporation of ^{14}C-leucine into isolated synaptosomes derived from within the cerebral cortex or structures of the limbic system, were inhibited by 26 to 32%, respectively, by 5-HT (3×10^{-7} M).

Inasmuch as previous behavioral studies have indicated a specificity of the retrograde amnesic effect of 5-HT that could not be demonstrated with related molecules, it was of some interest to consider whether such specificity for an inhibitory effect upon cerebral protein synthesis also obtained. Groups of CF-1S male mice were injected intrahippocampally with either 0.9% NaCl, 2 μg of 5-HT, or equimolar concentrations of either norepinephrine, N-acetyl-5-HT, or 5-methoxy-1-methyltryptamine in a volume of 5 μl of 0.9% NaCl. At 5 min post-treatment all animals were injected, that is, with ^{14}C-leucine (0.1 μC of 165 μC/μM/8, brought to a final concentration of 0.66 mM with the addition of "cold" leucine), and a 5-min labeling pulse was terminated by killing the mice and rapidly freezing the excised brain tissue from which the protein was precipitated and incorporated, and label counted. Several brain regions were isolated and with saline treatment as a baseline, changes in protein synthesis were estimated from differences between control and amine treatments in the ^{14}C-leucine incorporation into protein. These data have been summarized in Table 3. It is quite clear that only 5-HT treatment brought about a significant inhibition of synthesis for those regions measured except for the cerebellum. Consistent with the previously noted behavioral data, only intrahippocampal 5-HT significantly altered regional protein synthesis.

Several further studies were carried out to assess the effect of intrahippocampal 5-HT upon protein synthesis in subcellular fractions of the cerebral cortex of mice. Animals were treated with 2 μg of 5-HT, injected directly into the medial hippocampus as described before, and control animals were similarly injected with 0.9% NaCl. At 5 min following injection, ^{14}C-leucine was injected intracranially (0.1 C of 165 C/M/g with added nonlabeled leucine and a final concentration of

TABLE 3. *Percent inhibition of protein synthesis measured for several regions of the mouse brain after intrahippocampal amine injection*

Intrahippocampal treatment	Brain region			
	Cerebral cortex	Basal ganglia and diencephalon	Midbrain	Cerebellum
5-HT	50[a]	55[a]	46[a]	8
NE	5	0	0	0
N-Ac-5-HT	8	0	2	0
5-Me-1-MT	3	0	2	0

[a]$p < 0.01$.

TABLE 4. *The effect of 5-HT on the incorporation of ^{14}C-leucine into different subcellular fractions of the cerebral cortex*

Subcellular fraction	Incorporation of ^{14}C-leucine into protein (cpm/mg protein)		Percent inhibition
	Saline	5-HT	
Whole homogenate	144	116	19
Microsomes	264	213	19
Mitochondria	68	53	20
Synaptosomes	38	25	34
Soluble protein	154	121	21

0.66 mM). A 5-min labeling pulse was thereby given, obviating the possibility of both isotopic dilution or dilution through distribution to other tissues. Animals were killed by cervical dislocation, and the cerebral cortex was freed from underlying myelin and homogenized and fractionated to provide for several subcellular constituents. The results summarized in Table 4 indicate the regional as well as subcellular organal differences in the incorporation of ^{14}C-leucine into proteins as a consequence of 5-HT treatment. It may be noted that maximal inhibition occurred in the synaptosome fraction which, based upon enzymatic studies, appeared relatively free of microsomal contamination. This finding strongly supports the role of 5-HT in the mediation of an inhibitory event that closely parallels the behavioral effect observed for this amine. The similarity in the specificity, both behaviorally and biochemically, of 5-HT, and the significance of its synaptosomal effect suggest its potential import for modulation of memory consolidation events.

REFERENCES

Agranoff, B. W. (1969): Macromolecules and brain function. In: *Progress in Molecular and Subcellular Biology*, edited by F. E. Hahn, pp. 203–212. Springer-Verlag, New York.

Barondes, S. H. (1970): Cerebral protein synthesis inhibitors block long-term memory. *International Review of Biology*, 12:177–205.

Essman, W. B. (1968a): Electroshock induced retrograde amnesia and brain serotonin metabolism: Effect of several antidepressant compounds. *Psychopharmacologia*, 13:258–266.

Essman, W. B. (1968b): Changes in ECS-induced retrograde amnesia with DBMC: Behavioral and biochemical correlates of brain serotonin antagonism. *Physiology and Behavior*, 3:527–532.

Essman, W. B. (1970a): Some neurochemical correlates of altered memory consolidation. *Transactions of the New York Academy of Sciences*, 32:948–973.

Essman, W. B. (1970*b*): The role of biogenic amines in memory consolidation. In: *The Biology of Memory*, edited by G. Adam,   pp. 213–238. Akademiai Kiado Publishers, Budapest.

Essman, W. B. (1971*a*): Drug effects and learning and memory processes. In: *Advances in Pharmacology and Chemotherapy*, edited by S. Garattini, A. Goldin, F. Hawking, and I. J. Kopin, pp. 241–330. Academic Press, New York.

Essman, W. B. (1971*b*): Neurochemical changes associated with ECS and ECT. *Seminars in Psychiatry*, 4:67–70.

Essman, W. B. (1973*a*): Neuromolecular modulation of experimentally induced retrograde amnesia. *Confinia Neurologica*, 35:1–22.

Essman, W. B. (1973*b*): Age dependent effects of 5-hydroxytryptamine upon memory consolidation and protein synthesis. *Pharmacology, Biochemistry and Behavior*, 1:7–14.

Essman, W. B. (1973*c*): Effects of ECS on cerebral protein synthesis. In: *The Psychobiology of ECT*, edited by M. Fink, S. S. Kety, J. McGaugh, and T. Williams. V. H. Winston, Washington, D.C.

Essman, W. B. (1973*d*): *Neurochemistry of Cerebral Electroshock*. Spectrum Publishers, New York.

Essman, W. B., and Alpern, H. (1964): Single trial learning: Methodology and results with mice. *Psychological Reports*, 15:731–740.

Essman, W. B., Steinberg, M. I., and Golod, M. I. (1968): Alterations in the behavioral and biochemical effects of electroconvulsive shock with nicotine. *Psychonomic Science*, 12:107–108.

Flexner, L. B., and Flexner, J. B. (1966): Effect of acetoxycycloheximide and of an acetoxycycloheximide-puromycin mixture on cerebral protein synthesis and memory in mice. *Proceedings of the National Academy of Sciences*, U.S., 55:396–374.

Garattini, S., Valsecchi, A., and Valzelli, L. (1957): Variations in encephalic and intestinal serotonin after electrical shock. *Experientia*, 13:330.

Welch, A. S., and Welch, B. L. (1969): Solvent extraction method for simultaneous determination of norepinephrine, dopamine, serotonin, and 5-hydroxy-indoleacetic acid in a single mouse brain. *Analytical Biochemistry*, 30:161–179.

Whittaker, V. P. (1969): The synaptosome. In: *Handbook of Neurochemistry*, Vol. 2, edited by A. Lajtha. Plenum Press, New York.

Woolley, D. W. (1965): A method for demonstration of the effects of serotonin on learning ability. In: *Pharmacology of Conditioning, Learning and Retention*, edited by M. Ya. Mikhel'son and V. G. Longo. Pergamon Press, Oxford.

Advances in Biochemical Psychopharmacology, Vol. 11
Raven Press, New York © 1974

5-Hydroxytryptamine in Depression

C. M. B. Pare, Anne Trenchard, and Paul Turner

St. Bartholomew's Hospital, London, England

I. INTRODUCTION

There are two main ways in which a clinician can approach the biochemical basis for depression. The first is to investigate directly possible abnormalities in the patient, such as other authors in this volume have described, and then to assess whether these have a causal relationship to the depression or are a secondary manifestation.

Another way is to attempt to correlate the effects of different treatments with the patient's clinical response. One obvious example is to show that in patients dying from various terminal illnesses who were receiving monoamine oxidase inhibitors (MAOIs), brain amine levels as judged by 5-hydroxytryptamine (5-HT) and norepinephrine (NE) rise at about the time when an antidepressant effect would be expected (Maclean, Nicholson, Pare, and Stacey, 1965; Bevan-Jones, Pare, Nicholson, Price, and Stacey, 1972). Another example would be to demonstrate the potentiation of MAOIs by administration of tryptophan with the inference that a tryptophan derivative is important in the antidepressant effect of these drugs (Coppen, Shaw, and Farrell, 1963; Pare, 1963).

II. CLINICAL RESPONSE AND NONRESPONSE TO ANTIDEPRESSANTS

A more difficult task is to investigate the type of patient who responds to a particular antidepressant drug. An instance is the English Medical Research Council trial of imipramine, phenelzine, E.C.T., and placebo; it was found that imipramine and E.C.T. were much more effective than phenelzine in relieving depression. This does not mean, however, that phenelzine is ineffective in depression, but that in the rather low dosage used phenelzine was ineffective in those patients selected for the trial, that is, patients between the ages of 45 and 69 years with a short illness, who were ill enough to be admitted to a mental hospital and furthermore were the type of depressive suitable for E.C.T., which was one of the alternative treatments (Medical Research Council, 1965). The mass of depressed patients treated by psychiatrists do not fall into this category. The degree of depression is often moderate, and most patients can be treated as outpatients. The fact that the MAOIs are much more effective in this group of patients has led people to say that MAOIs are mild antidepressants and good only for mild depressions. However, in this great majority of depressed patients, we have always been impressed by the fact that one patient may respond dramatically to one type of antidepressant, say an MAOI, and show no improvement at all to a tricyclic, and vice versa, although there is a strong tendency for the patient to respond in the same way, improvement or no improvement, to drugs from the same antidepressant group. Furthermore, although there is a tendency for the more typical endogenous cases to respond to tricyclic antidepressants, clinically identical patients may respond quite differently to the same drug (Pare, 1965).

III. GENETICAL FACTORS AFFECTING THE RESPONSE TO ANTIDEPRESSANTS

By studying depressed patients in the same family, Angst and ourselves have suggested that this differing response might be due to different genetic types of depression breeding true in the same family, one type responding to MAOIs and another to tricyclic antidepressants (Angst, 1961, 1964; Pare, Rees, and Sainsbury, 1962; Pare and Mack, 1971).

Another explanation would be that patients metabolize the antidepressant drugs differently, some patients inactivating the drug quickly and thereby being less likely to respond clinically to a standard dose.

Alexanderson, Evans, and Sjöqvist (1969) demonstrated the genetic basis for this in the case of nortriptyline, and Braithwaite and his colleagues have suggested that in the case of amitriptyline, a therapeutic response depends on attaining a plasma level of amitriptyline and nortriptyline of approximately 120 μg/ml (Braithwaite, Goulding, Theano, Bailey, and Coppen, 1972). With regard to the MAOIs, Evans, Davidson, and Pratt (1965) first suggested that phenelzine might be acetylated slowly or rapidly on a genetic basis similarly to isoniazid, and recently Johnstone and Marsh (1973) showed that significantly greater improvement occurred in neurotic depressives who were slow acetylators of phenelzine as compared to rapid acetylators. This varied rate of metabolism of the antidepressant drug should show itself in a quantitative difference of tissue response to a standard dose of drug, and 2 years ago we pointed out a similar wide scatter of brain 5-HT levels after MAOIs had been given to patients dying from various terminal illnesses (Bevan-Jones et al., 1972). We did not use phenelzine, and the scatter was not bimodal.

Thus the response of a patient to an antidepressant drug may be related not only to the type of depressive illness but also to the rate of metabolism of the drugs and their varied biochemical actions. For these reasons we feel that an estimation of tissue response is important as a measure not only of drug activity, correlated with drug levels in the blood, but also of a possible abnormality in the target organ of the depressed patient.

IV. A MEASURE OF TISSUE RESPONSE TO TRICYCLIC ANTIDEPRESSANTS

We have started a research program to examine the uptake of biogenic amines in platelet-rich plasma from depressed patients receiving either chlorimipramine or protriptyline. Yates, Todrick, and Tait (1964) have shown *in vitro* that both of these tricyclic antidepressant drugs inhibit the uptake of 5-HT in human platelets. Lingjaerde (1971) has demonstrated that this uptake can also be reduced by lowering the incubation temperature from 37°C to 4°C. This indicates the involvement of an energy-dependent metabolic process. Preliminary experiments have shown that the uptake of dopamine in platelet-rich plasma is similarly temperature dependent. Norepinephrine, on the other hand, was taken up only poorly, and its rate of uptake was not influenced by the incubation temperature. This indicates that, under the experimental conditions used, it enters the platelets by physical diffusion only. Research has been concentrated, therefore, on the uptake of dopamine and 5-HT. This is dose related, but 5-HT is taken up more readily than dopamine. When tritiated solutions containing 5 $\mu C/\mu$mole of amine were used, it was found that it required about ten times the concentration of dopamine and an incubation time of 60 min instead of 15 min to achieve comparable uptake levels to 5-HT.

Subjects selected for this study are depressed patients who have never been treated with antidepressants and whose other medication is limited to diazepam, usually for sleep. They are randomly allocated to receive either chlorimipramine or protriptyline. Chlorimipramine is given in doses of 75 mg at night for the first week and then 25 mg t.d.s. and 75 mg at night thereafter. Protriptyline is given 15 mg at night for the first week and afterward 5 mg t.d.s. and 15 mg at night.

To date amine uptake studies have been carried out on six patients. Three patients received chlorimipramine and three patients protriptyline. It should be emphasized that the results obtained so far are only preliminary, and their validity has yet to be established by a more extensive study.

The uptake in platelet-rich plasma from untreated depressed patients was low for both amines in five of six patients when compared with the uptake in platelet-rich plasma from normal, healthy volunteers. During the first 2 weeks of treatment, plasma samples from two of the three patients on chlorimipramine showed a further reduction in 5-HT uptake, but there was little change in the uptake of dopamine into platelets. Platelet-rich plasma samples from patients on protriptyline did not show any further reduction in amine uptake. In fact, an increase in uptake of both amines was observed on the 14th day in one patient and on the 20th and 24th days, respectively, for the other two patients. This increase appeared to be transient, since it was not maintained in two of the patients. No further blood samples were obtainable from the other patient.

ACKNOWLEDGMENTS

This research is being supported by a generous grant from the Mental Health Research Fund.

REFERENCES

Alexanderson, B., Evans, D. A. P., and Sjöqvist, F. (1969): Steady state plasma levels of nortriptyline in twins: Influence of genetic factors and drug therapy. *British Medical Journal*, 4:764–768.

Angst, J. (1961): A clinical analysis of the effects of Tofranil in depression. Longitudinal and follow up studies. Treatment of blood relations. *Psychopharmacologia*, 2:381–407.

Angst, J. (1964): Antidepressiver Effekt und genetische Faktoren. *Arzneimittel-Forschnung*, 14:496–500.

Bevan-Jones, B., Pare, C. M. B., Nicholson, W. J. N., Price, K., and Stacey, R. S. (1972): Brain amine concentrations after monoamine oxidase inhibitor administration. *British Medical Journal*, 1:17–19.

Braithwaite, R. A., Goulding, R., Theano, G., Bailey, J., and Coppen, A. (1972): Plasma concentration of amitriptyline and clinical response. *Lancet*, 1:1297–1300.

Coppen, A. J., Shaw, D. M., and Farrell, J. P. (1963): Potentiation of the antidepressive effect of a monoamine oxidase inhibitor by tryptophan. *Lancet*, 1:79–81.

Evans, D. A. P., Davidson, K., and Pratt, R. T. C. (1965): Influence of acetylator phenotype on the effects of treating depression with phenelzine. *Clinical Pharmacology and Therapeutics*, 6:430–435.

Johnstone, E. C., and Marsh, W. (1973): Acetylator status and response to phenelzine in depressed patients. *Lancet*, 1:567–570.

Lingjaerde, O. (1971): Uptake of serotonin into blood platelets *in vitro*. 1. The effects of chloride. *Acta Physiologica Scandinavica*, 81:75–83.

Maclean, R., Nicholson, W. J. N., Pare, C. M. B., and Stacey, R. S. (1965): Effect of monoamine oxidase inhibitors on the concentrations of 5-hydroxytryptamine in the human brain. *Lancet*, 2:205–208.

Medical Research Council (1965): Report by Clinical Psychiatry Committee. *British Medical Journal*, 1:881–886.

Pare, C. M. B. (1963): Potentiation of monoamine oxidase inhibitors by tryptophan. *Lancet*, 2:527–528.

Pare, C. M. B. (1965): Some clinical aspects of antidepressant drugs. In: *The Scientific Basis of Drug Therapy in Psychiatry*, edited by J. Marks and C. M. B. Pare. Pergamon Press, London.

Pare, C. M. B., and Mack, J. W. (1971): Differentiation of two genetically specific types of depression by the response to antidepressants. *Journal of Medical Genetics*, 8:306–309.

Pare, C. M. B., Rees, L., and Sainsbury, M. J. (1962): Differentiation of two genetically specific types of depression by the response to antidepressants. *Lancet*, 2:1340–1343.

Yates, C. M., Todrick, A., Tait, A. C. (1964): Effect of imipramine and some analogues on the uptake of 5-HT by human blood platelets *in vitro*. *Journal of Pharmacy and Pharmacology*, 16:460–463.

Advances in Biochemical Psychopharmacology, Vol. 11
Raven Press, New York © 1974

Serotonin and Behavioral Inhibition

Larry Stein and C. David Wise

Wyeth Laboratories, Philadelphia, Pennsylvania 19101

More than 15 years ago, Brodie and Shore (1957) suggested that norepinephrine and serotonin may serve as chemical mediators of physiologically antagonistic systems in the brain for the control of a number of different central functions. Recent evidence suggests that norepinephrine and serotonin indeed exert opposing effects on temperature regulation, sleep, sexual behavior, and aggressive behavior (see Feldberg and Myers, 1964; Jouvet and Pujol, *This Volume*; Meyerson, Carrer, and Eliasson, *This Volume*; Gessa and Tagliamonte, *This Volume*; Valzelli, *This Volume*; Everett 1974). Here we review evidence which suggests that norepinephrine and serotonin also exert opposing effects on operant or goal-directed behavior (Stein, 1968; Wise and Stein, 1970; Wise, Berger, and Stein, 1970, 1973). Specifically, we propose that norepinephrine neurons comprise an important component of a behaviorally facilitatory reward system and that serotonin neurons form part of a behaviorally suppressant punishment system.

Recent histochemical work suggests that noradrenergic neurons may be organized into two ascending systems: a dorsal pathway originating in the locus coeruleus, which mainly innervates the cerebral cortex and hippocampus, and a ventral pathway originating in the reticular formation of the lower brainstem which mainly innervates the hypothalamus and ventral parts of the limbic system (Fuxe, Hökfelt, and Ungerstedt, 1970). Both pathways appear to mediate rewarding effects (Crow, Spear, and Arbuthnott, 1972; Ritter and Stein, 1972, 1973; Arbuthnott, Fuxe, and Ungerstedt, 1971), but their differential distribution suggests different functions: the ventral branch may regulate mainly motivational activities, whereas the dorsal branch may regulate mainly cognitive activities (Stein and Wise, 1971).

The cells of the serotonergic punishment system originate in the raphe nuclei and distribute extensively in the central gray of the midbrain; in addition, serotonin fibers ascend in the medial forebrain bundle and distribute in the forebrain in roughly parallel fashion to those of the noradrenergic system (Fuxe, 1974). Both systems richly innervate periventricular regions of the brain; however, serotonin terminals tend to be concentrated in the central gray of the midbrain, whereas noradrenergic terminals tend to be concentrated in hypothalamic and limbic forebrain regions that surround the third and lateral ventricles (Fuxe, 1965; Aghajanian and Bloom, 1967).

To test the idea that goal-directed behavior is reciprocally regulated by norepinephrine and serotonin, we manipulated the activities of these substances in the brain by various means and measured the effects of these changes in two behavioral tests. One emphasized facilitation of behavior by reward (brain self-stimulation), the other, suppression of behavior by punishment (passive avoidance or "conflict").

I. SELF-STIMULATION OF THE BRAIN

The self-stimulation test of Olds and Milner (1954) was used to measure the effects of central injections of norepinephrine and serotonin on reward function. The norepinephrine antagonists, phentolamine and propranolol, were also studied in order to assess the relative importance of α- and β-noradrenergic activity in the regulation of self-stimulation. Bipolar platinum electrodes for brain stimulation were stereotaxically implanted in the medial forebrain bundle, and cannulas for injection of solutions were implanted in the lateral ventricle on the side opposite the electrode in all rats. The details of method are described by Wise et al. (1973).

Intraventricular administration of norepinephrine facilitated self-stimulation of the lateral hypothalamus, whereas intraventricular administration of serotonin suppressed it (Fig. 1). The facilitatory effect of norepinephrine was dose-related, with maximum effect occurring about 10 to 15 min after an intraventricular injection of 5 or 10 μg. On the other hand, serotonin had approximately the same suppressant effect at 5 and 10 μg, with peak action within 2 to 8 min. In related experiments, similar doses of dopamine were generally ineffective, although in some cases suppression of self-stimulation was observed (Stein and Wise, 1974).

Phentolamine decreased the rate of self-stimulation (variable-interval reinforcement schedule) within 3 min after the intraventricular injection. Dose-related suppressant effects were obtained within the range, 20 to 100 μg (Table 1). In contrast, 50 and 100 μg of propranolol had relatively minor effects. The suppressive effect of phentolamine on self-stimulation was partially reversed by the norepinephrine-releasing agent, amphetamine (Wise et al., 1973).

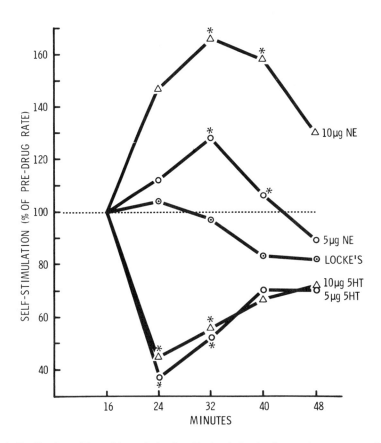

FIG. 1. Facilitation of lateral hypothalamic self-stimulation by L-norepinephrine HCl (NE) and suppression by serotonin HCl (5-HT). Intraventricular injections were made 16 min after the start of the test. Data are expressed as a percent of the self-stimulation rate in the second 8-min test period (8 to 16 min after start of test). The curves are obtained by averaging percent scores for the group of 11 rats. Starred points differ significantly from Locke's control at same time point at 0.05 level or beyond.

TABLE 1. *Effects of phentolamine and propranolol on self-stimulation*

Dose	Self-stimulations (% of control)	
(μg)	Phentolamine	Propranolol
20	73.1 ± 15.2	—
50	59.7 ± 16.1	90.8 ± 12.8
100	17.0 ± 6.4	76.2 ± 29.7

Intraventricular injections were made 15 min after the start of the 75-min test. Scores indicate rates during the first 15 min after phentolamine or propranolol as percentages of the rate prior to injection.

II. PASSIVE AVOIDANCE (CONFLICT)

The "conflict" test of Geller and Seifter (1960) was used to measure the influence of monoamines on the behavioral inhibition induced by punishment. In this test, hungry rats perform a lever-press response to obtain a sweetened milk reward. The reinforcement schedule consists of punishment and nonpunishment components alternating every 18 min. On the nonpunishment schedule (15 min), a response is rewarded with sweetened milk at infrequent and variable intervals—on the average, once every 2 min. On the punishment schedule (3 min), signaled by a tone, every response is rewarded with milk, but is also punished with a brief electrical shock to the feet. The rate of response in the tone period may be regulated by adjustment of shock intensity, and any degree of behavioral suppression may be obtained in well-trained animals. After several weeks of training, the animals were implanted with a permanently indwelling cannula in the lateral ventricle.

The effects of intraventricular injections of the monoamine transmitters on punished behavior are illustrated in Fig. 2. L-Norepinephrine caused large dose-related increases in the rate of punished responses; indeed, the anxiety-reducing activity of norepinephrine administered intraventricularly compares favorably with that of the benzodiazepines administered systemically (Stein, Wise, and Berger, 1973). Neurochemical specificity is suggested, since D-norepinephrine and dopamine produced only negligible effects.

Serotonin had complex and apparently triphasic effects in the conflict test (Wise et al., 1973). Doses of 5 to 20 μg lead to an initial phase of intense behavioral suppression that lasts for about 15 min, a longer secondard phase of normal response (Fig. 2G) or, frequently, behavioral facilitation (including release of punished behavior), and finally a prolonged period of behavioral suppression. At high doses of serotonin (25 to 80 μg), these delayed suppressant effects are regularly observed for as long as 2 days (Table 2). Also interesting are the behavioral effects of combined administrations of serotonin and norepinephrine (Fig. 2H). The facilitatory effects of norepinephrine predominated throughout the test and even seemed to be prolonged by serotonin.

Consistent with these findings, Graeff and Schoenfeld (1970) observed large increases in the punished response rates of pigeons after intramuscular administrations of the serotonin antagonists, methysergide and 2-bromo-D-lysergic acid (BOL); according to these authors, "this effect was of the same magnitude as that produced by chlordiazepoxide, diazepam, and nitrazepam." We also have obtained strong antipunishment effects with methysergide in the rat conflict test (Stein et al., 1973).

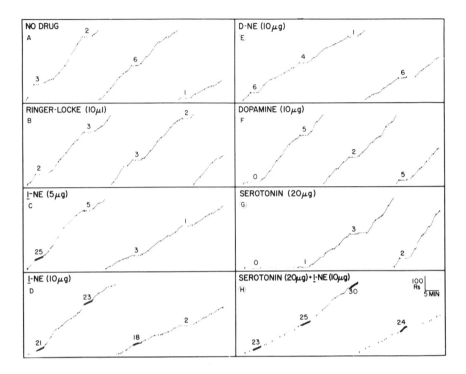

FIG. 2. Effects of intraventricular administration of various monoamine transmitters (as hydrochloride salts) on punished and nonpunished behavior in the same rat. Note especially dose-related antipunishment effect of L-norepinephrine (L-NE) (compare panels B–D), relative inactivity of D-norepinephrine (D-NE), and dopamine (panels E and F), initial suppressant action of serotonin (panel G), and paradoxical potentiation of L-norepinephrine antipunishment effect by serotonin (compare panels D and H).

Similarly, several groups have reported large releases of punishment-suppressed behavior after administration of the serotonin synthesis inhibitor, p-chlorphenylalanine (PCPA) (Tenen, 1967; Robichaud and Sledge, 1969; Stevens and Fechter, 1969; Geller and Blum, 1970). The time courses of behavioral disinhibition and serotonin depletion after PCPA coincide closely. In confirmation of the report of Geller and Blum (1970), we observed that intraperitoneal administration of the serotonin precursor, 5-hydroxytryptophan (5-HTP), reversed the disinhibitory effect of PCPA (Wise et al., 1973). Specificity of the 5-HTP reversal was suggested by tests with the catecholamine precursor, dihydroxyphenylalanine (DOPA). The rate of punished behavior in the PCPA-treated animals was increased rather than decreased by DOPA, both directly after injection and on the following day.

On the other hand, elevation of brain serotonin levels by combined

TABLE 2. *Effects of serotonin and phentolamine on punished and unpunished behavior (passive avoidance test)*[a]

		Punished responses ± SE			Unpunished responses ± SE		
Dose (μg)	N	Drug day	1 Day post	2 Days post	Drug day	1 Day post	2 Days post
Serotonin							
20 and 5	9	108.0 ± 32.6	53.3 ± 11.5[b]	61.0 ± 11.1[b]	163.3 ± 31.7	98.3 ± 12.0	111.2 ± 8.5
20 and 10	8	107.3 ± 18.9	59.1 ± 12.8[b]	69.7 ± 15.2[b]	147.6 ± 19.5	79.8 ± 13.8	87.1 ± 12.2
40 and 40	9	84.2 ± 19.2[c]	48.1 ± 9.8[b]	61.4 ± 16.2[b]	135.8 ± 30.4	92.7 ± 25.9	94.0 ± 23.5
Phentolamine							
5	5	59.9 ± 15.8[b]	80.5 ± 26.4	167.6 ± 41.3[b]	111.6 ± 26.6	96.5 ± 6.0	112.0 ± 12.5
10	5	9.2 ± 5.4[b]	91.5 ± 20.9	58.1 ± 14.8[b]	30.9 ± 19.5	84.1 ± 17.3	95.4 ± 9.6

[a]Serotonin was dosed twice (20 min and 5 min) and phentolamine was dosed once (10 min) before start of test. Data are expressed as percent of control performance on day before drug.

[b]Significantly different from control (*p* < 0.05).

[c]In eight of nine cases, punished behavior was suppressed. Excluding the single deviant case reduces the mean to 65.9 ± 6.3% (*p* < 0.01).

administration of 5-HTP and a monoamine oxidase inhibitor causes marked suppression of food-rewarded behavior in the pigeon (Aprison and Ferster, 1961). Furthermore, the long-lasting, centrally active serotonin agonist, α-methyltryptamine, strongly suppresses punished and nonpunished behaviors in the pigeon (Graeff and Schoenfeld, 1970) and in the rat (Stein et al., 1973).

The foregoing data are consistent with the idea that a serotonin system in the brain mediates the suppressive effects of punishment. If so, and if the antipunishment activity of benzodiazepine tranquilizers is mediated via a reduction of serotonin activity (Wise, Berger, and Stein, 1972; Stein et al., 1973), then it might be possible to demonstrate (a) suppression of behavior by direct activation of serotonin systems, and (b) antagonism of such directly induced behavioral suppression by administration of benzodiazepines.

As a first test of these ideas, minute quantities (about 5 μg) of crystalline carbachol were applied via permanently indwelling cannulas

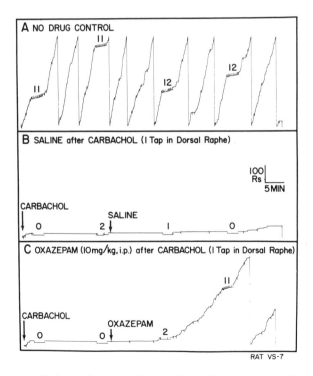

FIG. 3. Suppression of behavior in the conflict test by application of crystalline carbachol (1 tap = about 5 μg) to the ventral border of the dorsal raphe, and reversal of carbachol suppression by systemically injected oxazepam.

to the dorsal raphe region of five rats with stable rates of punished and unpunished responses in the conflict test. The dorsal raphe nucleus was selected for initial study because, as one of the major serotonin cell groups in the midbrain, it is the site of origin of many of the ascending serotonin fibers to the diencephalon and forebrain; furthermore, this nucleus consists almost entirely of serotonin cell bodies (Fuxe et al., 1970). Crystalline carbachol was used for chemical stimulation because powerful behavioral suppressive effects are readily obtained by direct application of this substance to the medial hypothalamic region (Margules and Stein, 1967).

In each of three cases, application of carbachol to the dorsal raphe nucleus caused marked suppression of both punished and unpunished behaviors within 1 to 2 min after treatment (Fig. 3). In all of these cases, oxazepam reversed the suppressive effects of raphe stimulation on both types of behavior. In two rats, the tip of the cannula penetrated into the reticular formation, about 1 mm lateral or caudal to the dorsal raphe; in these cases, carbachol had a much smaller or negligible suppressive effect on behavior.

III. DISCUSSION

Self-stimulation, a behavior which depends on activation of the reward mechanism, was facilitated by norepinephrine and suppressed both by serotonin and by the α-noradrenergic antagonist, phentolamine. Furthermore, inhibition of serotonin synthesis by PCPA has been reported to facilitate self-stimulation (Poschel and Ninteman, 1971). Passive avoidance, a behavior which depends on activation of the punishment mechanism, was antagonized by norepinephrine, PCPA, and the serotonin receptor blocker, methysergide; conversely, passive avoidance was facilitated by phentolamine and the serotonin agonist, α-methyltryptamine. Serotonin itself had complex triphasic effects in the conflict test. Intense suppression was observed both immediately after the intraventricular injection and later during a prolonged third phase, but a period of normal response or even behavioral disinhibition intervened between the phases of suppression. The phase of disinhibition is puzzling, but it might have an explanation, in part, in the observation of Aghajanian and Haigler (1974) that the firing rates of serotonin cells are suppressed by the direct application of serotonin, if it may be assumed that serotonin injected in the lateral ventricle could diffuse into the raphe nuclei within 10 to 20 min.

These results are consistent with the view that an α-noradrenergic system in the brain facilitates goal-directed behavior and a serotonergic system suppresses it (Wise et al., 1973). The two systems would seem to act reciprocally in such regulation, since parallel behavioral effects are obtained on the one hand by α-noradrenergic activation and serotonergic blockade, and on the other by serotonergic activation and α-noradrenergic blockade. The inputs to the noradrenergic and serotonergic systems have not been established experimentally, but anatomical considerations (Nauta, 1963) and physiological evidence (Olds, 1970) make it plausible to suggest that the α-noradrenergic system may be activated by rewarding signals and the serotonergic system by punishing signals. If so, it then would be appropriate to designate the α-noradrenergic system as a reward mechanism and the serotonergic system as a punishment mechanism.

IV. POSSIBLE CLINICAL SIGNIFICANCE

It may be worthwhile to speculate briefly about some clinical implications of our findings. We already have assumed that elation results from increased activity of the noradrenergic reward mechanism and depression from decreased noradrenergic reward activity (Stein, 1962). We have suggested further that the chronic impairment of goal-directed thinking and behavior in schizophrenia may result from a deterioration of central noradrenergic pathways (Stein and Wise, 1971). With regard to the punishment system, we have assumed that anxiety may result from an increase, and relief from a decrease, in its serotonergic activity (Wise et al., 1970). Since the norepinephrine and serotonin systems are assumed to act as reciprocal regulators, reduction of noradrenergic activity in mental depression would permit the serotonin system to predominate. This might explain the anxiety that is nearly always observed in the depressed patient. It may be of interest that such assumption of a *relative* increase of serotonergic activity in depression contradicts recent proposals that depression is associated with a central serotonin deficit. Absolute serotonin levels in depression could eventually decrease somewhat as a result of the lowered activity in the opposed noradrenergic system, but the balance of serotonin and norepinephrine activities would be the determining factor.

ACKNOWLEDGMENTS

We wish to thank Alfred T. Shropshire, Herman Morris, and Nicholas S. Buonato for their excellent technical assistance.

REFERENCES

Aghajanian, G. K., and Bloom, F. E. (1967): Localization of tritiated serotonin in rat brain by electron-microscopic autoradiography. *Journal of Pharmacology and Experimental Therapeutics*, 156:23–30.

Aghajanian, G. K., and Haigler, H. J. (1974): Mode of action of LSD on serotonergic neurons. In: *Advances in Biochemical Psychopharmacology*, Vol. 10, edited by E. Costa, G. L. Gessa, and M. Sandler. Raven Press, New York.

Aprison, M. H., and Ferster, C. B. (1961): Neurochemical correlates of behavior: II. Correlation of brain monoamine oxidase activity with behavioral changes after iproniazid and 5-hydroxytryptophan. *Journal of Neurochemistry*, 6:350–357.

Arbuthnott, G., Fuxe, K., and Ungerstedt, U. (1971): Central catecholamine turnover and self-stimulation behavior. *Brain Research*. 27:406–413.

Brodie, B. B., and Shore, P. A. (1957): A concept for a role of serotonin and norepinephrine as chemical mediators in the brain. *Annals of the New York Academy of Sciences*, 66:631–642.

Crow, T. J., Spear, P. J., and Arbuthnott, G. W. (1972): Intracranial self-stimulation with electrodes in the region of the locus coeruleus. *Brain Research*, 36:275–287.

Everett, G. M. (1974): Effect of 5-hydroxytryptophan on brain levels of dopamine, norepinephrine, and serotonin in mice. In: *Advances in Biochemical Psychopharmacology*, Vol. 10, edited by E. Costa, G. L. Gessa, and M. Sandler. Raven Press, New York.

Feldberg, W., and Myers, R. D. (1964): Effects on temperature of amines injected into the cerebral ventricles. A new concept of temperature regulation. *Journal of Physiology*, 173:226–237.

Fuxe, K. (1965): The distribution of monoamine nerve terminals in the central nervous system. *Acta Physiologica Scandinavica* (64, Suppl.), 247:37–102.

Fuxe, K. (1974): Morphological and functional studies on 5-HT neurons using intracerebral injections of dihydroxytryptamines. This Volume.

Fuxe, K., Hökfelt, T., and Ungerstedt, U. (1970): Morphological and functional aspects of central monoamine neurons. *International Review of Neurobiology*, 13:93–126.

Geller, I., and Blum, K. (1970): The effects of 5-HT on *para*-chlorophenylalanine (*p*-CPA) attenuation of "conflict" behavior. *European Journal of Pharmacology*, 9:319–324.

Geller, I., and Seifter, J. (1960): The effects of meprobamate, barbiturates, D-amphetamine and promazine on experimentally induced conflict in the rat. *Psychopharmacologia* (Berlin), 1:482–492.

Graeff, F. G., and Schoenfeld, R. I. (1970): Tryptaminergic mechanisms in punished and nonpunished behavior. *Journal of Pharmacology and Experimental Therapeutics*, 173:277–283.

Margules, D. L., and Stein, L. (1967): Neuroleptics vs. tranquilizers: Evidence from animal studies of mode and site of action. In: *Neuropsychopharmacology*, edited by H. Brill, J. O. Cole, P. Deniker, H. Hippius, and P. B. Bradley, pp. 108–120. Excerpta Medica Foundation, Amsterdam.

Nauta, W. J. H. (1963): Central nervous organization and the endocrine nervous system. In: *Advances in Neuroendocrinology*, edited by A. V. Nalbandov, pp. 5–21. University of Illinois Press, Urbana.

Olds, J. (1970): The behavior of hippocampal neurons during conditioning experiments. In: *The Neural Control of Behavior*, edited by R. Whalen, pp. 257–293. Academic Press, New York.

Olds, J., and Milner, P. (1954): Positive reinforcement produced by electrical stimulation of septal area and other regions. *Journal of Comparative and Physiological Psychology*, 47:419–427.

Poschel, B. P. H., and Ninteman, F. W. (1971): Intracranial reward and the forebrain's serotonergic mechanism: Studies employing *para*-chlorophenylalanine and *para*-chloroamphetamine. *Physiology and Behavior*, 7:39–46.

Ritter, S., and Stein, L. (1972): Self-stimulation of the locus coeruleus. *Federation Proceedings*, 31:820.

Ritter, S., and Stein, L. (1973): Self-stimulation of noradrenergic cell groups in the locus coeruleus of the rat. *Journal of Comparative and Physiological Psychology*, 85:443–452.

Robichaud, R. C., and Sledge, K. L. (1969): The effects of p-chlorophenylalanine on experimentally induced conflict in the rat. *Life Sciences*, 8:965–969.

Stein, L. (1962): Effects and interactions of imipramine, chlorpromazine, reserpine, and amphetamine on self-stimulation: Possible neurophysiological basis of depression. In: *Recent Advances in Biological Psychiatry*, Vol. 4, edited by J. Wortis, pp. 228–308. Plenum Press, New York.

Stein, L. (1968): Chemistry of reward and punishment. In: *Psychopharmacology, A Review of Progress: 1957–1967*, edited by D. H. Efron, pp. 105–123. U.S. Government Printing Office, Washington, D.C.

Stein, L., and Wise, C. D. (1971): Possible etiology of schizophrenia: Progressive damage to the noradrenergic reward system by 6-hydroxydopamine. *Science*, 171:1032–1036.

Stein, L., and Wise, C. D. (1974): Amphetamine and noradrenergic reward pathways. *Pharmacological Reviews (in press)*.

Stein, L., Wise, C. D., and Berger, B. D. (1973): Antianxiety action of benzodiazepines: Decrease in activity of serotonin neurons in the punishment system. In: *The Benzodiazepines*, edited by S. Garattini, E. Mussini, and L. O. Randall, pp. 299–326. Raven Press, New York.

Stevens, D. A., and Fechter, L. D. (1969): The effects of p-chlorophenylalanine, a depletor of brain serotonin, on behavior: II. Retardation of passive avoidance learning. *Life Science*, 8:379–385.

Tenen, S. S. (1967): The effects of p-chlorophenylalanine, a serotonin depletor, on avoidance acquistion, pain sensitivity, and related behavior in the rat. *Psychopharmacologia* (Berlin), 10:204–219.

Wise, C. D., Berger, B. D., and Stein, L. (1970): Serotonin: A possible mediator of behavioral suppression induced by anxiety. *Diseases of the Nervous System* (GWAN Suppl.), 31:34–37.

Wise, C. D., Berger, B. D., and Stein, L. (1972): Anxiety reducing activity by reduction of serotonin turnover in the brain. *Science*, 177:180–183.

Wise, C. D., Berger, B. D., and Stein, L. (1973): Evidence of α-noradrenergic reward receptors and serotonergic punishment receptors in the rat brain. *Biological Psychiatry*, 6:3–21.

Wise, C. D., and Stein, L. (1970): Amphetamine: Facilitation of behavior by augmented release of norepinephrine from the medial forebrain bundle. In: *Amphetamines and Related Compounds*, edited by E. Costa and S. Garattini, pp. 463–485. Raven Press, New York.

Advances in Biochemical Psychopharmacology, Vol. 11
Raven Press, New York © 1974

The "Kynurenine Shunt" and Depression

Alfonso Mangoni

Department of Psychiatry,
University of Cagliari Medical School,
Cagliari, Italy

I. INTRODUCTION

To use the term "shunt" in reference to a major pathway of tryptophan metabolism may not sound appropriate if one is unaware that up until a few years ago this pathway tended to be ignored in studies on depression. The first observation correlating certain changes in the kynurenine pathway with depression, to my knowledge, came from our laboratory (Cazzullo, Mangoni, and Mascherpa, 1966). An increased output of xanthurenic acid in depressed patients compared with manic and control subjects was found in these experiments, and it was assumed that it was due to a raised activity of tryptophan pyrrolase. The following year Richter (1967) ascribed the reduced levels of indoleamines in depression to increased activity of tryptophan pyrrolase following raised level of plasma corticoids (Gibbons and McHugh, 1962). These observations have been confirmed by tracer experiments (Rubin, 1967).

Weil-Malherbe (1972), in a recent review on the biochemistry of affective disorders, has stated that one of the few unequivocal observations in depression is "the change in the activity of tryptophan pyrrolase which probably accounts, at least in part, for some of the changes of indole metabolism and serotonin synthesis that have been reported." We have carried out further studies on the kynurenine pathway in depression to try to answer two main questions:

Is pyrrolase really responsible for the xanthurenic acid increase, or does the raised output of this metabolite depend on decreased activity of kynureninase?

What happens to other metabolites of the kynurenine pathway in depression?

II. RESULTS AND DISCUSSION

The study we devised to answer these questions includes a statistical evaluation of the relationship between severity of depressive symptoms and output of xanthurenic acid and N'-methylnicotinamide.

The methods we adopted and some of the results obtained have been reported previously (Cazzullo et al., 1966; Cazzullo, Mangoni, Bozzoli, and Freni, 1972). Clinical symptoms were assessed using a Wittenborn Rating Scale in conjunction with determinations of the two metabolites before and after treatment with amitriptyline. Some of the patients were classified as unipolar depressives, some as endo-reactive depressives, others as being in the depressed phase of manic-depressive psychosis. Pyridoxine hydrochloride was administered to one group of patients, but not to another.

The results of these experiments are summarized in Table 1 and show the correlation coefficient (r) between depressive symptoms scores measured with the Wittenborn Scale and xanthurenic acid and N'-methylnicotinamide in a urinary sample collected during a standard diet with or without added pyridoxine hydrochloride.

They point to a positive correlation between xanthurenic acid and depression; that is, there is a higher output of xanthurenic acid in patients with higher depression scores in the pyridoxine-fed group, confirming our earlier observation (Cazzullo et al., 1966) under similar experimental conditions (pyridoxine-supplemented diet). When no pyridoxine was added to the diet, a statistically significant correlation

TABLE 1. *Correlation coefficient (r) between depressive symptoms score and urinary output of xanthurenic acid (XA) and N'-methylnicotinamide (NMN)*

	XA		NMN	
	Before amitriptyline treatment	After amitriptyline treatment	Before amitriptyline treatment	After amitriptyline treatment
Pyridoxine 150 mg daily	0.896 $p < 0.05$	-0.158 N.S.	0.476 N.S.	0.540 N.S.
No pyridoxine added	0.049 N.S.	0.263 N.S.	-0.721 $p < 0.001$	0.149 N.S.

Xanthurenic acid and N'-methylnicotinamide were measured in 8-hr urine samples collected during a standard diet with or without added pyridoxine hydrochloride. Depressive symptoms were scored with a 52-item Wittenborn Rating Scale in conjunction with the determination of tryptophan metabolites.

between xanthurenic acid output and depressive symptom score was not observed. This finding tends to rule out the possibility of the increased xanthurenic acid excretion in depressed patients being due to decreased activity of pyridoxal phosphate-dependent kynureninase, and therefore gives support to the view that it stems from increased activity of tryptophan pyrrolase.

Another observation of interest is the negative correlation between N'-methylnicotinamide output and depression; that is, there was a lower urinary excretion of N'-methylnicotinamide in patients with higher depressive scores in the group for whom no pyridoxine hydrochloride had been added to the diet. This finding indicates that the N'-methylnicotinamide decrease in depression, which is in the opposite direction from the change in xanthurenic acid, may be pyridoxal phosphate-dependent, perhaps secondary to decreased activity of kynureninase.

We do not know whether nicotinamide is also decreased in depressed patients. If it is, it might be relevant to the development of some depressive symptoms because of the role of the nicotinamide mononucleotides in the metabolism of nerve cells and in mental function (Lehmann, 1972). Further experiments are also needed on methylation in depressed subjects and on renal clearance of N'-methylnicotinamide.

In parallel with these clinical studies, we have carried out a series of experiments on the effect of some antidepressant drugs on tryptophan pyrrolase activity of rat liver. The experimental procedures and some of the findings have been reported (Mangoni and Paracchi, 1965; Mangoni, Cabibbe, and Paracchi, 1968; Paracchi, Lanzara, Bernetti, and Mangoni, 1966). To different groups of animals having a standard initial body weight we administered daily, for 4 weeks, imipramine, 6 mg/kg body weight, tranylcypromine, 5 mg/kg body weight, and amitriptyline, 5 mg/kg body weight. This amount is weight for weight about twice the therapeutic dose given to patients. At the end of the experimental period, pyrrolase activity was assayed. L-Tryptophan (1000 mg/kg) was administered intraperitoneally to one group of the animals treated with antidepressant drugs and to untreated controls 5 hr before sacrifice. No major differences were observed as to the weight of whole liver among the various groups of rats.

The results, which are reported in Fig. 1, show an inhibitory effect of all these drugs on tryptophan pyrrolase activity, particularly evident in imipramine- and tranylcypromine-treated animals ($p < 0.01$) and less marked in rats treated with amitriptyline ($p < 0.05$).

A further experiment was carried out as follows: One group of rats was adrenalectomized with sham-operated controls. All, apart from a control group, had received either imipramine or tranylcypromine intraperitoneally for 4 weeks at the same dosage as in the experiment

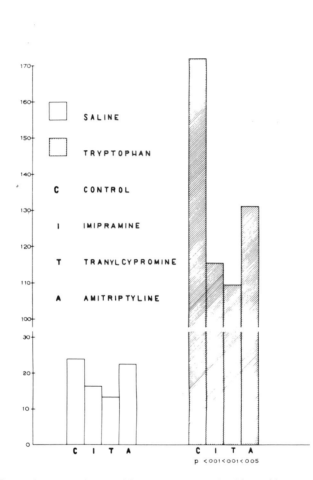

LIVER   TRYPTOPHAN   PYRROLASE
μ MOLES KYNURENINE /HOUR/ WHOLE LIVER

SALINE

TRYPTOPHAN

C   CONTROL

I   IMIPRAMINE

T   TRANYLCYPROMINE

A   AMITRIPTYLINE

p <001<001<005

FIG. 1. Tryptophan pyrrolase activity in rats treated with antidepressant drugs. Drugs administered daily for 4 weeks: imipramine, 6 mg/kg body weight; tranylcypromine, 5 mg/kg body weight; amitriptyline, 5 mg/kg body weight. Five hours before sacrifice, L-tryptophan, 1000 mg/kg body weight, was administered intraperitoneally to some animals.

described above. To a group of adrenalectomized animals, the following hormones were administered daily, in addition to imipramine or tranylcypromine: aldosterone, 0.20 mg/kg body weight intraperitoneally; hydrocortisone, 2.50 mg/kg body weight intraperitoneally; and methylandrostendiol, 6.25 mg/kg body weight subcutaneously. After 4 weeks, the animals were sacrificed and liver tryptophan pyrrolase activity assayed. The results are shown in Fig. 2. The most relevant conclusion deriving from these experiments is that the fall of pyrrolase activity induced by imipramine and tranylcypromine is sustained by the presence of adrenal hormones in intact animals. In fact, in adrenal-ectomized rats, antidepressant drug inhibition of pyrrolase does not occur apart, presumably, from hydrocortisone-treated animals.

Considering that in depressed subjects, plasma cortisol levels lie above normal values (Gibbons and McHugh, 1962; Sachar, Hellman, Roffwarg, Halpern, Fukushima, and Gallagher, 1973), we might reasonably assume that the therapeutic effect of imipramine and tranylcypromine might be due, at least in part, to their inhibitory effect on tryptophan pyrrolase, so that higher concentrations of free tryptophan are available for serotonin synthesis.

Thus, if a depressed mood is the consequence of a diminished concentration of 5-hydroxytryptamine at active brain sites secondary

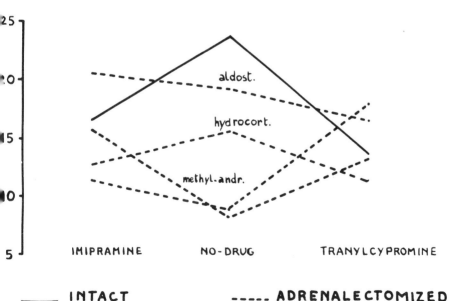

FIG. 2. Tryptophan pyrrolase activity in rats treated with antidepressant drugs and adrenocortical hormones. Imipramine, 6 mg/kg body weight, and tranylcypromine, 5 mg/kg body weight, administered daily for 4 weeks. Aldosterone, 0.20 mg/kg body weight, hydrocortisone, 2.50 mg/kg body weight, and methylandrostendiol, 6.25 mg/kg body weight, administered daily for 4 weeks to some adrenalectomized animals.

to a raised activity of tryptophan pyrrolase in the liver, the observation may be connected with the metabolic effect of some widely distributed antidepressant drugs on this "kynurenine shunt" enzyme apart from their better-known effects on oxidation and reuptake of 5-hydroxytryptamine.

REFERENCES

Cazzullo, C. L., Mangoni, A., Bozzoli, A., and Freni, S. (1972): Some metabolic parameters in depressive illness. *Proceedings of the Symposium on Depression, World Psychiatric Association*, Madrid.

Cazzullo, C. L., Mangoni, A., and Mascherpa, G. (1966): Tryptophan metabolism in affective psychoses. *British Journal of Psychiatry*, 112:157–162.

Gibbons, J. L., and McHugh, P. R. (1962): Plasma cortisol in depressive illness. *Journal of Psychiatric Research*, 1:162–171.

Lehmann, J. (1972): Mental and neuromuscular symptoms in tryptophan deficiency. *Acta Psychiatrica Scandinavica, Suppl. 237*, 5–28.

Mangoni, A., Cabibbe, F., and Paracchi, G. (1968): Effetto della fenilalanina sulla via metabolica triptofano-acido nicotinico. *Biochimica e Biologia Sperimentale*, 7:149–152.

Mangoni, A., and Paracchi, G. (1965): L'attività triptofano-pirrolasica in ratti trattati con farmaci antidepressivi. *Bollettino della Società* Italiana di Biologia Sperimentale, 41:1552–1554.

Paracchi, G., Lanzara, D., Bernetti, M. G., and Mangoni, A. (1966): Effect of antidepressant drugs on tryptophan metabolism. *Excerpta Medica International Congress Series No. 150*, pp. 1891–1892.

Richter, D. (1967): Tryptophan metabolism in mental illness. In: *Amines and Schizophrenia*, edited by H. E. Himwich, S. S. Kety, and J. R. Smythies, pp. 167–179. Pergamon Press, Oxford.

Rubin, R. T. (1967): Adrenal cortical activity changes in manic-depressive illness. Influence on intermediary metabolism of tryptophan. *Archives of General Psychiatry*, 17:671–679.

Sachar, E. J., Hellman, L., Roffwarg, H. P., Halpern, F. S., Fukushima, D. K., and Gallagher, T. F. (1973): Disrupted 24-hour pattern of cortisol secretion in psychotic depression. *Archives of General Psychiatry*, 28:19–24.

Weil-Malherbe, H. (1972): The biochemistry of affective disorders. In: *Handbook of Neurochemistry*, Vol. VII, edited by A. Lajtha, pp. 371–416. Plenum Préss, New York.

Advances in Biochemical Psychopharmacology, Vol. 11
Raven Press, New York © 1974

N,N-Dimethyltryptamine — A Possible Relationship to Schizophrenia?

Richard Jed Wyatt,* J. Christian Gillin,* Jonathan Kaplan,*
Richard Stillman,* Lewis Mandel,** H. S. Ahn,**
W. J. A. VandenHeuvel,** and R. W. Walker**

*Laboratory of Clinical Psychopharmacology, National Institute
of Mental Health, St Elizabeths Hospital, Washington, D.C. 20032, and
**Merck Institute for Therapeutic Research, Rahway, New Jersey 07065

I. INTRODUCTION

Murphy and Wyatt (1972) recently demonstrated that platelet monoamine oxidase (MAO) activity is low in some schizophrenic patients compared with patients with affective disorders and normals. Subsequently Wyatt, Murphy, Belmaker, Cohen, Donnelly, and Pollin (1973b) found that platelet MAO activity was highly correlated in pairs of monozygotic twins discordant for schizophrenia, indicating that the enzyme activity was, at least in part, genetically determined. These data suggest that low platelet MAO activity is present in some individuals with a high susceptibility to schizophrenia.

Information is lacking as to whether abnormal MAO activity is present in parts of the schizophrenic's body other than platelets and whether it is causative or secondary to the schizophrenic illness. Nevertheless, some speculations can be made regarding biochemical defects which might be associated with low MAO activity. Our laboratory has been particularly interested in the possible role of N,N-dimethyltryptamine (DMT) in schizophrenia (Fig. 1). DMT is a known hallucinogen. An enzyme capable of forming DMT from tryptamine has recently been described in human blood (Wyatt, Saavedra, and Axelrod, 1973c), lung (Mandel, Ahn, VandenHeuvel, and Walker, 1972), and brain (Mandell and Morgan, 1971; Saavedra and Axelrod, 1972b). Since tryptamine is normally degraded by MAO, abnormally low levels of MAO could predispose to elevated levels of tryptamine, the immediate precursor of DMT.

Biochemical theories of schizophrenia often assume that the clinical symptoms of schizophrenia are caused by an offending chemical

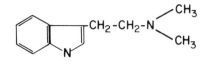

DIMETHYLTRYPTAMINE
(DMT)

FIG. 1. Chemical structure of dimethyltryptamine (DMT).

substance. Furthermore, it is assumed that this substance could produce similar symptoms if it could be isolated or synthesized and administered to normal subjects. Based upon this reasoning, the following criteria (after Hollister, 1968) must be met before such an agent would be considered an endogenous hallucinogen causative of schizophrenia:

1. The agent must be capable of mimicking clinical aspects of schizophrenia.
2. The agent must be found in man.
3. The precursor of the agent should be found in man (see 4).
4. The agent should be synthesized in man (it remains a possibility that the agent could be of dietary origin).
5. The agent must be differentially synthesized or metabolized in schizophrenics.
6. Tolerance to the agent should not develop: Slow tolerance to the agent might be acceptable for acute schizophrenia but not for the chronic illness. (This assumes that the agent does not produce irreversible damage.)
7. Neuroleptic drugs must be capable of antagonizing the synthesis, increasing the metabolism, or antagonizing the effect of the agent.

II. CLINICAL EXPERIENCE WITH DMT

In this chapter we shall try to summarize how well DMT fits these criteria, where it fails, and where further research efforts are necessary.

In 1956, Dr. Stephen Szara (see also Sai-Halasz, Brunecker, and Szara, 1958) synthesized and gave DMT to a group of 20 normal volunteers. He noted that the effects of DM T (0.7 to 2 mg/kg) were similar to those produced by mescaline and LSD. Oral dosages were without effect. The minimally effective dosage was 0.2 mg/kg and the optimal, above which there was no increase in symptoms, was 0.7 to 1.0 mg/kg. The principal symptoms were visual illusions and hallucinations, distortion of spatial perception and of body image, disturbances of speech, and euphoria. He noted mydriasis and elevation of blood pressure as well as athetoid and compulsive movements. The most striking aspect of this "model psychosis" was its rapidity of onset (3 to 5 min) after injection and the brevity of duration (less than 1 hr). Arnold and Hofman (1957), administering DMT to psychologists who had previous LSD experience, found that DMT produced less intense hallucinations, with less movement and color, than were remembered for LSD. These results were confirmed in 1963 by Rosenberg, Isbell, and Miner in five human subjects (0.75 to 1 mg/kg). Turner and Merlis (1959) found intranasal (0.5 to 20 mg) and oral (up to 350 mg) DMT had no psychotomimetic effect in a group of schizophrenic subjects. Intravenous DMT injections of up to 25 mg were without effect except in one schizophrenic patient who saw nurses as paper dolls and walls as crumbling paper. Saline injections did not produce this effect. Turner and Merlis also gave 5 mg intramuscularly for 1 week without psychic changes. Intramuscular doses below 25 mg were without effect, but above 25 mg psychic changes were present. The patients became restless and withdrew from personal contact out of fear. The patients gave the appearance of hallucinating. There were periods of perplexity and of irrelevant or incoherent speech. Sai-Halasz (1962, 1963) gave DMT to 40 normal volunteers. Fifteen of these were then first given the 5-HT antagonist, 1-methyl-D-lysergic acid butanolamide, followed by DMT. The serotonin antagonist greatly potentiated the effect of DMT. Subsequently he found that pretreatment with a monoamine oxidase inhibitor (iproniazid) before DMT was given (0.65 to 0.83 mg/kg) produced an attenuated effect with illusions and hallucinations without color. After 14 to 20 min, nothing was of interest to the subjects even though the hallucinations had worn off unusually rapidly. The subjects were "emotionally bleak." Dosages of DMT below 0.65 mg/kg had no hallucinatory effect but produced the same bleakness.

Despite dramatic effects produced by DMT, we have been able to find no other published studies of this substance in man—most investigators having chosen to use longer acting but perhaps less physiological agents. Because of the potential importance of understanding how DMT works in man, we have initiated a study of the effects of DMT in a group of medically and psychologically healthy male volunteers, all of whom have had extensive previous experience with hallucinogens. To date five subjects have been studied. Each was administered an intramuscular dosage of DMT (0.7 mg/kg) and blood pressure, pulse rate, and pupillary diameter were measured at 5-min intervals. In addition, several behavioral and subjective tests were given.

In agreement with previous studies, we found that DMT is a short-acting, potent hallucinogenic drug with a rapid onset. DMT blood concentrations, autonomic and subjective measures demonstrate this and follow a very similar time course. Figure 2 shows that there is a

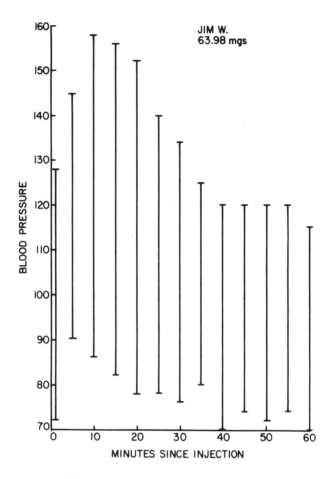

FIG. 2. Measurements of diastolic and systolic blood pressure in a normal subject receiving 64 mg of DMT.

rapid and parallel increase in both diastolic and systolic blood pressure in a normal subject given DMT. This is also true for the pulse rate.

Subjects were asked to rate how high they were, on a scale of 0 to 10, with a rating of 10 as "high as you have ever been in your life." In this particular subject, who had taken LSD over 50 times, and in three of the four others, the DMT experience was rated as "more intense than any dose of LSD they had ever taken."

Figure 3 shows the subjective effects of DMT on subject J.B. It can be seen that DMT produces an intense experience, with the first rush perceived within 3 min of the injection. Almost immediately, the subjects became completely absorbed in a very complex experience which so far has defied their attempts to verbalize completely, and

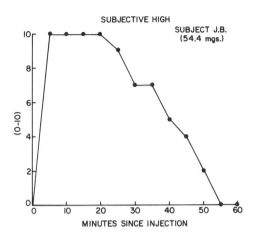

FIG. 3. Measurement of subjective "high" in a subject receiving 54 mg of DMT. The "high" is measured on a subjective 0 to 10 scale, with the rating of 10 as "high," that is, "as 'high' as you have ever been in your life."

which cannot be described at all while it is happening.

The DMT experience was described as being associated with an overload of thoughts and perceptions—so compelling that the subjects were forced to attend to them—neglecting interactions with the experimenters. They had problems with concentration. Their extreme difficulty with communication occasionally was associated with resentment. This necessitated the use of "yes" and "no" questionnaires, since open-ended questions led simply to perseveration and no answers.

The perceptions were largely visual, primarily with patterns and changing shapes. Colors seemed brighter and more beautiful than usual. There were sensations that the surroundings were moving. Some subjects reported that the experimenter looked different, became a yellow color, or broke into cubist abstracts. Only rarely were formed visual hallucinations noted.

There was a feeling among the subjects that they were generating thoughts faster than normal, and in two of the five sadness was reported. All subjects were at times euphoric. Four subjects reported feeling both calm and excited at the same time.

Three of the subjects had dysphoric reactions. One of these reported severe anxiety about losing control, one became very suspicious, and one frankly paranoid. Both of the latter felt, to different degrees, that the experimenters were controlling their minds. One demanded, "You started all this, now stop it." Inexplicably, as soon as the drug effect wore off (30 to 60 min), the subjects remembered the experience only as being pleasurable. This behavior, while at times somewhat similar to that of schizophrenia, always took place with the realization that it was inappropriate. The subjects slid in and out of contact with reality.

Comment: DMT in normals was originally described as producing frank visual hallucinations. In our experience this has not been the case—DMT produced some illusions or distortions of overwhelming interest to subjects such that they ignored all outside communication. The subjects, however, did on rare occasions see objects that were not present. With effort, they could communicate, and when they did, it was relevant and clear. It is our experience that much of the schizophrenic communication, such as "Get away from me," or "No" is also relevant and clear. Nevertheless, schizophrenic thought disorder was not observed in our subjects. The presence of suspiciousness (frank paranoia in one subject), however, might have been indistinguishable from acute paranoid schizophrenia if it had been sustained. Of course, all our subjects were experienced with psychotomimetic drugs and, unlike schizophrenics, knew the source of the altered state of consciousness.

The schizophrenics of Turner and Merlis (1959) given DMT had an increase of intensiveness of their schizophrenic symptomatology. Since a number of drugs, including methylphenidate (Janowsky, El-Yousef, Davis, and Sekerke, 1973), produce similar exacerbations of the illness, it is at the moment difficult to draw any conclusions from this data relevant to the involvement of DMT in schizophrenia.

III. DMT CONCENTRATIONS IN MAN

Several authors have reported the presence of DMT in human tissue (Franzen and Gross, 1965; Gross and Franzen, 1965; Tanimukai, Ginter, Spaide, Bueno, and Himwich, 1970; Rosengarten, Szemis, Piotrowski, Romaszewska, Matsumoto, Stencka, and Jus, 1970; Narasimhachari, Heller, Spaide, Haskovec, Meltzer, Strahilevitz, and Himwich, 1971). An important advance in this pursuit was the development of a gas chromatographic-mass spectrometric isotope dilution assay for DMT by Walker, Ahn, Mandel, and VandenHeuvel (1972). A year later Walker, Ahn, Albers-Schonberg, Mandel, and VandenHeuvel (1973) measured DMT in human plasma using this methodology and reported that DMT, when present, varied between 0.5 to 2.0 ng/ml. Analysis of plasma samples from patients who were chronic alcoholics, acute and chronic schizophrenics, and depressives showed no variation from these low levels (Wyatt, Mandel, Ahn, Walker, and VandenHeuvel, 1973a).

IV. SYNTHESIS OF DMT IN MAN

DMT is formed by the enzymatic addition of two methyl groups to the free amino group of tryptamine from S-adenosyl methionine (Fig. 4). Tryptamine itself is the product of decarboxylation of the essential amino acid tryptophan.

There has been a controversy over whether tryptamine is present in mammalian brain (Hess and Doepfner, 1961; Green and Sawyer, 1960; Eccleston, Ashcroft, Crawford, and Loose, 1966), but it now seems clear that it is normally present in human brain (Martin, Sloan, Christian, and Clements, 1972; Saavedra and Axelrod, 1972a,b). It also seems evident that under conditions of high tryptophan loading, with or without monoamine oxidase inhibitors, that tryptamine can be endogenously synthesized *in vivo* in rat brain (Saavedra and Axelrod, 1972a). The K_m, however, for L-amino acid decarboxylase toward tryptophan is 1.4×10^{-2} M (Ichiyama, Nakamura, Nishizuka, and Hayaishi, 1970), and does not favor the *in vivo* synthesis of tryptamine under normal circumstances. Nevertheless, inhibition of monoamine oxidase by parenteral administration of iproniazid does increase brain tryptamine, suggesting that tryptamine is normally synthesized in the body. (More direct studies, which are able to exclude the possibility of tryptamine being synthesized by bacteria in the gastrointestinal system, seem necessary.)

Whether tryptamine is exogenously or endogenously formed, it *is* present in human tissue (Saavedra and Axelrod, 1972b), including plasma (0.5 to 2 ng/ml) (Saavedra—personal communication), and its principal metabolite, indoleacetic acid, has been consistently found in human spinal fluid (Bertilsson and Palmer, 1972). Since tryptamine itself crosses the blood-brain barrier, it could be converted in the brain to DMT by the enzymes described by Mandell and Morgan (1971) and

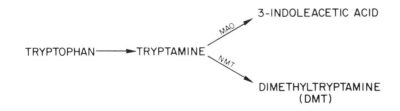

FIG. 4. Metabolism of tryptamine. MAO = monoamine oxidase, NMT = N-methyltransferase.

by Saavedra and Axelrod (1973). Tryptamine urinary excretion may be high in some schizophrenics (Brune and Himwich, 1962). High tryptamine concentrations favor its conversion to DMT.

DMT, however, does not have to be formed in the brain to be active there. Mandel et al. (1972) described very active enzymatic conversion of tryptamine to N-monomethyl tryptamine and of N-monomethyl tryptamine to DMT in the human lung. The lung also appears to accumulate high concentrations of tryptamine (Saavedra and Axelrod, 1973). Since the lung is the last filtering organ of blood on its way to the brain, DMT formed in the lung would be shunted directly to the brain. An enzyme capable of forming DMT has also been described in human plasma (Wyatt et al., 1973*b*), serum (Narasimhachari, Plant, and Himwich, 1972), platelets (Wyatt et al., 1973*c*), and red blood cells (Wyatt et al., 1973*c;* Rosengarten, Meller, and Friedhoff, 1973).

Comments: To date there has been no demonstration of *in vivo* DMT synthesis in man. Saavedra and Axelrod (1972*b*), however, using ^{14}C-tryptamine injected intracisternally, were able to demonstrate its conversion to N-methyltryptamine and DMT in rat brain. Similar experiments are possible in man.

V. DMT METABOLISM

The implication of the rapid disappearance of DMT's psychological effects is that the drug is rapidly metabolized in the body (Szara, 1956). Erspamer (1955), studying rats, noted that DMT appeared to be broken down in the urine to 3-indoleacetic acid (3-IAA). Szara (1956) identified 8.6% of an original DMT dosage as free 3-IAA in the urine of normals. Alkalinization of the urine brought the recovered product up to 25% of the administered DMT. DMT itself could not be found in the urine. In 1959, Szara and Axelrod identified 6-hydroxy DMT-N-oxide as an *in vivo* metabolite of rats.

Szara (1961) and Szara and Hearst (1962) proposed that the behavioral effects of DMT are due to a metabolite because they isolated a high-potency substance (not DMT) in the urine of rats administered DMT. Using chromatographic techniques, this compound was identified as 6-hydroxy-DMT. Subsequently, however, Kalir and Szara (1963) found that 6-fluoro-N,N-diethyltryptamine (N,N-diethyltryptamine is a longer acting hallucinogen than DMT) produced no hallucinogenic effects in man. In addition, Rosenberg et al. (1963) administered 6-hydroxy-DMT, DMT, and placebo in a single-blind study in equal dosages to five volunteers. The 6-hydroxy compound failed to produce signs or symptoms that might be considered peripherally (changes in blood pressure, kneejerk threshold, and pupil size) or centrally (the presence of hallucinations and perceptual distortions) associated with the hallucinogen DMT.

Using a fluorescent technique, Cohen and Vogel (1972) studied the disposition of DMT in rat tissues. After injecting DMT intraperitoneally, it was found to peak within 5 min, remain at a plateau for another 5 min, and have a sharp decrease thereafter. No DMT was detectable after 30 min. The highest concentrations of DMT were achieved in the liver, with the brain and plasma having one-sixth and one-seventeenth the liver concentration, respectively. There was a high brain-to-plasma ratio (5.4:1), suggesting that DMT penetrates the brain with ease.

In studies in our laboratories using gas-liquid chromatography and mass spectrometry, mice given 15 mg/kg of DMT intraperitoneally (Fig. 5) had peak DMT concentrations at 10 min, which fell within 30 min. The peak concentration of 4500 ng/g tissue was in the liver, with the brain and lung having about two-thirds that concentration.

In a group of normal volunteers given 0.7 mg/kg of DMT, we have found a large range in peak blood concentrations. Figure 6 demonstrates one subject at the extreme high end of the range. The time course, however, of the peaks and troughs is very similar, no matter

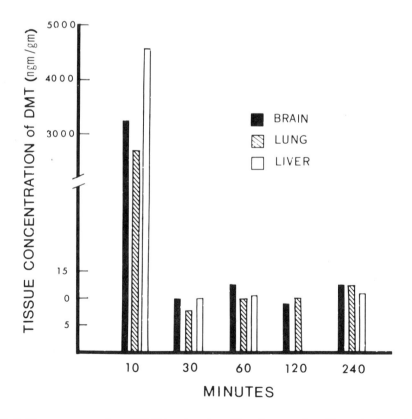

FIG. 5. Tissue concentration of DMT in mice, given 15 mg/kg, over time. Two animals used for each point.

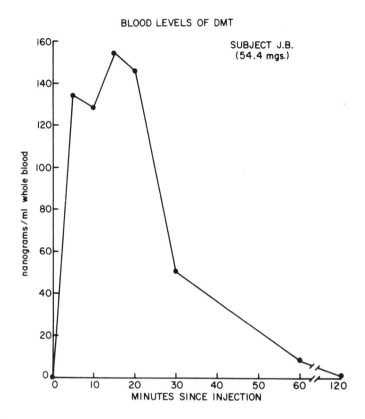

FIG. 6. Concentrations of DMT in whole blood over time after intramuscular injection of 54 mg in one normal subject. Subject had highest blood concentration of anyone studied.

what the peak blood concentration. In subject J.B., blood peak concentrations occurred at 15 min, about 5 min after the peak behavioral and autonomic change. The subsequent decay corresponds with the subjective effects. The peak blood concentration, calculated for total body blood, represents about 2% of the injected dosage.

Urinary excretion over 24 hr in the same individual was 36 μg, representing 0.067% of the injected dosage with 97% of that being excreted in the first 5.5 hr of collection: DMT, a nonpolar substance, quickly reaches tissues with high lipid concentrations such as the brain, and is found in low concentrations in urine, which is a highly polar solvent. In fact, after an injection of 0.7 mg/kg, only about 7/10,000 of the total dosage was found in a 24-hr urine. In addition, it is clear from the animal studies that DMT is not sequestered in the liver and brain, but is metabolized very quickly. While there is some evidence that metabolism occurs through hydroxylation and oxidation, which of these represents the principal metabolic route is not known. DMT is reported to be a poor substrate for monoamine oxidase, and may even inhibit this enzyme.

Comments: What is clear is that the normal person has excellent mechanisms for dealing with this highly toxic substance. It is also clear that looking for abnormal concentrations of DMT in the blood and urine of schizophrenic patients may be unrewarding due to DMT's rapid metabolism. Urine seems to be a very poor place to look for DMT, with blood only a little better. If DMT is produced and released in spurts rather than constantly, frequent blood sampling from the same person might be more rewarding than occasional samples. Demonstrating an increased quantity of DMT in schizophrenic patients (if present) will require attempts at determination of turnover rates or finding a specific stable metabolite if one exists. Prior to resorting to these measures, it will be necessary to determine if exogenous DMT is cleared from the blood in schizophrenics as rapidly as it is in controls.

VI. FAILURE OF DMT TO PRODUCE TOLERANCE

One of the chief arguments against a DMT-like compound being involved in a psychosis is the well-known development of tolerance to LSD and mescaline. Normal humans as well as schizophrenic patients and animals have the expected reaction to these drugs when they are first given them, but when dosages are given over consecutive days the effects usually disappear by the third day. Bridger, Stoff, and Mandel (1973), using a shuttle-avoidance test, have recently found that while tolerance develops to the inhibitory effects of mescaline, the excitatory effects produced by mescaline do not decrease with time.

Cole and Pieper (1973) trained five squirrel monkeys for food reinforcement on fixed-ratio schedules. A DMT dosage (up to 2 mg/kg) which was found to disrupt the operant behavior was administered once daily for 36 to 38 consecutive days. No tolerance developed using this measure.

Kovacie and Domino (personal communication) demonstrated that rats given DMT, 10 mg/kg, once a day for 14 days do not become tolerant in fixed-ratio 4 schedule; however, when the same dosage of DMT was given every 2 hr for 14 days, there at first appeared a hypersensitivity, but subsequently three of nine rats became tolerant to the drug.

Gillin, Cannon, Magyar, Schwartz, and Wyatt (1973) gave DMT (3 mg/kg), LSD (150 to 200 μg/kg) and saline in a double-blind fashion to five cats implanted with cortical EEG, electrooculogram, electromyogram, and lateral geniculate body electrodes. Drug injections were both given twice daily for from 7 to 15 consecutive days and every 2 hr for

24 hr. Neurological and autonomic effects of LSD were mild compared to those of DMT, and became attenuated with repeated administration. High-voltage slow waves produced by both drugs did not occur beyond the fourth or fifth day with LSD. DMT, on the other hand, produced sustained neurologic, autonomic, and EEG effects for as long as it was given. Even when DMT was given every 2 hr for 24 hr, no tolerance was seen. The typical DMT response began within 5 to 15 min of the injection and lasted 1 to 2 hr. Spontaneous movements decreased, pupils dilated, were unresponsive to light, and abnormalities of posture were evident. There was a wide-based stance, particularly of the hindlimbs. The cats assumed unusual fixed postures, such as hanging for long periods by the forelimbs when placed on the edge of a table. Pontine-geniculate-occipital ("PGO spikes") monophasic waves, said to be present during waking with LSD administration (Stern, Morgan, and Bronzin, 1972), were not present with DMT administration. In this experiment it was our impression that the DMT effect became greater with time.

In still another species (male albino mice), we attempted to determine whether tolerance developed to a DMT-produced impairment in a conditioned avoidance to an electric shock. Five groups of animals were trained to avoid an electric shock by running from a start box with an electrified grid. Each group of mice was given four injections at 90-min intervals ranging from all saline (group 1), DMT (15 mg/kg, i.p.), and three saline injections to four DMT injections (group V). DMT significantly interfered with established avoidance in all mice receiving DMT, no matter how many DMT injections had preceded it (no tolerance developed).

Chronic administration of DMT, 15 mg/kg, twice a day continued to disrupt avoidance for a 10-day period, also indicating a failure to develop tolerance to DMT. In the experiment with mice there was no evidence of increased sensitivity to repeated DMT administration.

Comment: Failure to develop tolerance to DMT has been found in four species of mammals. Even repeated dosage over very short time periods does not produce tolerance. An initial hint of reverse tolerance or hypersensitivity in our cat experiments was not seen in similar studies in mice, but was in Kovacie and Domino's rat experiments. Almost continuous DMT administration to their rats produced tolerance in three of nine animals.

Ultimately, tests will have to be made to determine whether tolerance to DMT develops in normal and schizophrenic individuals.

VII. EFFECTS OF ANTIPSYCHOTIC DRUGS ON DMT

While there are a fairly large number of studies demonstrating the

ability of antipsychotic drugs to block the effects of other hallucinogens, data for DMT are sparse. Teixeira, Bueno, and Do Lago (1968)
gave chlorpromazine (1.5 mg/kg) intramuscularly for 27 days and
found that *in vitro* red blood cell uptake of DMT was diminished 11%.
In our laboratory we have demonstrated that pretreating mice for 3
days with 2.5 mg/kg of chlorpromazine partially blocks a DMT-produced depression of motoric activity.

VIII. DISCUSSION

DMT is one of a number of substances which have been from time to
time considered as a cause of schizophrenia. Our laboratory is
interested in DMT because an increase in DMT synthesis is consistent
with a deficiency of monoamine oxidase. While a deficiency of
monoamine oxidase has been found in some patients with schizophrenia, to date only blood platelets have been found deficient, and
whether the deficiency is more generalized remains unknown. Nevertheless, an examination of whether DMT could fit a number of required
criteria for serious consideration as a schizophrenogen seems warranted.
DMT does appear to fit at least several of the criteria. In single
dosages it is capable of producing at least some schizophrenic-like
symptomatology. It and its precursor can be found in human tissue,
and the enzymes for DMT synthesis are present in man. In several
species tolerance to repeated administration has not been found to
develop. The most significant piece of negative evidence is the failure to
find elevated DMT plasma concentrations in schizophrenic individuals,
but in retrospect the extremely short *in vivo* life of DMT suggests that
it might be metabolized too quickly to be able to use blood
concentrations to differentiate schizophrenics and normals. Urine
appears to provide an even poorer measurement. Even those criteria
that are met at this point are met only in a small number of
experiments. It would seem that DMT remains a candidate for the
causative agent for at least some schizophrenias.

REFERENCES

Arnold, O. H., and Hofman, G. (1957): Zur Psychopathologie des Dimethyltrypta-
min: ein weiterer Beitrag zur Pharmakopsychiatrie. *Wien Zeitschrift für
Nervenheink*, 13:438–445.
Bertillsson, L., and Palmer, L. (1972): Indole-3-acetic acid in human cerebrospinal
fluid: Identification and quantitative determination by mass fragmentography.
Science, 177:74–76.
Bridger, W. H., Mandel, I. J., and Stoff, D. M. (1973): Mescaline: No tolerance to
excitatory effects. *Biological Psychiatry*, 7:129–133.
Brune, G. G., and Himwich, H. E. (1962). Effects of methionine loading on the
behavior of schizophrenic patients. *Journal of Nervous and Mental Disease*,
134:447–450.

Cohen, I., and Vogel, W. H. (1972): Determination and physiological disposition of dimethyltryptamine and diethyltryptamine in rat brain, liver and plasma. *Biochemical Pharmacology*, 21:1214–1216.

Cole, J. M., and Pieper, W. A. (1973): The effects of N,N-dimethyltryptamine on operant behavior in squirrel monkeys. *Psychopharmacologia* (Berlin), 29:107–112.

Eccleston, D., Ashcroft, G. W., Crawford, J. B., and Loose, R. (1966): Some observations on the estimation of tryptamine in the tissues. *Journal of Neurochemistry*, 13:93–101.

Erspamer, V. (1955): Observations of the fate of indolealkylamines in the organism. *Journal of Physiology*, 127:118–133.

Franzen, F., and Gross, H. (1965): Tryptamine, N,N-dimethyltryptamine, N,N-dimethyl-5-hydroxytryptamine and 5-methoxytryptamine in human blood and urine. *Nature*, 206:1052.

Gillin, J. C., Cannon, E., Magyar, R., Schwartz, M., and Wyatt, R. J. (1973): Failure of DMT to evoke tolerance in cats. *Biological Psychiatry*, 7:213–219.

Green, H., and Sawyer, J. L. (1960): Correlation of tryptamine induced convulsions in rats with brain tryptamine concentration. *Proceedings of the Society for Experimental Biology and Medicine (New York)*, 104:153–155.

Gross, H., and Franzen, F. (1965): Zur Bestimmung korporeigener Amine in biologischen Substraten. I Mitteilung: Bestimmung von N,N-Dimethyltryptamin in Blut und Harn. *Zeitschrift für Klinische Chemie*, 3:99–102.

Hess, S. M., and Doepfner, W. (1961): Behavioral effects and brain amine content in rats. *Archives Internationales de Pharmacodynamie et de Thérapie (Gand)*, 134:89–99.

Hollister, L. E. (1968): *Chemical Psychoses-LSD and Related Drugs*. Charles Thomas, Springfield, Ill.

Ichiyama, A., Nakamura, S., Nishizuka, Y., and Hayaishi, O. (1970): Enzymatic studies on biosynthesis of serotonin in mammalian brain. *Journal of Biological Chemistry*, 245:1699–1709.

Janowsky, D. S., El-Yousef, M. K., Davis, J. M., and Sekerke, H. J. (1973): Provocation of schizophrenic symptoms by intravenous administration of methylphenidate. *Archives of General Psychiatry*, 28:185–191.

Kalir, A., and Szara, S. (1963): Synthesis and pharmacological activity of fluorinated tryptamine derivatives. *Journal of Medical Chemistry*, 6:716–719.

Mandel, L. R., Ahn, H. S., VandenHeuvel, W. J., and Walker, R. W. (1972): Indoleamine-N-methyl transferase in human lung. *Biochemical Pharmacology*, 21:1197–1200.

Mandell, A. J., and Morgan, M. (1971): Indole(ethyl)amine N-methyltransferase in human brain. *Nature New Biology*, 230:85–87.

Martin, W. R., Sloan, J. W., Christian, S. T., and Clements, T. H. (1972): Brain levels of tryptamine. *Psychopharmacologia* (Berlin), 24:331–346.

Murphy, D. L., and Wyatt, R. J. (1972): Reduced monoamine oxidase activity in blood platelets from schizophrenic patients. *Nature*, 238:225.

Narasimhachari, N., Heller, B., Spaide, J., Haskovec, L., Meltzer, H., Strahilevitz, M., and Himwich, H. E. (1971): N,N-Dimethylated indoleamines in blood. *Biological Psychiatry*, 3:21–23.

Narasimhachari, N., Plaut, J. M., and Himwich, H. E. (1972): Indolethylamine-N-methyltransferase in serum samples of schizophrenics and normal controls. *Life Sciences II*, Part II:221–227.

Rosenberg, D. E., Isbell, H., and Miner, E. J. (1963): Comparison of a placebo, N-dimethyltryptamine, and 6-hydroxy-N-dimethyltryptamine in man. *Psychopharmacologia*, 4:39–42.

Rosengarten, H., Meller, E., and Friedhoff, A. J. (1973): Indolethylamine-N-methyl transferase in human blood cells. *Abstract in Fourth Meeting of American Society for Neurochemistry*, Columbus, Ohio.

Rosengarten, H., Szemis, A., Piotrowski, A. Romaszewska, K., Matsumoto, H., Stencka, K., and Jus, A. (1970): N,N-Dimethyltryptamine and bufotenine in the urine of patients with chronic and acute schizophrenic psychosis. *Psychiatria Polska*, 4:519–521.

Saavedra, J. M., and Axelrod, J. (1972a): A specific and sensitive enzymatic assay for tryptamine in tissues. *Journal of Pharmacology and Experimental Therapeutics*, 182:363–369.

Saavedra, J. M., and Axelrod, J. (1972b): Psychotomimetic N-methylated tryptamine: Formation in brain *in vivo* and *in vitro*. *Science*, 24:1365–1366.

Saavedra, J. M., and Axelrod, J. (1973): Effect of drugs on the tryptamine content of rat tissues. *Journal of Pharmacology and Experimental Therapeutics*, 185:523–529.

Sai-Halasz, A. (1962): The effect of antiserotonin on the experimental psychosis induced by dimethyltryptamine. *Experientia* (Basel),18:137–138.

Sai-Halasz, A. (1963): The effect of MAO inhibition on experimental psychosis induced by dimethyltryptamine. *Psychopharmacologia*, 4:385–388.

Sai-Halasz, A., Brunecker, G., and Szara, S. (1958): Dimethyltryptamine: Ein neues Psychoticum. *Psychiatry Neurology* (Basel), 135:285–301.

Stern, W. C., Morgan, P. J., and Bronzino, J. D. (1972): LSD: Effects on sleep patterns and spiking activity in the lateral geniculate nucleus. *Brain Research*, 41:199–204.

Szara, S. (1956): Dimethyltryptamine—Its metabolism in man; the relation of its psychotic effect to the serotonin metabolism. *Experientia*, 12:441–442.

Szara, S. (1961): Hallucinogenic effects and metabolism of tryptamine derivatives in man. *Federation Proceedings*, 20:885–888.

Szara, S., and Axelrod, J. (1959): Hydroxylation and N-dimethylation of N,N-dimethyltryptamine. *Experientia*, 15:216–217.

Szara, S., and Hearst, E. (1962): The 6-hydroxylation of tryptamine derivatives: A way of producing metabolites. *Annals of the New York Academy of Sciences*, 96:134–141.

Tanimukai, H., Ginter, R., Spaide, J., Bueno, J. R., and Himwich, H. E. (1970): Detection of psychotomimetic N,N-dimethylated indoleamines in the urine of four schizophrenic patients. *British Journal of Psychiatry*, 117:421–430.

Teixeira, A. R., Bueno, J. R., and Do Lago, G. P. (1968): Uptake of serotonin (5-hydroxy-tryptamine) and N,N-dimethyltryptamine (DMT) by dog's blood cells "in vitro": Influence of chlorpromazine. *Journal Brasileiro de Psyquiatria*, 17:53–58.

Turner, W. J., and Merlis, S. (1959): Effect of some indolealkylamines in man. *American Medical Association Archives of Neurology and Psychiatry*, 81:121–129.

Walker, R. W., Ahn, H. S., Albers-Schonberg, G., Mandel, L. R., and VandenHeuvel, W. J. (1973): Gas chromatographic-mass spectrometric isotope dilution assay for N,N-dimethyltryptamine in human plasma. *Biochemical Medicine*, 8:105–113.

Walker, R. W., Ahn, H. S., Mandel, L. R., and VandenHeuvel, W. J. (1972) Identification of N,N-dimethyltryptamine as the product of an *in vitro* enzymatic methylation. *Analytical Biochemistry*, 47:228–234.

Wyatt, R. J., Mandel, L. R., Ahn, H. S., Walker, R. W., and VandenHeuvel, W. J. (1973a): Gas-chromatographic-mass spectrometric isotope dilution determination of N,N-dimethyltryptamine concentrations in normals and psychiatric patients. *Psychopharmacologia*, 31:265–270.

Wyatt, R. J., Murphy, D. L., Belmaker, R., Cohen, S., Donnelly, C. H., and Pollin, W. (1973b): Reduced monoamine oxidase activity in platelets: A possible genetic marker for vulnerability to schizophrenia. *Science*, 173:916–918.

Wyatt, R. J., Saavedra, J. M., and Axelrod, J. (1973c): A dimethyltryptamine (DMT) forming enzyme in human blood. *American Journal of Psychiatry*, 130:754–760.

Advances in Biochemical Psychopharmacology, Vol. 11
Raven Press, New York © 1974

Melatonin: Effects on Brain Function

F. Antón-Tay

Department of Endocrinology, Biomedicas UNAM, Mexico City 20, Mexico

I. INTRODUCTION

Although our knowledge of the effects of melatonin administration is far from complete, it seems certain that this hormone has pronounced effects on the activity of the CNS. The administration of melatonin both to experimental animals and human subjects results in changes in neuroendocrine function, behavior, electroencephalographic activity, and some aspects of brain metabolism. Moreover, melatonin has been reported to have beneficial effects on Parkinson's disease and temporal lobe epilepsy.

Despite the interesting implications of these actions, there is little information regarding its precise mechanism of action on brain activity. A tentative model developed to integrate available information about melatonin effects on the CNS is therefore presented here.

II. MELATONIN EFFECTS ON BRAIN ACTIVITY

Early evidence showed that melatonin exerts an important effect upon diverse endocrine functions (Wurtman, Axelrod, and Kelly, 1968; Reiter and Fraschini, 1969; Wurtman and Antón-Tay, 1969). However, from more recent results, these effects are thought to be mediated via endocrine areas in the hypothalamus (see review by Reiter, 1973). Evidence that other aspects of brain function and structure are modified by melatonin administration is accumulating. In recent years we have been interested in studying metabolic, electrophysiological, and clinical effects of melatonin. Since the proposed model is based largely on these studies, we shall review them briefly.

A. Metabolic Effects

After melatonin administration there is a wide range of changes in brain metabolic activity. Initially we observed an almost twofold increase in levels of γ-aminobutyric acid (GABA) in hypothalamus and cerebral cortex following intraperitoneal administration of melatonin (50 μg/kg) to rabbits. Conversely, with larger doses (250 μg/kg), a decrease in GABA concentration was found (Antón-Tay, 1971).

It was later reported that serotonin (5-HT) levels are also modified by melatonin. Serotonin concentration increased in whole rat brain 60 min after administration of 500 μg/kg of melatonin (Antón-Tay, Chou, Antón, and Wurtman, 1968). The timing of 5-HT changes differed according to the region studied. Twenty minutes after hormone administration, 5-HT content was significantly decreased, by 14%, in the cerebral cortex. After 60 min, 5-HT levels were still decreased, while there was a significant increase in the midbrain hypothalamus. 5-HT level was still rising in the midbrain after 180 min, while cortical and hypothalamic levels returned to control values. Larger doses of melatonin appeared to be less effective in eliciting these changes.

Pyridoxal phosphokinase activity in brains of rats treated intraperitoneally with 200 μg/kg of melatonin showed an almost sixfold increase 45 min after hormone injection (Antón-Tay, Sepúlveda, and González, 1970). Enzyme activity remained high even after 180 min, peaking before the maximal changes observed for 5-HT and GABA.

Pyridoxal phosphokinase catalyzes the formation of pyridoxal phosphate, a prosthetic group of aromatic L-amino acid decarboxylase, and of glutamic acid decarboxylase, the synthesizing enzymes of 5-HT and GABA, respectively. It has therefore been suggested that the effects

TABLE 1. *Changes in cAMP content of various brain regions in the rat after melatonin administration*

Region	cAMP (pM/mg of wet tissue)	
	Control	Melatonin
Midbrain	2.33 ± 0.23	1.62 ± 0.24^{a}
Cerebellum	5.40 ± 0.33	8.50 ± 0.68^{b}
Cerebral cortex	3.21 ± 0.39	3.31 ± 0.25

Groups of five rats received 200 μg of melatonin intraperitoneally. The animals were killed 15 min later and the brain immediately frozen. Each brain was dissected and assayed for cAMP. Each sample was assayed in triplicate by radioimmunoassay. Data are presented as mean ± SE.

[a] $p < 0.01$.
[b] $p < 0.001$.

of melatonin on GABA and 5-HT are mediated by modifications on pyridoxal phosphokinase activity.

Recently, evidence was obtained indicating that melatonin also modifies the activity of adenylcyclase (Ortega, Antón-Tay, Esparza, and Cancino, 1974). Brain 3',5'-adenosine monophosphate (cAMP) levels were assayed by radioimmunoassay after intraperitoneal administration of melatonin to rats. The effect of melatonin on cAMP levels varied according to the region studied (Table 1); the midbrain showed a 31% fall in contrast to the cerebellum, which showed a 57% increase, while concentrations in cerebral cortex remained unchanged. Maximal changes in cAMP levels were detected 10 min after melatonin administration. The decreased level in midbrain was apparent for 40 min.

B. Electrophysiological Effects

Behavioral and electroencephalographic studies have shown that melatonin administration to cats and chickens results in EEG desynchronization and sleep (Marczynski, Yamaguchi, Ling, and Grodzinska, 1964; Hishikawa, Cramer, and Kuhlo, 1969; Barchas, Da Costa, and Spector, 1967).

In human subjects melatonin also elicited EEG and behavioral changes. Thus, the administration of 1.25 μg/kg of melatonin to healthy volunteers was followed by slight EEG deactivation and sleep during the next 20 min (Antón-Tay, Díaz, and Fernández-Guardiola, 1971). EEG monitoring during the following 2 hr showed an increase in alpha rhythm which appeared to have greater amplitude. On questioning after these experiments, most subjects referred to a sensation of well-being, comfort, and elation.

The effects of melatonin on sleep have recently been studied in greater detail (Fernández-Guardiola, Contreras, Calvo, Brailowsky, Solis, and Antón-Tay (*Unpublished Results*). Six healthy volunteers received melatonin or placebo in a double-blind trial, either 250 mg in carbowax 400 orally four times daily over a period of 6 days or carbowax 400 alone. Between test and placebo, subjects were allowed a rest period of 7 days. Hypnograms were constructed from all-night polygraphic recordings for each subject during placebo, melatonin, or habituation periods. The most prominent changes were enhancement of phase II, shortening of phase IV, and increase of the cholinergic components of sleep (galvanic skin response, heart rate, and number of rapid eye movements).

Some information about the electrical activity of deep brain structures is also available. A group of three patients, suffering from temporal lobe epilepsy, and wearing deep electrodes in both temporal lobes for diagnostic and therapeutic purposes, were given a single dose of melatonin (1.25 μg/kg) (Antón-Tay et al., 1971). The hormone produced deactivation of the percutaneous EEG, a progressive decrease in the amplitude of temporal lobe electrical activity, and a tendency for the amygdaloid nucleus to be synchronized with the cerebral cortex. In these patients, the typical theta-hypocampal activity was mixed with short trains of 15 to 25 cps. Concurrently with these changes, the frequency of paroxysmal graphoelements decreased in the 5 hr following injection. These results suggest that melatonin has a depressing effect on the recorded subcortical limbic structures (Antón-Tay et al., 1971).

Additional information has been obtained of melatonin effects on subcortical electrical activity in cats. The multineuronal unit activity of the red nucleus was increased after intraperitoneal administration of small doses of melatonin (5 μg/kg). The increased firing of red nucleus neurons started 10 min after administration and lasted 30 min (Fernández-Guardiola, Brailowsky, Calvo, Solis, and Antón-Tay (*Unpublished Results*). These results suggest that the effects are mediated through a multineuronal pathway, that is, the cerebello-rubro-dentate pathway.

C. Therapeutic Studies

Some possible therapeutic applications of melatonin based upon its metabolic effects were suggested in 1971 (Antón-Tay et al., 1971) and investigated thereafter by several groups.

In an initial study, melatonin was found to be effective in ameliorating the clinical signs of two patients with Parkinson's disease (Antón-Tay et al., 1971). This study was later extended and confirmed in two groups of patients: The first consisted of five patients to whom the hormone was given orally during 4 weeks at a dose of 1.2 g daily in an open trial; in the second the same dose was given in a random double-blind, cross-over trial (Antón-Tay and Fernández-Guardiola, 1973). During the first weeks of treatment, subjective improvement was reported by all patients in both groups. Objective changes were observable during the second week of treatment and consisted mainly in improvement of daily task performance. At the end of the fourth week all patients showed amelioration in their clinical picture. According to Webster's scale, patients in the first group showed 36% improvement, while those in the second group showed 20%. Two weeks after the end of the trial, all patients showed a progressive deterioration. Partial results of these studies have been published elsewhere (Antón-Tay and Fernández-Guardiola, 1973).

The beneficial effects of melatonin in Parkinson's disease have not

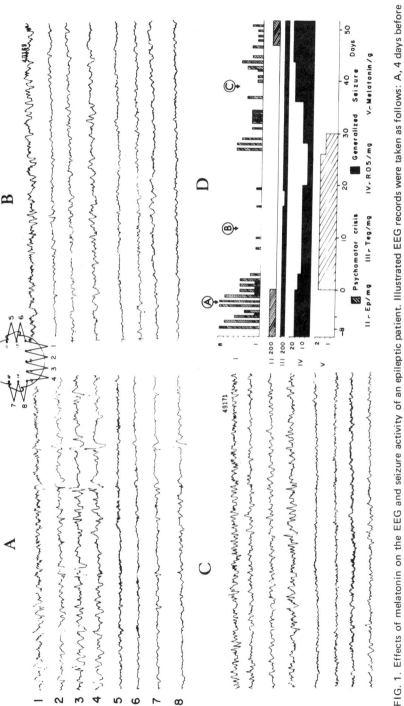

FIG. 1. Effects of melatonin on the EEG and seizure activity of an epileptic patient. Illustrated EEG records were taken as follows: A, 4 days before the administration of melatonin; B, the 11th day of administration; and C, 9 days after withdrawal of the hormone. D shows the number of seizures per day (I); the dosage of diphenylhydantoinate; (III): carbamezapin; (III): Ro5-4023 (IV); and melatonin (V). The patient was a chronic temporal lobe epileptic with intractable generalized seizures (3 to 8 per day). During his habitual treatment the EEG (A) showed a parietotemporal left focus of repetitive spikes. When melatonin was given (2 g daily) diphenylhydantoinate was discontinued and the dose of Ro5-4023 reduced. During this period, the number of seizures decreased and the EEG showed diffuse theta activity and decrease of paroxysmal grapho-elements. (B) After melatonin was discontinued, the number of seizures increased and the EEG recordings again showed a similar pattern of spiking to that before melatonin treatment.

been confirmed by other authors. Cotzias, Papavasiliou, Ginos, Steck, and Duby (1971) reported that one patient treated for 7 weeks with an oral daily dose of 1.3 g showed control of tremor and somnolence but no changes in the other clinical features. More recently, Shaw, Stern, and Sandler (1973) reported that doses up to 1 g of melatonin were ineffective in modifying the clinical picture in a group of four parkinsonian patients in whom the only noticeable effect was sedation. These contradictory results are probably due to differences in method of administration; since melatonin is practically insoluble in biological fluids, it must be given in an appropriate vehicle in order to be absorbed by the gastrointestinal tract. In our initial studies, we used hormone solubilized in 2% ethanol for intravenous administration and in elixir for oral administration. However, since ethanol is active on the CNS, it was difficult to eliminate nonspecific effects. Recently, we have found that carbowax 400 is a more suitable solvent for melatonin. Furthermore, it is inactive when administered orally. Even though we have not used this vehicle for melatonin administration to parkinsonians as yet, evidence that melatonin reaches the brain when given in this vehicle was obtained in the study on sleep mentioned above (Fernández-Guardiola, Contreras, Calvo, Brailowsky, Solis and Antón-Tay, *Unpublished Results*) and in a new group of epileptic patients (*vide infra*).

In order to evaluate the results obtained earlier in epileptics, a group of six patients with intractable epilepsy of various etiologies were given orally 2 g of melatonin dissolved in carbowax 400 in an open trial over 30 days (Foster, Fernández-Guardiola, and Antón-Tay, *Unpublished Results*). In all patients a reduction in spiking activity was observed. The patients also showed a decrease in frequency of seizures during the treatment, but needed other anticonvulsants (Fig. 1). These preliminary results confirm previous data and show that melatonin acts on inhibitory subcortical mechanisms.

III. PROPOSED SCHEME FOR THE ACTION OF MELATONIN ON MAMMAL BRAIN

A scheme for the action of melatonin on the CNS of mammals is proposed in Fig. 2. It has been postulated (Robison, Butcher, and Sutherland, 1968) that hormones exert their effects in peripheral organs by modifying the activity of enzymes in close relationship to the receptor, mainly adenylcyclase. This enzyme catalyzes the conversion of ATP to cAMP which, acting as a "second messenger," is responsible for the intracellular metabolic effects of the hormone. The activity of adenylcyclase in the CNS is higher than in any other tissue (Sutherland, Rall, and Menon, 1962; De Robertis, Rodríguez de Lores Arnaiz, and Alberici, 1967; Weiss and Costa, 1968), and much evidence indicates

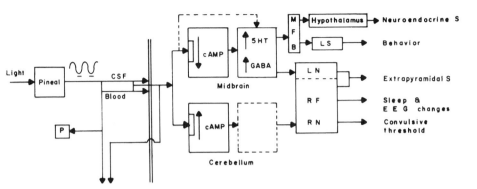

FIG. 2. Schematic diagram of melatonin effects on CNS. CSF, cerebrospinal fluid; MFB, medial forebrain bundle; LN, locus niger; LS, limbic system; P, peripheral tissues; RF, reticular formation; RN; red nucleus. For description see text.

that it is related to receptor sites (Robison et al., 1967). The formation of cAMP in the CNS appears to be regulated by at least two interacting factors: biogenic amines, such as catecholamines, serotonin, and histamine, and membrane depolarization level (Shimizu, Crevelin, and Daly, 1970). The changes in brain cAMP observed after melatonin administration suggest that the hormone acts on at least two different kinds of receptor, located in midbrain and cerebellum. The midbrain receptor responds to melatonin by a fall in cAMP level while that in the cerebellum produces an increase of nucleotide. Although the physiological role of cAMP in the CNS is not clear, it is interesting to note that the sequence of metabolic changes induced by melatonin is such that we might assume a causal relationship between them. Thus, after melatonin administration maximal changes in cAMP level appear after 10 min, in pyridoxal phosphokinase activity after 45 min, and in midbrain 5-HT concentration after 60 min.

It is interesting to examine the effect of melatonin on brain cAMP from the phylogenetic point of view. In lower vertebrates, the skin darkening effect of β-MSH is thought to be mediated by increased cAMP levels in melanophores (Abe, Robison, Liddle, Butcher, Nicholson, and Baird, 1969). Melatonin inhibits skin darkening in response to MSH, and also inhibits the MSH-induced rise of cAMP. Thus, it seems acceptable to think that melatonin acts on ectodermic structures through the same mechanism. In lower vertebrates the target organ is the skin, while in mammals it is the brain.

Some melatonin effects may be explicable by different mechanisms; that is, melatonin may compete with 5-HT at receptor sites in the telencephalon, causing an increased synthesis of the monoamine in the midbrain. Moreover, it is possible that melatonin also acts on other brain systems. In this context the hypothalamus is particularly interesting, since it shows a rich uptake of melatonin (Antón-Tay and Wurtman, 1969) and hypothalamic levels of melatonin remain unchanged after pinealectomy (Green, Koslow, Spano, and Costa, 1972).

The differential actions of melatonin on midbrain and cerebellum provide an explanation for various effects observed after its administration. Since serotonergic neurons of the midbrain have axons projecting to the hypothalamus and to the limbic system through the medial forebrain bundle, the neuroendocrine and behavioral effects of melatonin are probably mediated by activation of these serotonergic pathways. Similarly, changes in sleep pattern and convulsive threshold observed after melatonin administration may be related to the activation of neural pathways that leave the cerebellum to enter the red nucleus and the reticular formation.

It has been reported that electrical stimulation of the cerebellum depresses convulsive activity (Snider and Cooke, 1954; Dow, Fernández-Guardiola, and Manni, 1962), so that the action of melatonin upon this structure is likely to trigger the same mechanism. Moreover, the increased neuronal activity of the red nucleus is followed by depression of the spinal monosynaptic reflex through the rubro-spinal tract (Fernández-Guardiola, Muñzo-Martizez, and Velazco, 1964), a mechanism that might account for the effects of the hormone on the tremor of Parkinson's disease.

IV. SUMMARY

Melatonin effects on the CNS are probably mediated by changes in levels of cAMP. According to the kind of response (increase or decrease in cAMP concentration), two different receptors at least might exist in the midbrain and the cerebellum, respectively. By acting at these sites, it is possible for melatonin to modulate the activity of various neural pathways.

ACKNOWLEDGMENTS

Published and unpublished studies described in this paper were supported by grants from the Ford Foundation, 710-0099, and Patronato Lucio Laguete and FORGUE.

REFERENCES

Abe, K., Robison, G. A., Liddle, G. W., Butcher, R. W., Nicholson, W. E., and Baird, C. E. (1969): Role of cyclic AMP in mediating the effects of MSH,

norepinephrine and melatonin on frog skin color. *Endocrinology*, 85:674–682.
Antón-Tay, F. (1971): Pineal-brain relationships. In: *The Pineal Gland*, edited by G. E. W. Wolstenholme and J. Knight, pp. 213–227. Churchill, London.
Antón-Tay, F., Chou, C., Antón, S., and Wurtman, R. J. (1968): Brain serotonin concentration: Elevation following intraperitoneal administration of melatonin. *Science*, 162:277–278.
Antón-Tay, F., Díaz, J. L., and Fernández-Guardiola, A. (1971): On the effect of melatonin upon human brain. Its possible therapeutic implications. *Life Sciences*, Part I, 10:841–850.
Antón-Tay, F., and Fernández-Guardiola, A. (1973): Changes in human brain activity after melatonin administration. (in preparation).
Antón-Tay, F., Sepúlveda, J., and González, S. (1970): Increase of brain pyridoxal phosphokinase activity following melatonin administration. *Life Sciences*, Part II, 9:1283–1288.
Antón-Tay, F., and Wurtman, R. J. (1969): Regional uptake of ^{3}H-melatonin from blood or cerebrospinal fluid by rat brain. *Nature*, 221:474–475.
Barchas, J., Da Costa, F., and Spector, S. (1967): Acute pharmacology of melatonin. *Nature*, 214:919–920.
Cotzias, G. C., Papavasiliou, P. S., Ginos, J., Steck, A., and Duby, S. (1971): Metabolic modification of Parkinson's disease and of chronic manganese poisoning. In: *Annual Review of Medicine*, Vol. 22, edited by A. C. Degraff and W. P. Creger, pp. 305–326. Annual Reviews, Inc., Palo Alto, Calif.
De Robertis, E., Rodriquez de Lores Arnaiz, G., and Alberici, M. (1967): Subcellular distribution of adenyl cyclase and cyclic phosphodiesterase in rat brain cortex. *Journal of Biological Chemistry*, 242:3487–3493.
Dow, R. S., Fernández-Guardiola, A., and Manni, E. (1962): The influence of the cerebellum on experimental epilepsy. *Electroencephalography and Clinical Neurophysiology*, 14:383–398.
Fernández-Guardiola, A., Muñóz-Martínez, E., and Velazco, M. (1964): Facilitatory and inhibitory influences on spinal activity during the cortical tonic-clinic postdischarge. *Boletín del Instituto de Estudios Médicos y Biológicos*, 22:205–215.
Green, A. R., Koslow, S. H., Spano, P. F., and Costa, E. (1972): Identification of melatonin and 5-methoxy-tryptamine in rat hypothalamus by gas chromatography-mass spectrometry. In: *Proceedings of the Fifth International Congress on Pharmacology*, p. 87.
Hishikawa, Y., Cramer, H., and Kuhlo, W. (1969): Natural and melatonin-induced sleep in young chickens—A behavioral and electrographic study. *Experimental Brain Research*, 7:84–94.
Marczynski, T. J., Yamaguchi, N., Ling, G. M., and Grodzinska, L. (1964): Sleep induced by the administration of melatonin (5-methoxy-N-acetyltryptamine) to the hypothalamus in unrestrained cats. *Experientia*, 20:435–436.
Ortega, B. G., Antón-Tay, F., Esparza, N., and Cancino, F. M. (1974): Melatonin: Effect on cAMP concentration in the brain. In: *Proceeding of the Fourth International Meeting of the International Society for Neurochemistry*. Tokyo, Japan (*in press*).
Reiter, R. J. (1973): Comparative physiology: Pineal gland. In: *Annual Review of Physiology*, Vol. 35, edited by V. E. Hall, A. C. Giese, and R. R. Sonnenshein, pp. 305-328. Annual Reviews, Inc., Palo Alto, Calif.
Reiter, R. J., and Fraschini, F. (1969): Endocrine aspects of the mammalian pineal gland: A review. *Neuroendocrinology*, 5:219-255.
Robison, G. A., Butcher, R. W., and Sutherland, E. W. (1968): Cyclic AMP. In: *Annual Reviews of Biochemistry*, Vol. 37, edited by E. E. Snell, P. D. Boyer, A. Meister, and R. L. Sinsheimer, pp. 149–176. Annual Reviews, Inc., Palo Alto, Calif.
Shaw, K. M., Stern, G. M., and Sandler, M. (1973): Melatonin and parkinsonism.

Lancet, 1:271.

Shimizu, H., Creveling, C. R., and Daly, J. W. (1970): Effect of membrane depolarization and biogenic amines on the formation of cyclic AMP in incubated brain slices. In: *Advances in Biochemical Psychopharmacology*, Vol. 3, edited by E. Costa and P. Greengard, pp. 135—154. Raven Press, New York.

Snider, R. S., and Cooke, P. M. (1954): Cerebral seizures as influenced by cerebellar stimulation. *Transactions of the American Neurological Association*, 79:87—89.

Sutherland, E. W., Rall, T. W., and Menon, T. (1962): Adenyl cyclase. I. Distribution, preparation and properties. *Journal of Biological Chemistry*, 237:1220—1227.

Weiss, B., and Costa, E. (1968): Regional and subcellular distribution of adenyl cyclase and 3',5'-cyclic nucleotide phosphodiesterase in brain and pineal gland. *Biochemical Pharmacology*, 17:2107—2116.

Wurtman, R. J., and Antón-Tay, F. (1969): The mammalian pineal as a neuroendocrine transducer. In: *Recent Progress in Hormone Research*, Vol. 25, edited by E. B. Astwood, pp. 493—514. Academic Press, New York and London.

Wurtman, R. J., Axelrod, J., and Kelly, D. E. (1968): *The Pineal*, edited by R. J. Wurtman, J. Axelrod, and D. E. Kelly. Academic Press, New York.

Advances in Biochemical Psychopharmacology, Vol. 11
Raven Press, New York © 1974

Plasma Tryptophan Binding and Depression

A. Coppen, E. G. Eccleston, and M. Peet

*Medical Research Council Neuropsychiatry Unit, Carshalton and
West Park Hospital, Epson, Surrey, England*

I. INTRODUCTION

We wish to report an investigation into the concentration of total acid-soluble and of ultrafiltrable ("free") tryptophan in the plasma of women suffering from a depressive illness. We have previously reported (Coppen, Eccleston, and Peet, 1972*b*) that although total plasma acid-soluble tryptophan is normal, the percentage bound to plasma protein is increased in women suffering from depression and, hence, the free tryptophan in plasma is considerably reduced. It is probable that free plasma tryptophan concentration determines the concentration of brain tryptophan (Tagliamonte, Biggio, and Gessa, 1971*a*); this, in turn, is one of the important factors influencing the rate of 5-hydroxy-tryptamine (5-HT) synthesis (Tagliamonte, Tagliamonte, Perez-Cruet, and Gessa, 1971*b*; Fernstrom and Wurtman, 1971), which many investigators suggest may be reduced in depressive illness (Coppen, 1972; Coppen, Prange, Whybrow, and Noguera, 1972*c*; Lapin and Oxenkrug, 1969; van Praag, 1970). Our previous finding (Coppen, Brooksbank, and Peet, 1972*a*) of a decreased concentration of tryptophan in the lumbar cerebrospinal fluid of depressive patients also supports the notion that the concentration of tryptophan may be decreased in the brains of depressive patients.

In this chapter we present further data on a larger sample of depressive female patients and female controls, and we also present evidence on the changes in total acid-soluble and free plasma tryptophan occurring after recovery, both in patients not receiving drugs when they were tested and in a group of patients receiving lithium carbonate prophylactically.

II. METHODS

Twenty-four female patients suffering from a depressive illness were tested. They did not have any history of manic illness. The patients were tested after they had been in hospital for at least a week and none had received any medication, apart from night sedation, for at least this period. Control subjects were female volunteers who had no known history of psychiatric disorder and who were not receiving any medication. The controls were selected so that their age range was similar to that of the depressive patients. Two other groups of patients were examined. A group of 11 depressive female patients, who at the time of testing were not receiving any treatment or drugs and who had been well for at least 2 weeks, were examined after clinical recovery. Six of these patients had been tested when ill, then treated by electroconvulsive therapy, and retested. A group of 11 recovered patients receiving prophylactic lithium were also tested. These patients had been well for several months. Thirty milliliters of blood was obtained from both patients and controls at 9 a.m. The blood was collected into calcium heparin tubes, centrifuged immediately, and the plasma pipetted at once into plain plastic tubes. The plasma was ultrafiltered on the same day and samples of the ultrafiltrate and plasma were deep-frozen (Eccleston, 1973). Free and total acid-soluble tryptophan were then estimated by a modification of the method of Denckla and Dewey (1967). All subjects had been fasting overnight before the blood was taken.

III. RESULTS

The results are shown in Tables 1 and 2 and in Figs. 1 and 2. There was no significant difference between the means of the ages of the control subjects and the depressive patients.

Total acid-soluble plasma tryptophan: There was no significant difference between the mean total acid-soluble plasma tryptophan concentration of the control subjects and patients tested when depressed or after recovery. The recovered patients who were being treated by lithium, however, had a significantly decreased mean total acid-soluble plasma tryptophan concentration.

Free plasma tryptophan: The mean free plasma tryptophan concentration was very significantly decreased in patients tested when depressed. This increased in patients tested after recovery to a level which was significantly higher both in the patients who had recovered and were not on drugs, and in those who were receiving lithium when tested. There was no significant difference in mean free tryptophan levels between these two groups of recovered patients. Both recovered groups had lower free plasma tryptophan levels than the control group, and when the two recovered groups were combined, their mean was significantly lower ($p < 0.01$) than the control group. In the six

TABLE 1. *Plasma total and free tryptophan concentration in female depressive patients and control subjects*

	N	Age (yrs) mean	Plasma Tryptophan Total (μg/ml) mean	Free (μg/ml) mean	Free as percent of total mean
Control subjects	26	48.6	12.4	1.39	11.2
Depressed patients	24	55.3	12.4	0.80[a]	6.5[a]
Recovered depressives (not on treatment)	11	56.3	11.5	1.17	10.2
Recovered depressives (on lithium)	11	52.5	10.5[b]	1.13	10.9

[a]Depressed patients significantly less ($p < 0.001$) than controls and recovered depressives (both untreated and on lithium).
[b]Recovered depressives (on lithium) less than control subjects ($p < 0.01$).

patients who were tested both before and after recovery, the free plasma tryptophan significantly increased on the second occasion (Table 2). There was no correlation between free plasma tryptophan and age in the control subjects ($r = -0.16$, not significant), but there was a small positive correlation in the depressive patients ($r = 0.46$, $p < 0.05$). There was a significant positive correlation between total acid-soluble and free tryptophan concentration in the control group (r

TABLE 2. *Plasma total and free tryptophan concentration in six patients tested before and after clinical recovery*

	N	Total (μg/ml) mean	SE	Free (μg/ml) mean	SE	Free as percent of total mean	SE
Depressed	6	10.65	0.67	0.66	0.10	6.32	1.12
After recovery	6	11.55	0.73	1.22[a]	0.10	10.58[b]	0.72

Recovered greater than depressed [a]$p < 0.01$, [b]$p < 0.005$.

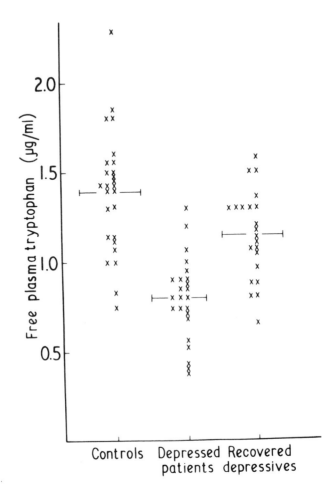

FIG. 1. Free tryptophan concentration in the plasma of female depressive patients, recovered depressives, and control subjects.

= 0.56, $p < 0.01$) but not in the depressive patients ($r = 0.38$, not significant).

Free tryptophan as percent of total acid-soluble plasma tryptophan: This also was very significantly lower in the patients suffering from depression, but rose almost to normal in both recovered groups. There was no significant difference in percent free tryptophan between the recovered groups and the control group. There was no correlation between percent free tryptophan and age in either the control or depressive group.

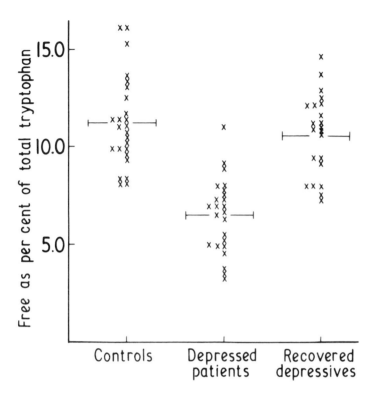

FIG. 2. Free tryptophan as percent of total tryptophan in the plasma of female depressive patients, recovered depressives, and control subjects.

IV. DISCUSSION

The concentration of total acid-soluble tryptophan in the plasma of female depressive patients tested at 9 a.m. was normal and did not change with clinical recovery. These observations are similar to those reported by Birkmayer and Linauer (1970). There was a very marked reduction in free tryptophan in the plasma of female depressive patients tested when clinically depressed. This increased very significantly after clinical recovery, but did not quite attain normal levels. The amount of free tryptophan, expressed as a percent of total acid-soluble trypto-phan, was also very significantly reduced in depressive patients, and this returned to normal levels after clinical recovery. These differences in both free tryptophan and the percentage that is free are very marked and, indeed, there is little overlap in the values obtained from depressive women and normal subjects.

There is little known about physiological factors which determine the amount of free tryptophan in plasma. It is of significance that if the binding of tryptophan to protein is diminished by drugs, such as salicylates (Tagliamonte et al., 1971a; Korf, van Praag and Sebens, 1972), the level of tryptophan in the brain is increased while total plasma tryptophan is unchanged or even decreased. The cause of the change in binding in the depressive patients is unknown. The patients were not on drugs, they had been on a full hospital diet for at least a week, and all subjects were tested after overnight fasting and at a standardized time. The concentration of plasma free fatty acids, which has been demonstrated to affect protein binding of tryptophan (Curzon, Friedel, and Knott, 1973; Lipsett, Madras, Wurtman, and Munro, 1973) is now being measured in these patients.

There is considerable evidence that there is some deficiency in brain 5-hydroxytryptamine (5-HT) in depressive illness (Coppen, 1972; Coppen et al., 1972c; Lapin and Oxenkrug, 1969; van Praag, 1970). This is based on postmortem examination of the brains of depressive subjects for 5-HT and for its metabolite 5-hydroxyindoleacetic acid (5-HIAA) (Shaw, Camps, and Eccleston, 1967; Bourne, Bunney, Colburn, Davis, Davis, Shaw, and Coppen, 1968: Pare, Yeung, Price, and Stacey, 1969). There is also evidence of decreased cerebrospinal fluid (CSF) levels of 5-HIAA (Coppen et al., 1972c; Ashcroft, Crawford, Eccleston, Sharman, MacDougall, Stanton, and Burns, 1966; Dencker, Malm, Roos, and Werdinius, 1966), and diminished increase of CSF 5-HIAA after the administration of probenecid (Sjöstrom and Roos, 1970; van Praag, Korf, and Puite, 1970) in these patients. Tryptophan concentration in cerebrospinal fluid has been found to be decreased in depressive illness (Coppen et al., 1972a). These findings could all be related to a decrease in brain tryptophan which, in turn, could be due to the very considerable decrease in plasma free tryptophan, since it is the free tryptophan only that is available for transfer to the central nervous system.

There is now good evidence that tryptophan will enhance the antidepressant action of monoamine oxidase inhibitors (Ayuso-Gutierrez and Lopez-Ibor, 1971; Coppen, Shaw, and Farrell, 1963; Glassman and Platman, 1969; Pare, 1963), and there are reports that tryptophan by itself is an antidepressant (Coppen, Whybrow, Noguera, Maggs, and Prange, 1972d; Broadhurst, 1970), although this had not been confirmed by all investigators (Carroll, 1971). It is not known what changes occur in the free tryptophan levels of patients fed large doses of tryptophan over a period of time, although the immediate effect is to increase the free tryptophan concentration (*unpublished observation*). Lithium has been reported to increase tryptophan levels and the synthesis of 5-HT in the brain (Tagliamonte, Tagliamonte, Perez-Cruet, Stern, and Gessa, 1971c). However, in this investigation we found no difference in plasma free tryptophan levels between patients maintained on lithium and recovered patients not on treatment after recovery.

It is known that a number of the antirheumatic drugs will increase the concentration of plasma free tryptophan. It is therefore of interest that there is a report that the antirheumatic drugs have an antidepressant action in rheumatic patients, and that the decrease in depression is correlated with the increase in plasma free tryptophan (Aylward and Maddock, 1973).

V. SUMMARY

Plasma acid-soluble total and free tryptophan concentrations was measured in 26 control women; in a group of 24 women suffering from a depressive illness; in 11 patients after recovery, not on drugs; and in 11 recovered patients on prophylactic lithium. It was found that the group of depressive women had very significantly reduced plasma levels of free tryptophan, but that their levels increased significantly after recovery although not returning fully to normal. This increase in free tryptophan after recovery was not influenced by the administration of lithium. Total acid-soluble plasma tryptophan was normal in depressive illness both before and after clinical recovery.

REFERENCES

Ashcroft, G. W., Crawford, T. B. B., Eccleston, D., Sharman, D. F., MacDougall, E. J., Stanton, J. B., and Burns, J. F. (1966): 5-Hydroxyindole compounds in the cerebrospinal fluid of patients with psychiatric or neurological disorders. *Lancet*, ii:1049–1052.

Aylward, M., and Maddock, M. (1973). Plasma-tryptophan levels in depression. *Lancet*, i:936.

Ayuso-Gutierrez, J. L., and Lopez-Ibor Alino, J. J. (1971): Tryptophan and an MAOI (Nialamide) in the treatment of depression. *International Pharmacopsychiatry*, 6:92–97.

Birkmayer, W., and Linauer, W. (1970): Storung des Tyrosin- und Tryptophanmetabolismus bei Depression. *Archiv für Psychiatrie und Nervenkrankheiten*, 213:377–387.

Bourne, H. R., Bunney, W. E., Colburn, R. W., Davis, J. M., Davis, J. N., Shaw, D. N., and Coppen, A. J. (1968): Noradrenaline, 5-hydroxytryptamine and 5-hydroxyindoleacetic acid in hindbrains of suicidal patients. *Lancet*, ii:805–808.

Broadhurst, A. D. (1970): Tryptophan in the treatment of depression. *Lancet*, i:1392.

Carroll, B. J. (1971): Monoamine precursors in the treatment of depression. *Clinical Pharmacology and Therapeutics*, 12:743–761.

Coppen, A. (1972): Indoleamines and affective disorders. *Journal of Psychiatric Research*, 9:163–171.

Coppen, A., Brooksbank, B. W. L., and Peet, M. (1972a): Tryptophan concentration in the cerebrospinal fluid of depressive patients. Lancet i:1393.
Coppen, A., Eccleston, E. G., and Peet, M. (1972b): Total and free tryptophan concentration in the plasma of depressive patients. Lancet ii:1415.
Coppen, A., Prange, A. J., Whybrow, P. C., and Noguera, R. (1972c): Abnormalities of indoleamines in affective disorders. Archives of General Psychiatry, 26:474–478.
Coppen, A., Shaw, D. M., and Farrell, J. P. (1963): Potentiation of the antidepressive effect of a monoamine-oxidase inhibitor by tryptophan. Lancet i:79–81.
Coppen, A., Whybrow, P. C., Noguera, R., Maggs, R., and Prange, A. J. (1972d): The comparative antidepressant value of L-tryptophan and imipramine with and without attempted potentiation by liothyronine. Archives of General Psychiatry, 26:234–241.
Curzon, G., Friedel, J., and Knott, P. J. (1973): The effect of fatty acids on the binding of tryptophan to plasma protein. Nature, 242:198–200.
Dencker, S. J. Malm, V., Roos, B.-E., and Werdinius, B. (1966): Acid monoamine metabolites in CSF in mental depression and mania. Journal of Neurochemistry, 13:1545–1548.
Denckla, W. D., and Dewey, H. K. (1967): The determination of tryptophan in plasma, liver and urine. Journal of Laboratory and Clinical Medicine, 69:160–169.
Eccleston, E. G. (1973): To be published.
Fernstrom, J. D., and Wurtman, R. J. (1971): Brain serotonin content: physiological dependence on plasma tryptophan levels. Science, 173:149–152.
Glassman, A. H., and Platman, S. R. (1969): Potentiation of a monoamine oxidase inhibitor, by tryptophan. Journal of Psychiatric Research, 7:83–88.
Korf, J., van Praag, H. M., and Sebens, J. B. (1972): Serum tryptophan decreased, brain tryptophan increased and brain serotonin synthesis unchanged after probenecid loading. Brain Research, 42:239–242.
Lapin, I. P., and Oxenkrug, G. F. (1969): Intensification of the central serotoninergic processes as a possible determinant of the thymoleptic effect. Lancet, i:132–136.
Lipsett, D., Madras, B. K., Wurtman, R. J., and Munro, H. N. (1973): Serum tryptophan level after carbohydrate ingestion: Selective decline in non-albumin-bound tryptophan coincident with reduction of serum free fatty acids. Life Sciences II, 12:57–64.
Pare, C. M. B. (1963): Potentiation of monoamine oxidase inhibitor by tryptophan. Lancet, ii:527.
Pare, C. M. B., Yeung, D. P. H., Price, K., and Stacey, R. S. (1969): 5-Hydroxytryptamine, noradrenaline, and dopamine in brain stem, hypothalamus and caudate nucleus of controls and patients committing suicide by coal-gas poisoning. Lancet, ii:133–135.
Shaw, D. M., Camps, F. E., and Eccleston, E. G.,(1967): 5-Hydroxytryptamine in the hindbrain of depressive suicides. British Journal of Psychiatry, 113:1407–1411.
Sjöstrom, R., and Roos, B.-E. (1970): Measurement of 5-HIAA and HVA in manic depressive patients after probenecid application. Read before the Seventh Congress of the Collegium Internationale Neuropsychopharmacologium, Prague.
Tagliamonte, A., Biggio, G., and Gessa, G. L. (1971a): Possible role of "free" plasma tryptophan in controlling brain tryptophan concentrations. Rivista di Farmacologia e Terapia, 11:251–255.
Tagliamonte, A., Tagliamonte, P., Perez-Cruet, J., and Gessa, G. L. (1971b): Increase of brain tryptophan caused by drugs which stimulate serotonin synthesis. Nature New Biology, 229:125–126.
Tagliamonte, A., Tagliamonte, P., Perez-Cruet, J., Stern, S., and Gessa, G. L. (1971c): Effect of psychotropic drugs on tryptophan concentration in the rat brain. Journal of Pharmacology and Experimental Therapeutics, 177:475–480.
van Pragg, H. M. (1970): Indoleamines and the central nervous system. A sounding

of their clinical significance. *Psychiatria, Neurologia, Neurochirurgia*, 73:9–36.
van Praag, H. M., Korf, J., and Puite, J. (1970): 5-Hydroxyindole levels in the CSF
of depressive patients treated with probenecid. *Nature*, 225:1259.

Advances in Biochemical Psychopharmacology, Vol. 11
Raven Press, New York © 1974

Mood and Neuronal Functions: A Modified Amine Hypothesis for the Etiology of Affective Illness

G. W. Ashcroft and A. I. M. Glen

*MRC Brain Metabolism Unit, University Department of Pharmacology,
Edinburgh, Scotland*

I. INTRODUCTION

We would like to report the results of investigations carried out by a number of workers in our unit in Edinburgh, which will show that there is no consistent relation between level of 5-hydroxyindoleacetic acid (5-HIAA) and mood state in patients with affective illness. As a result of these experiments, we have been led to develop a modification of the amine hypothesis for affective illness as it has been previously stated (Ashcroft et al., 1972); that is, we cannot now support the simple hypothesis that depression can occur when the levels of biological amines at reactive sites are reduced, and we shall outline a modification of the amine hypothesis which we believe may fit the facts more closely and which is susceptible to further testing.

II. CEREBROSPINAL FLUID STUDIES IN MAN

In 1960, Ashcroft and Sharman found that 5-HIAA was low in the cerebrospinal fluid (CSF) of patients with depression. Subsequent research has, on the whole, confirmed the findings. Recently, however, we have investigated 5-HIAA in diagnostic subgroups of patients with affective illness (Ashcroft et al., 1973a). Table 1 shows the results of these investigations. We confirmed that 5-HIAA is low in recurrent unipolar depressive illness, but we found normal values in bipolar depression and mania. The findings for the dopamine metabolite, homovanillic acid (HVA), are similar. We found no significant change in 5-HIAA following recovery in any of the groups of patients. Clearly, the CSF levels of these amine metabolites are not related to mood.

It has been argued by previous speakers, that low levels of 5-HIAA in

TABLE 1. *Mean ± SD concentrations (ng/ml) of 5-HIAA and HVA in lumbar CSF of psychiatric patients both before and after treatment and of controls*

	Unipolar depression	Bipolar depression	Bipolar mania	Control subjects
5-HIAA before treatment	$10 \pm 4(11)^a$	$18 \pm 8(9)$	$15 \pm 6(11)$	$16 \pm 8(30)$
5-HIAA after treatment	$12 \pm 3(10)$	$17 \pm 11(6)$	$8 \pm 4(4)$	
Change on treatment (paired difference)	None	None	None	
HVA before treatment	$20 \pm 9(11)^b$	$34 \pm 16(9)$	$35 \pm 15(11)$	$41 \pm 23(31)$
HVA after treatment	$23 \pm 10(9)^a$	$28 \pm 14(7)$	$17 \pm 4(6)^a$	
Change on treatment (paired difference)	None	Fall^a	None	

Number of observations are shown in parentheses.
$^a p < 0.05$ ⎱ comparison with controls; student's t test.
$^b p < 0.01$ ⎰

the unipolar depressive represent a decrease in turnover of 5-hydroxy-tryptamine (5-HT), and it has been suggested that there may be a deficiency of tryptophan in some patients. Unfortunately, we cannot substantiate this. Table 2, however, shows the results of experiments (Ashcroft et al., 1973*b*) designed to detect a deficiency in the activity of the rate-limiting enzyme for the synthesis of 5-HT, tryptophan-5-

TABLE 2. *Mean ± concentrations of 5-HIAA and tryptophan in the lumbar CSF of psychiatric patients 8 hr after the oral administration of L-tryptophan (50 mg/kg) as compared with controls treated similarly*

	Unipolar depression	Bipolar depression	Bipolar mania	Controls
5-HIAA (ng/ml)	$32 \pm 13 (10)$	$26 \pm 14 (5)$	$27 \pm 7(4)$	$29 \pm 14 (13)$
Tryptophan (ng/ml)	$2.4 \pm 1.0 (10)$	$3.3 \pm 1.0 (5)$	$2.1 \pm 0.5 (3)$	$2.8 \pm 1.4 (12)$

The concentrations of 5-HIAA and tryptophan in any of the psychiatric groups of patients do not differ significantly ($p < 0.05$) from those of the control group.

hydroxylase. In these experiments a loading dose of 50 mg/kg was given by mouth, and the concentration of 5-HIAA and of tryptophan in the CSF of patients and controls assessed 8 hr later. There was no significant difference between the subgroups of patients and controls. The results suggested that there was no defect in the activity of tryptophan-5-hydroxylase or indeed in the transport of tryptophan. The findings suggest that in unipolar depressive illness there is a change in functional release of 5-HT without a change in the capacity for synthesis. For the bipolar depressive and for the manics, however, an alternative hypothesis seems necessary.

III. ANIMAL STUDIES

One possibility we have considered is that there may be an alteration in the sensitivity of the receptor to the released transmitter. If, for example, the sensitivity of the receptor falls, there would be a decrease in the firing rate of the postsynaptic cell. Conversely, an abnormal rise in the sensitivity of the receptor would result in an increase in firing of the postsynaptic cell in response to a normal release transmitter. Ungerstedt (1971) has shown that administration of 6-hydroxy-dopamine in rats destroys the dopaminergic systems in rat brain and that this is followed by supersensitivity of postsynaptic receptors to apomorphine and L-DOPA. Recently, in our unit in Edinburgh, Eccleston and his colleages have lesioned the norepinephrine system in rats (Eccleston, 1973) and have found that on the lesioned side the increase in receptor sensitivity was accompanied by an increase in cyclic-AMP in response to norepinephrine. Thus there is the possibility of developing an index of receptor sensitivity. No such methods have as yet been developed for the 5-HT systems, but the work of Aghajanian and his colleagues on the effects of LSD on 5-HT receptors (Foote, Sheard, and Aghajanian, 1969; Aghajanian, 1972) suggests that there may well be similar alteration in receptor sensitivity in this system.

Are there any clinical observations to support such a hypothesis for bipolar depression and mania?

IV. MODIFIED AMINE HYPOTHESIS

We have suggested that changes in exploratory and stereotyped behavior patterns are seen in both depression and mania (Ashcroft et al., 1972). In depression there is diminished exploration in terms of both physical and mental activity, with an increase in stereotyped behavior. In mania, stereotyped behavior is often marked. The patient may rummage through drawers for hours at a time, apparently without purpose. It is of interest that in animals an increase in stereotyped behavior results from stimulation of dopamine systems (Iverson, 1971). Simpler forms of stereotypes are the faciobuccal dyskinesias which have been described in other conditions, such as parkinsonian patients treated with L-DOPA. These dyskinesias often occur independently of drug treatment in manic-depressives, but are identical with those seen in amphetamine addicts and similar to movements produced by amphetamine in animals, where they can be shown to be caused by dopamine stimulation (Ungerstedt, 1971). An interesting observation is that some hyperkinetic children are improved when treated with small doses of amphetamine over a long period of time (Bradley, 1937), suggesting that prolonged administration of amphetamine may depress the sensitivity of the receptor.

We are therefore suggesting that "mood" is a complex behavior pattern and that a change in mood may indicate a change in a number of neuronal systems and possibly a change in the sensitivity of postsynaptic receptors in these systems.

The conclusion we could draw in relation to depression is that both low output of amine transmitters (low output depression) and low sensitivity of amine receptors (low sensitivity depression) may give rise to a similar clinical picture. Our preliminary results suggest that the clinical division into unipolar and bipolar depression might parallel such a classification.

A similar subdivision into high output and high sensitivity mania does not yet emerge from the clinical data, but certainly this is susceptible to further testing.

REFERENCES

Aghajanian, G. K. (1972): Influence of drugs on the firing of serotonin-containing neurones in brain. *Federation Proceedings*, 31:91–96.

Ashcroft, G. W., and Sharman, D. F. (1960): 5-Hydroxyindols in human cerebrospinal fluid. *Nature*, 186:1050–1051.

Ashcroft, G. W., Eccelston, D., Murray, L. G., Glen, A. I. M., Crawford, T. B. B., Pullar, I. A., Shields, P. J., Walter, D. S., Blackburn, I. M., Connechan, J., and Lonergan, M. (1972): Modified amine hypothesis for the aetiology of affective illness. *Lancet* i:573–577.

Ashcroft, G. W., Blackburn, I. M., Eccleston, D., Glen, A. I. M., Hartley, W., Kinloch, N. E., Lonergan, M., Murray, L. G., and Pullar, I. A. (1973a): Changes on recovery in the concentrations of tryptophan and the biogenic amine metabolites in the CSF of patients with affective illness. *Psychological Medicine* 3:319–325.

Ashcroft, G. W., Crawford, T. B. B., Cundall, R. L., Davidson, D. L., Dobson, J., Dow, R. C., Eccleston, D., Loose, R. W., and Pullar, I. A. (1973b): 5-Hydroxytryptamine metabolism in affective illness. The effect of tryptophan administration. *Psychological Medicine*, 3:326–332.

Bradley, C. (1937): The behavior of children receiving Benzedrine. *American Journal of Psychiatry*, 94:577–585.

Eccleston, D. (1973): Adenosine 3′:5′ cyclic monophosphate and affective disorders: Animal models. *Biochemical Society Special Publication*, 1:121–126.

Foote, W. E., Sheard, M. H., and Aghajanian, G. K. (1969): Comparison of effects of LSD and amphetamines on midbrain raphe units. *Nature*, 222:567–569.

Iversen, S. D. (1971): The effect of surgical lesions to frontal cortex and substantia nigra on amphetamine responses in rats. *Brain Research*, 31:295–311.

Ungerstedt, U. (1971): Post synaptic super sensitivity after 6-hydroxy dopamine induced degeneration of the nigro striatal dopamine system. *Acta Physiologica Scandanavica*, Supplement 367:69–93.

Advances in Biochemical Psychopharmacology, Vol. 11
Raven Press, New York © 1974

Brain Serotonin, Affective Illness, and Antidepressant Drugs: Cerebrospinal Fluid Studies with Probenecid

Frederick K. Goodwin and
Robert M. Post

*NIMH Intramural Program, NIH, Bethesda,
Maryland 20014*

I. INTRODUCTION

Recently we have witnessed a renewed interest in the possible role of brain serotonin in the pathogenesis of various mental disorders (for reviews see Barchas and Usdin, 1973). For example, in relation to the affective disorders a serotonin deficiency has been hypothesized to underlie some forms of depression (Lapin and Oxenkrug, 1969) or both depression and mania (Coppen, 1967); reports of antidepressant or antimanic effect of serotonin precursors have been cited in support of this hypothesis (Coppen, 1967; van Praag and Korf, 1971), but these findings have not been confirmed by others (Carroll, 1971; Bunney, Brodie, Murphy, and Goodwin, 1971). The findings of decreased brainstem concentrations of serotonin (Shaw, Camps, and Eccleston, 1967) or its major metabolite, 5-hydroxyindoleacetic acid (5-HIAA) (Bourne, Bunney, Colburn, Davis, Davis, Shaw, and Coppen, 1968; Pare, Young, Price, and Stacey, 1969), tend to support the hypothesis, but the paucity of data and the pitfalls of postmortem studies limit the usefulness of these studies.

Investigations of 5-HIAA levels in the cerebrospinal fluid (CSF) of patients provide the major body of direct data relevant to the possible role of central serotonin in affective illness. Some studies have reported a decrease in CSF concentration of 5-HIAA in depressed patients as compared to controls (Ashcroft, Crawford, Eccleston, Sharman, MacDougall, Stanton, and Binns, 1966; Dencker, Malm, Roos, and Werdinius, 1966; van Praag, Korf, and Puite, 1970; Coppen, Prange, Whybrow, and Noguera, 1972), while others have found no significant differences from controls (Bowers, Henninger, and Gerbode, 1969; Roos and Sjöström, 1969; Papeschi and McClure, 1971; Goodwin and

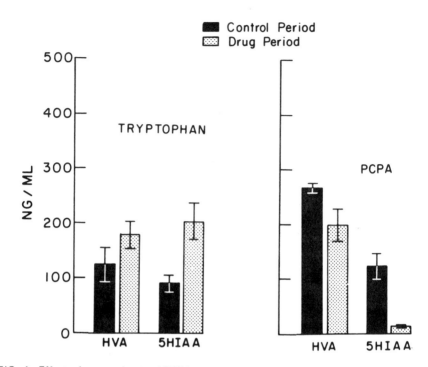

FIG. 1. Effect of tryptophan and PCPA on probenecid-induced accumulation of HVA and 5-HIAA. The average dose per day at time of LP and duration of drug trial is as follows: L-tryptophan, 8.7 g, 12.5 days, $n = 6$ (5-HIAA increase $p < 0.01$, paired t); PCPA, 3.8 g, 20 days, $n = 2$ (5-HIAA decrease $p < 0.001$, group t). Higher initial HVA values in patients subsequently given PCPA is related to small n and that some were manic during the trial.

Post, 1973). The data from these studies are summarized in Table 1. The interpretation of the data on CSF 5-HIAA levels is confounded by several factors, including the following: (1) A large portion of the 5-HIAA released from brain into the ventricular system is removed from the CSF by an active transport system resulting in a 3 to 1 ventricular-lumbar concentration gradient for 5-HIAA (Moir, Ashcroft, Crawford, Eccleston, and Guldberg, 1970); (2) the spinal cord itself contains serotonergic neurons; thus it has been shown that a substantial portion of the 5-HIAA in the lumbar CSF probably originates from serotonin in the cord (Bulat and Zivkovic, 1971; Post, Goodwin, Gordon, and Watkin, 1973).

In an attempt to overcome these difficulties and to provide an estimate of the dynamics ("turnover") of central serotonin, high doses of probenecid have been administered to patients in conjunction with CSF 5-HIAA determinations. The rationale for the use of probenecid in the study of serotonin turnover in the central nervous system is based

TABLE 1. *Results of studies of CSF amine metabolites in affective illness*

	Control		Depression		Main	
	N	Mean ± SD	N	Mean ± SD	N	Mean ± SD
		CSF 5-HIAA				
		(ng/ml)				
Ashcroft et al., 1966	21	19.1 ± 4.4	24	11.1 ± 3.9	4	18.7 ± 5.4
Dencker et al., 1966	34	30 (median)	14	10 (median)	6	10 (median)
Fotherby et al., 1963	6	16.6 ± 9.4	11	12.2 ± 8.2		
Coppen et al., 1972	20	42.3 ± 14	31	19.8 ± 8.5	18	19.7 ± 6.8
Roos and Sjöström, 1969	26	29 ± 7	17	31 ± 8	19	36 ± 9
Bowers et al., 1969	18	43.5 ± 16.8	8	34.0 ± 11.5	8	42.0 ± 10.3
van Praag and Korf, 1971	11	40 ± 24	14	17 ± 17		
Papeschi and McClure, 1971	10	28 ± 3	12	22 ± 2		
Gottfries et al., 1971	60	32.4 ± 10.4				
McLeod and McLeod, 1972	12	32.6 ± 11.4	25	20.5 ± 12.1		
Mendels et al., 1972			12	12.9 ± 6.0	4	17.1 ± 14.6
Goodwin et al., 1973a	29	27.3 ± 1.6	55	25.5 ± 1.3	16	28.7 ± 2.5

on the finding that probenecid inhibits the transport of 5-HIAA out of CSF and that the rate of elevation of this metabolite gives a turnover value which is comparable to that obtained by other methods (Neff, Tozer, and Brodie, 1967). Furthermore, the inhibition of 5-HIAA removal results in drastically altering the conditions responsible for the steep ventricular-lumbar concentration gradient and, in effect, results in the lumbar CSF becoming a better reflection of ventricular CSF. Thus, as the total lumbar concentration of 5-HIAA increases (from above), the percentage which represents the cord contribution decreases. From these considerations it would seem reasonable to assume that a measure of CSF 5-HIAA following probenecid is considerably better as a reflection of brain 5-HIAA (and therefore brain serotonin) than is a baseline measure.

The present study first examines the alterations in the probenecid-induced accumulation of 5-HIAA produced by a serotonin precursor and synthesis inhibitor in order to provide evidence for the validity of this technique in the estimation of CNS serotonin turnover. Then differences in the probenecid-induced accumulation of 5-HIAA are examined in normal controls and in patients with affective illness divided according to diagnostic subgroups, severity of various symptoms, and response to treatment. Finally, two tricyclic antidepressants, imipramine (IMI) and amitriptyline (AMI), are examined for their effects on the probenecid-induced accumulation of 5-HIAA.

II. METHODS

Forty patients with affective illness and ten normal controls were studied on a metabolic research unit at the NIMH specifically designed for the collection of behavioral and biochemical data on a longitudinal basis. Subjects were limited to those who volunteered after a careful explanation of the procedures involved. The depressed patients had symptoms which ranged from moderately severe to severe, and included psychomotor retardation or agitation, anorexia, weight loss, sleep disturbance, and depressive thought content often of psychotic proportions. They were all free of extenuating medical problems, and their physical condition was such that they could tolerate the requirements of a research protocol. The diagnostic criteria and behavioral rating system were as previously described (Kotin and Goodwin, 1972).

All subjects were drug free for at least 2 weeks prior to the procedure. During the study, diet was limited in foods which might affect indoleamine or catecholamine metabolism. The initial lumbar puncture (LP) was performed at 9 a.m. with the patients in the lateral decubitus position. The second LP was done at 3 p.m. on the following day; during the 18 hr preceding the second LP, probenecid (100 mg/kg) was administered in divided doses (9 p.m., 2 a.m., 7 a.m., and 12 noon). Patients were kept at bedrest for 8 hr prior to the initial LP and throughout the time of probenecid administration. For the studies focused on drug effects, the entire probenecid procedure was repeated in an identical fashion so that placebo and drug data could be compared in the same patient. The serotonin precursor, L-tryptophan (average dose, 8.7 g), and the synthesis inhibitor, parachlorophenylalanine (PCPA, average dose 3.8 g), were administered during clinical trials conducted for the purpose of evaluating their therapeutic effectiveness in depression and mania. Other drugs, including aspirin and sedatives, were carefully avoided throughout the study period. (In order to have some data for direct comparison with the previously studied control group, a small number of depressed patients were given an LP after only 9 hr of probenecid. In this group the dose averaged 115 mg/kg.)

LP's were nontraumatic; the samples were collected in 20 mg of ascorbic acid per 10 ml of fluid and stored at $-20°C$ until analyzed. 5-HIAA was determined on the first 8 ml of CSF obtained according to modifications of the fluorimetric method of Ashcroft and Sharman (1962). Except as otherwise indicated, statistical comparisons were done by Student's t test for grouped or paired data.

III. RESULTS

A. Effects of L-Tryptophan and PCPA

If the probenecid-induced accumulation of 5-HIAA in the CSF does in fact reflect the turnover (and therefore also the synthesis) of serotonin in the CNS, then this accumulation should be increased by compounds which increase synthesis and decreased by inhibitors of synthesis. L-Tryptophan, the amino acid precursor of serotonin, caused a large increase in 5-HIAA accumulation, while not significantly affecting accumulation of homovanillic acid (HVA), the dopamine metabolite. On the other hand, PCPA, a specific inhibitor of serotonin synthesis (Koe and Weissman, 1966), virtually eliminated the probenecid-induced accumulation of 5-HIAA while HVA accumulation was not affected (Fig. 1).*

B. Probenecid-Induced Alterations in 5-HIAA in Different Groups

Figure 2 illustrates the effect of probenecid on CSF 5-HIAA in two groups of patients compared to normal controls. A number of patient variables were examined for possible effects on the probenecid 5-HIAA data (Table 2). Although a wide age range existed in our patient population, there was no correlation between age or sex and the probenecid-induced accumulation of 5-HIAA. In our initial studies, using a fixed dose of probenecid, a significant negative correlation was found between patient weight and 5-HIAA accumulation on probenecid (Tamarkin, Goodwin, and Axelrod, 1970). As illustrated in Table 2, when the drug is administered on a milligram per kilogram basis there is no significant relationship between amine metabolite accumulation and patient weight or absolute dose level of probenecid. Because the dose was not available in units smaller than 250 mg, the 100 mg/kg represented an approximation. The actual dose range was from 85 to 118 mg/kg, allowing for the small but nonsignificant positive correlation between dose per kilogram and metabolite accumulation. The marginal nature of this correlation is consistent with earlier work

---

*A recent study by Sjöström (1972) did not find 5-HIAA decreases with probenecid, but the PCPA dose was much smaller and the duration of the trial shorter than those reported here.

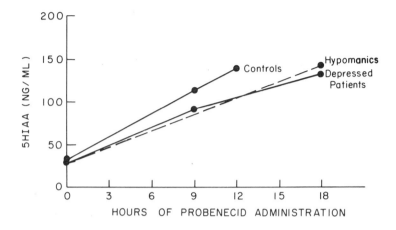

FIG. 2. Accumulation of 5-HIAA after probenecid in different diagnostic groups. The number of patients in each group are: controls, n = 29 baseline, n = 8 at 9 hr; depressed, n = 58 baseline, n = 6 at 9 hr, n = 40 at 18 hr; manic, n = 17 baseline, n = 18 at 18 hr.

showing that above 90 mg/kg the relationship between probenecid dose per kilogram and metabolite accumulation falls off sharply (Tamarkin et al., 1970).

For comparisons with the control group, only the data from the smaller number of depressed patients studied at 9 hr can be used. 5-HIAA levels in the depressed patients are lower than in the controls at the 9-hr point in spite of the fact that the depressed patients received about 20% more probenecid on a milligram/kilogram basis. With the small number of patients studied at this time, the difference does not reach statistical significance. The 5-HIAA levels at the 18-hr point are virtually identical in the depressed and manic groups.

TABLE 2. *Correlations between CSF 5-HIAA*
after probenecid and patient variables
(pearsons r)

	5-HIAA
Patient age	–0.08
Patient weight	0.00
Probenecid dose (g)	0.08
Probenecid dose (mg/kg)	0.13
Sleep	0.17
Depression ratings	–0.22

The depressed patients were classified as bipolar or unipolar based on the presence or absence of a prior history of mania as documented by hospitalization; patients with a history of recurrent depression interspersed with periods of abnormal mood elevation but not mania were separately classified as bipolar II according to previously described criteria (Dunner, Gershon, and Goodwin, 1974). There were no unipolar-bipolar differences in 5-HIAA accumulation. Patients were divided into three subgroups according to their mean depression ratings for the two days preceding the first LP. There was a slight but nonsignificant trend for the patients with less severe depressive symptoms to have higher levels of 5-HIAA following probenecid. The degree of psychomotor agitation and/or retardation did not significantly affect the probenecid-induced accumulation of 5-HIAA (Table 3).

The patients were subdivided according to their subsequent antidepressant response to either lithium or tricyclic antidepressants employing criteria previously described (Goodwin, Murphy, Dunner, and Bunney, 1972). The CSF amine metabolite data for these groups is summarized in Table 4. When compared to lithium nonresponders,

TABLE 3. *CSF levels of 5-HIAA after probenecid in subgroups of depressed patients*

	5-HIAA (ng/ml)
Diagnostic subgroups	
Bipolar I (12)	144 ± 7
Bipolar II (12)	128 ± 11
Unipolar (16)	127 ± 13
Level of depression[a]	
Moderate (7)	151 ± 4
Moderately	
severe (20)	125 ± 10
Severe (13)	134 ± 13
Degree of agitation[a]	
Mild (24)	133 ± 7
Moderate to	
severe (16)	131 ± 12
Degree of retardation[a]	
Mild (30)	135 ± 7
Moderate to	
severe (10)	123 ± 15

Number of patients in parentheses.
[a]See text

TABLE 4. *CSF levels of 5-HIAA after probenecid in*
depressed patients: Relationship to
subsequent drug response

		5-HIAA (ng/ml)
Lithium carbonate		
Responders	(13)	119 ± 10
Nonresponders	(10)	161 ± 8^{a}
Tricyclic antidepressants		
Responders	(12)	144 ± 16
Nonresponders	(7)	127 ± 6

Number of patients in parentheses
[a]Significantly different from lithium responders ($p < 0.02$, Student's t test).

those depressed patients who showed a therapeutic response to lithium
had significantly lower levels of 5-HIAA ($p < 0.02$) after probenecid.
On the other hand, the tricyclic responders tended to have somewhat
higher levels of 5-HIAA after probenecid when compared to nonresponders, although this difference was not significant.

A frequency distribution plot of the data indicates an apparent
subgroup of ten patients with very low 5-HIAA accumulation—a mean
postprobenecid 5-HIAA of 72 ng/ml, compared to 141 ng/ml for the
entire patient group. As yet we have not been able to differentiate this
apparent subgroup clinically.

C. The Effects of Tricyclic Antidepressants

Sixteen patients who had probenecid studies while on placebo were
restudied at an average of 25 days into a full course of either
imipramine (nine patients) or amitriptyline (seven patients) at doses
ranging from 150 to 350 mg/day. The data are summarized in Fig. 3. The
tricyclic antidepressants reduced baseline 5-HIAA levels by 20% ($p <$
0.05), while the extent of accumulation on probenecid was reduced by
35% ($p < 0.01$) compared to the same patients studied while drug free.
There were no differences in the effect of IMI compared with AMI on
the pattern of CSF amine metabolites. The one patient who did not
have a reduced level of 5-HIAA during tricyclic treatment had been on
the highest doses (350 mg/day) of IMI for the longest duration (41
days). However, this patient's ratio of 5-HIAA to HVA accumulation
with probenecid did decrease substantially during tricyclic administration. Patients who were studied after more than 3 weeks of IMI and
AMI treatment had smaller average reductions in probenecid-induced
accumulations of 5-HIAA (−23 ng/ml) compared to those who were

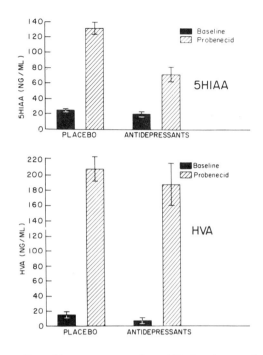

FIG. 3. Effects of tricyclic antidepressants on probenecid-induced accumulation of 5-HIAA and HVA in depressed patients. 5-HIAA levels in 16 patients treated with amitriptyline or imipramine are lower than when the same patients are untreated. $p<0.05$, baseline; $p<0.01$, probenecid, paired t.

studied with less than 3 weeks of treatment (-71 ng/ml) ($p < 0.05$). The relationship of tricyclic dose to the magnitude of the decrease in 5-HIAA accumulation with treatment was not significant ($r = -0.36$).

Although, as expected, the depression ratings were lower in the majority of patients on IMI and AMI treatment, the decreases in probenecid-induced levels of 5-HIAA did not appear to be related to clinical improvement. The decrease in 5-HIAA with treatment was not significantly correlated with degree of improvement in depression ratings ($r = -0.33$). Patients who showed no improvement had at least as large a reduction in 5-HIAA during treatment as those who showed substantial clinical change.

The 5-HIAA/HVA ratio following probenecid decreased significantly with tricyclic treatment ($p < 0.05$, paired t on the square root of the ratios). The change in ratio indicates that decreases in 5-HIAA were not due to nonspecific transport effects or reduction in probenecid effectiveness during IMI and AMI administration, since such factors would alter HVA accumulation in the same direction. Patients received the same average mg/kg dose of probenecid during both the medication-free LP (100.9 ± 2.3 mg/kg) and the LP on tricyclics (100.2 ± 2.6 mg/kg). The reduction in 5-HIAA was evident in the two patients who received their drug-free LP's 2 weeks or more *after* tricyclic treatment, suggesting that the reduction in 5-HIAA during tricyclic administration was not a function of sequence and was not related to factors (such as stress) which might be associated with having a repeat probenecid procedure.

IV. DISCUSSION

How adequately does the probenecid-induced accumulation of 5-HIAA in the lumbar CSF of patients reflect alterations in the metabolism of turnover of serotonin in the brain? We have detailed elsewhere the potential difficulties in the interpretation of lumbar CSF amine metabolite data (Goodwin and Post, 1973; Post et al., 1973); these include the concentration gradient for 5-HIAA from relatively high levels in brain ventricular fluid to relatively low levels in lumbar fluid, the transport time required for amine metabolites formed in brain to reach the lumbar CSF, and finally the possible contribution of local spinal cord serotonin metabolism to the CSF 5-HIAA levels.

The concentration gradient is maintained by the active removal of 5-HIAA as it moves down the cord from ventricle to lumbar sac (Moir et al., 1970). High doses of probenecid reduce the gradient by inhibiting this removal process, allowing amine metabolite levels in the lumbar CSF to reflect ventricular levels more truly. The lumbar CSF levels of 5-HIAA found after probenecid in the present study approximate the levels found in human ventricular fluid (Gordon, *Personal Communication*, 1972).

The effects of a serotonin precursor and synthesis inhibitor on the probenecid-induced accumulation of 5-HIAA further support the validity of these data as a reflection of brain serotonin metabolism. As we noted above, some of the 5-HIAA in the lumbar CSF of patients appears to have its origin in the spinal cord (Bulat and Zivkovic, 1971; Curzon, Gumpert, and Sharpe, 1971; Post et al., 1973). Nevertheless, 5-HIAA levels after probenecid should represent a relatively greater contribution from brain by virtue of the decreased removal of this metabolite along the ventricular-lumbar axis.

Previous reports of a high correlation between plasma (or CSF) probenecid levels and amine metabolite accumulation (Korf and van Praag, 1971; Sjöström, 1972) have raised the possibility that differences in probenecid absorption or metabolism could account for some of the observed differences in amine metabolite accumulation. This caution is relevant only when the accumulation of 5-HIAA and HVA are both altered in the same direction and to approximately the same degree, that is, when the 5-HIAA/HVA ratio is not altered (Bowers, 1972*a*). Sjöström has recently reported (1972) that the high correlation between CSF probenecid levels and 5-HIAA accumulation ($r = 0.86$) begins to drop off at CSF probenecid concentration above 4 to 5 μg/ml. Preliminary data indicate that the relatively high dose schedule employed in the present studies results in CSF probenecid levels well over 5 μg/ml, with plasma levels in the 300 to 400 μg/ml range. Recently, Perel, Levitt, and Dunner (1974) have demonstrated that at these high probenecid levels there is no correlation between CSF probenecid and metabolite accumulation. Thus, in our patients, variations in probenecid absorption are less likely to make a meaningful contribution to variations in amine metabolite accumulation.

In light of the above considerations it is not surprising that the probenecid technique has "uncovered" a number of alterations in amine metabolites which were not evident from baseline CSF metabolite levels. In patients with affective illness, 18 hr of probenecid administration (100 mg/kg) increased levels of 5-HIAA 400% and HVA 1000% over baseline. At these higher levels of accumulation, contrary to the finding of Bowers (1972*a*), we do not find a significant correlation between baseline and probenecid levels of amine metabolites. Thus, the high-dose probenecid technique does not appear to be merely an amplification of baseline values, but may be reflecting a different parameter, such as amine turnover.

Our data on probenecid-induced accumulation of 5-HIAA in patients with affective illness versus normal controls, although not quite reaching statistical significance, is nevertheless not inconsistent with previous reports of decreased 5-HIAA accumulation in depressed patients on probenecid (van Praag and Korf, 1971; Sjöström and Roos, 1972). The possibility of biochemical heterogeneity in affective illness, with a "low-serotonin" subgroup, has previously been suggested (van Praag, Korf, Dols, and Schut, 1972). Our data are also suggestive of a low-serotonin subgroup, although a larger number of patients will have to be studied before this can be stated with confidence. It is of interest that this low-probenecid–5-HIAA subgroup does not overlap with another subgroup of depressed patients who have very low CSF levels of MHPG (3-methoxy-4-hydroxy-phenylglycol—the major metabolite of norepinephrine in brain). Although we have been unable to characterize the low 5-HIAA subgroup clinically, there is some preliminary evidence that they may respond differentially to various tricyclic antidepressants.

Our results with the tricyclic drugs would suggest that central nervous system turnover of serotonin is reduced during treatment with these antidepressants. These findings are consistent with those of Bowers (1972*b*) as well as those from animal studies demonstrating that

IMI decreases brain 5-HIAA with probenecid (Bruinvels, 1972), inhibits the uptake of tryptophan competitively (Bruinvels, 1972), decreases serotonin depletion after synthesis inhibition (Bruinvels, 1972; Corrodi and Fuxe, 1968, 1969), and retards the disappearance of labeled serotonin from brain (Schildkraut, Schanberg, Breese, Gordon, and Kopin, 1969)—all suggesting that IMI decreases brain serotonin turnover.

It is of interest that widely diverse treatments effective in depression—tricyclics, lithium (Goodwin, Post, and Murphy, 1973b), monoamine oxidase inhibitors (Bowers and Kupfer, 1971), and electroconvulsive therapy (Goodwin et al., 1973b)—tend to decrease 5-HIAA accumulation in the CSF. L-Tryptophan, which of course increases 5-HIAA accumulation, has been reported to have antidepressant properties by some investigators, but not by others (Carroll, 1971). The phenothiazines (Bowers, 1972c, Persson and Roos, 1969) and the butyrophenones (Persson and Roos, 1968; Sjöström and Roos, 1972), which are not considered as primary antidepressant agents, do not reduce 5-HIAA levels in the CSF after probenecid.

That the degree of clinical improvement with IMI and AMI treatment was unrelated to the reduction in 5-HIAA with tricyclics does not rule out the possibility that reduced serotonin turnover is related to the antidepressant effect of these drugs. The tricyclics and other antidepressant treatments may consistently reduce serotonin turnover while only some patients with a biochemically distinct illness or predisposition respond clinically. To date, however, we have not been able to separate tricyclic responders from nonresponders (IMI and AMI patients grouped together) on the basis of pretreatment levels of 5-HIAA in the CSF as has been recently suggested by Asberg, Bertilsson, Tuck, Cronholm, and Sjöqvist (1973).

It could appear paradoxical that antidepressant treatments might decrease central serotonin turnover in light of the evidence that some depressed patients already have abnormally low turnover of this amine. This paradox might be resolved if our measurement of amine turnover reflects presynaptic events. Thus, if postsynaptic serotonergic receptor activity were actually increased in some depressed patients, one might expect to find decreased serotonin turnover presynaptically, the result of feedback regulatory mechanisms (Carlsson, Snider, and Almgren, 1974; Corrodi and Fuxe, 1968; Schubert, Nybäck, and Sedvall, 1970; Sheard, Zolovick, and Aghajanian, 1972). If this were the situation, then antidepressants might work by a direct action on the presynaptic neuron (Bruindels, 1972), resulting in a further reduction in serotonin turnover, a reduction sufficient to overcome the excessive postsynaptic activity.

The probenecid technique provides a useful tool for the direct assessment of central amine neurotransmitter turnover in patients and, as such, appears to reflect presynaptic function. However, before we can adequately understand the functional meaning of alterations in presynaptic amine metabolism, we will need methods for the assessment of postsynaptic receptor function in man.

REFERENCES

Asberg, M., Bertilsson, L., Tuck, D., Cronholm, B., and Sjöqvist, F. (1973): Indoleamine metabolites in the cerebrospinal fluid of depressed patients before and during treatment with nortriptyline. *Clinical Pharmacology and Therapeutics,* 14:277−286.

Ashcroft, G. W., Crawford, T. B. B., Eccleston, D., Sharman, D. F., Mac Dougall, E. J., Stanton, J. B., and Binns, J. K. (1966): 5-Hydroxyindole compounds in the cerebrospinal fluid of patients with psychiatric or neurological disease. *Lancet,* II:1049−1052.

Ashcroft, G. W., and Sharman, D. F. (1962): Measurement of acid monoamine metabolites in human CSF. *British Journal of Pharmacology,* 19:153−160.

Barchas, J., and Usdin, E. (1973): *Serotonin and Behavior.* Academic Press, New York/London.

Bourne, H. R., Bunney, W. E., Colburn, R. W., Davis, J. N., Davis, J. M., Shaw, D. M., and Coppen, A. J. (1968): Noradrenaline, 5-hydroxytryptamine and 5-hydroxyindoleacetic acid in the hindbrains of suicidal patients. *Lancet,* II:805−808.

Bowers, M. B. (1972a): Cerebrospinal fluid 5-hydroxyindoleacetic acid (5HIAA) and homovanillic acid (HVA) following probenecid in unipolar depressives treated with amitriptyline. *Psychopharmacologia* (Berlin) 23:26−33.

Bowers, M. B., Jr. (1972b): Clinical measurements of central dopamine and 5-hydroxytryptamine metabolism: reliability and interpretation of cerbrospinal fluid acid monoamine metabolite measures. *Neuropharmacology,* 11:101−111.

Bowers, M. B., Jr. (1972c): Acute psychosis induced by psychomimetic drug abuse. II. Neurochemical findings. *Archives of General Psychiatry,* 27:440−442.

Bowers, M. B., Heninger, G. R., and Gerbode, F. A. (1969): Cerebrospinal fluid, 5-hydroxyindoleacetic acid and homovanillic acid in psychiatric patients. *International Journal of Neuropharmacology,* 8:255−262.

Bowers, M. B., and Kupfer, D. J. (1971): Central monoamine oxidase inhibition and REM sleep. *Brain Research,* 35:561−564.

Bruinvels, J. (1972): Inhibition of the biosynthesis of 5-hydroxytryptamine in rat brain by imipramine. *European Journal of Pharmacology,* 20:231−237.

Bulat, M., and Zivkovic, B. (1971): Origin of 5-hydroxyindoleacetic acid in the spinal fluid. *Science,* 173:738−740.

Bunney, W. E., Jr., Brodie, H. K. H., Murphy, D. L., and Goodwin, F. K. (1971): Studies of alpha-methyl-para-tyrosine, L-DOPA, and L-tryptophan in depression and mania. *American Journal of Psychiatry,* 127:48−57.

Carlsson, A., Snider, S. R., and Almgren, O. (1974): The neurogenic short-term control of catecholamine synthesis and release in the sympatho-adrenal system, as reflected in the levels of endogenous dopamine and β-hydroxylated catecholamines. *Journal of Neurochemistry (in press).*

Carroll, B. J. (1971): Monoamine precursors in the treatment of depression. *Clinical Pharmacology and Therapeutics,* 12:743−761.

Coppen, A. J. (1967): The biochemistry of affective disorder. *British Journal of Psychiatry,* 113:1237.

Coppen, A., Prange, A. J., Jr., Whybrow, P. C., and Noguera, R. (1972): Abnormalities of indoleamines in affective disorders. *Archives of General Psychiatry,* 26:474−478.

Corrodi, H., and Fuxe, K. (1968): The effect of imipramine on central monoamine neurons. *Journal of Pharmacy and Pharmacology*, 20:230—231.

Corrodi, H., and Fuxe, K. (1969): Decreased turnover in central 5-HT nerve terminals induced by antidepressant drugs of the imipramine type. *European Journal of Pharmacology*, 7:56—59.

Curzon, G., Gumpert, E. J. W., and Sharpe, D. M. (1971): Amine metabolites in the lumbar cerebrospinal fluid of humans with restricted flow of cerebrospinal fluid. *Nature New Biology*, 231:189—191.

Dencker, S. J., Malm, V., Roos, B.-E., and Werdinius, B. (1966): Acid monoamine metabolites of cerebrospinal fluid in mental depression and mania. *Journal of Neurochemistry*, 13:1545—1548.

Dunner, D. L., Gershon, E. S., and Goodwin, F. K. (1974): Heritable factors in the severity of affective illness. *Biological Psychiatry* (*in press*).

Fotherby, K., Ashcroft, G. W., Affleck, J. W., and Forrest, A. D. (1963): Studies on sodium transfer and 5-hydroxyindoles in depressive illness. *Journal of Neurology, Neurosurgery and Psychiatry*, 26:71—73.

Goodwin, F. K., Murphy, D. L., Dunner, D. L., and Bunney, W. E., Jr. (1972): Lithium response in unipolar vs. bipolar depression. *American Journal of Psychiatry*, 129:44—47.

Goodwin, F. K., and Post, R. M. (1973): The use of probenecid in high doses for the estimation of central serotonin turnover in patients. In: *Serotonin and Behavior*, edited by J. Barchas and E. Usdin, pp. 469—480. Academic Press, New York.

Goodwin, F. K., Post, R. M., Dunner, D. L., and Gordon, E. K. (1973a): Cerebrospinal fluid amine metabolites in affective illness: The probenecid technique. *American Journal of Psychiatry*, 130:73—79.

Goodwin, F. K., Post, R. M., and Murphy, D. L.(1973b):Cerebrospinal fluid amine metabolites and therapies for depression. Presented at the Annual Meeting, American Psychiatric Association, Honolulu, Hawaii.

Gottfries, C. G., Gottfries, I., Johansson, B., Olsson, R., Persson, T., Roos, B. E., and Sjöstrom, R. (1971): Acid monamine metabolites in human cerebrospinal fluid and their relations to age and sex. *International Journal of Neuropharmacology*, 10:665—672.

Koe, B. K., and Weissman, A. (1966): Para-chlorophenylalanine: A specific depletor of brain serotonin. *Journal of Pharmacology and Experimental Therapeutics*, 154:499—516.

Korf, J., and van Praag, H. M. (1971): Amine metabolism in the human brain: Further evaluation of the probenecid test. *Brain Research*, 35:221—230.

Kotin, J., and Goodwin, F. K. (1972): Depression during mania: Clinical observations and theoretical implications. *American Journal of Psychiatry*, 129:687—692.

Lapin, I. P. and Oxenkrug, G. F. (1969): Intensification of the central serotonergic processes as a possible determinant of thymoleptic effect. *Lancet*, i:132-136.

McLeod, W. R., and McLeod, M. (1972): Indoleamines and cerebrospinal fluid. In: *Depressive Illness: Some Research Studies*, edited by B. M. Davies, B. J. Carroll, and R. M. Mowbray. Charles C. Thomas, Springfield, Ill.

Mendels, J., Frazer, A., Fitzgerald, R. J., Ramsey, T. A., and Stokes, J. W. (1972): Biogenic amine metabolites in cerebrospinal fluid of depressed and manic patients. *Science*, 175:1380—1381.

Moir, A. T. B., Ashcroft, G. W., Crawford, T. B. B., Eccleston, D., and Guldberg, H. C. (1970): Central metabolites in cerebrospinal fluid as a biochemical approach to the brain. *Brain*, 93:357-368.

Neff, N. H., Tozer, T. N., and Brodie, B. B. (1967): Application of steady-state kinetics to studies of the transfer of 5-hydroxyindoleacetic acid from brain to plasma. *Journal of Pharmacology and Experimental Therapeutics*, 158:214—218.

Papeschi, R., and McClure, D. J. (1971): Homovanillic acid and 5-hydroxyindoleacetic acid in cerebrospinal fluid of depressed patients. *Archives of General Psychiatry*, 25:354—358.

Pare, C. M. B., Young, D. P. U., Price, K., and Stacey, R. S. (1969): 5-Hydroxytryptamine, noradrenaline and dopamine in brainstem, hypothalamus and caudate nucleus of controls and patients committing suicide by coal gas poisoning. *Lancet*, ii:133—135.

Perel, J., Levitt, M., and Dunner, D. L. (1974): Plasma and cerebrospinal fluid probenecid concentrations as related to accumulation of acidic biogenic amine metabolites in man. *Psychopharmacologia (in press)*.

Persson, T., and Roos, B.-E. (1968): Clinical and pharmacological effects of monoamine precursors of Haloperidol in chronic schizophrenia. *Nature*, 217:854.

Persson, T., and Roos, B.-E. (1969): Acid metabolites from monoamines in cerebrospinal fluid of chronic schizophrenics. *British Journal of Psychiatry*, 115:95—98.

Post, R. M., Goodwin, F. K., Gordon, E. K., and Watkin, D. M. (1973): Amine metabolites in human cerebrospinal fluid: Effects of cord transection and spinal fluid block. *Science*, 179:897—899.

Roos, B.-E., and Sjöström, R. (1969): 5-Hydroxyindoleacetic acid (and homovanillic acid) levels in the CSF after probenecid: Application in patients with manic-depressive psychosis. *Pharmacologia Clinica*, 1:153—155.

Schildkraut, J. J., Schanberg, S. M., Breese, G. R., Gordon, E., and Kopin, I. J. (1969): Effects of psychoactive drugs on serotonin metabolism. *Biochemical Pharmacology*, 18:1971—1978.

Schubert, J., Nyback, H., and Sedvall, G. (1970): Effect of antidepressant drugs on accumulation and disappearance of monoamines formed *in vivo* from labelled precursors in mouse brain. *Journal of Pharmacy and Pharmacology*, 22:136—139.

Shaw, D. M., Camps, F. E., and Eccleston, E. (1967): 5-Hydroxytryptamine in the hindbrains of depressive suicides. *British Journal of Psychiatry*, 113:1407—1411.

Sheard, M. H., Zolovick, A., and Aghajanian, G. K. (1972): Raphe neurons: Effect of tricyclic antidepressant drugs. *Brain Research*, 43:690—694.

Sjöström, R. (1972): Steady-state levels of probenecid and their relation to acid monoamine metabolites in human cerebrospinal fluid. *Psychopharmacologia*, 25:96—100.

Sjöström, R., and Roos, B.-E. (1972): 5-Hydroxyindoleacetic acid and homovanillic acid in cerebrospinal fluid in manic-depressive psychosis. *European Journal of Clinical Pharmacology*, 4:170—176.

Tamarkin, N. R., Goodwin, F. K., and Axelrod, J. (1970): Rapid elevation of biogenic amine metabolites in human CSF following probenecid. *Life Sciences*, 9:1397—1408.

van Praag, H. M., and Korf, J. (1971): Retarded depression and the dopamine metabolism. *Psychopharmacologia*, 19:199—203.

van Praag, H. M., Korf, J., Dols, L. C. W., and Schut, T. (1972): A pilot study of the predictive value of the probenecid test in application of 5-hydroxytryptophan as antidepressant. *Psychopharmacologia*, 25:14.

van Praag, H. M., Korf, J., and Puite, J. (1970): 5-Hydroxyindoleacetic acid levels in the cerebrospinal fluid of depressive patients treated with probenecid. *Nature*, 225:1259.

Advances in Biochemical Psychopharmacology, Vol. 11
Raven Press, New York © 1974

Toward a Biochemical Classification of Depression

H. M. van Praag

*Department of Biological Psychiatry, Psychiatric University
Clinic, Groningen, The Netherlands*

I. BASIC DATA AND PROBLEM DEFINITION

My argument will run as follows: I shall begin with two statements which will be formulated but not presented for discussion. On the basis of these statements I shall pose a question. My attempt to answer it will be the principal feature of my argument.

First statement: All known psychotropic drugs primarily exert an influence on impulse transmission in central synapses. The antidepressants are no exception to this rule. It is more specifically the monoaminergic synapse—the synapse in which a monoamine (MA) acts as transmitter—that is influenced by antidepressants. MAO inhibitors inactivate MAO and thus inhibit the intraneuronal degradation of MA. The result is a "leakage" of nonbound MA to the synaptic cleft. Tricyclic antidepressants block the "MA pump," that is, the transport of MA from the synaptic cleft back to the synaptic vesicles. The two types of antidepressants now in use are unrelated in their chemical structure; moreover, they influence central MA metabolism in different ways. Nevertheless they produce the same net effect: They increase the concentration of physiologically active MA in the synapse, thus probably increasing the activity of monoaminergic neuronal systems. There are indications that the therapeutic action of antidepressants is based on their MA-potentiating influence.

Second statement: Antidepressants can have a mood-improving effect in depressions. However, this statement is subject to at least two qualifications. To begin with, these agents are not equally effective in all types of depression. The syndrome known as endogenous depression provides an obvious indication of choice. But even within the limits of a given syndrome (and this is the second restriction), not all patients show improvement. The improvement rate attained with antidepressants in endogenous depression is about 60 to 65% (Klerman and Cole, 1965).

The same applies to other biological methods of treating depressions.

In the syndrome of endogenous depression, electroconvulsive therapy (ECT) is still the therapy with the highest chance of success, approximating 90%. Results obtained in other types of depression are substantially less favorable (Hordern, Burt, and Holt, 1965).

A very recent therapy is based on sleep deprivation. The patient is systematically deprived of sleep throughout several nights per week. Investigations made in our department have revealed a similar phenomenon: Some depressive patients show a strikingly favorable response to this therapy, even if only temporary, whereas other patients with similar symptoms show no response at all.

Questions: I have made two statements: Antidepressants increase the amount of MA at the central receptors; antidepressants are effective in some, but quite ineffective in other patients, even if they belong to the same diagnostic category. Three questions result from these statements:

1. Does a central MA deficiency occur in depressive patients?
2. If so, is this disorder present in only a proportion of the patients?
3. If so, can this explain the apparent selectivity of antidepressants in the sense that MA-deficient patients benefit from this type of therapy?

II. CENTRAL MA DEFICIENCY IN DEPRESSIVE PATIENTS

The answer to the first question is affirmative. There are phenomena which indicate that disorders of the metabolism of 5-hydroxytryptamine (5-HT), dopamine (DA), and/or norepinephrine (NE) can occur in the brain in depressive patients. I shall list the most important of these phenomena, confining myself to findings which directly concern the CNS.

A. Postmortem Studies

The 5-HT concentration found in the brainstem in suicide victims was lower than that found in a comparable control group. A decreased concentration of a given compound can indicate either decreased synthesis or increased degradation. Since not only the 5-HT concentration but also that of 5-hydroxyindoleacetic acid (5-HIAA, the principal 5-HT metabolite) proved to be diminished, decreased 5-HT synthesis must have been involved in the above-mentioned individuals. So far, three independent groups of investigators have carried out postmortem studies of suicide victims (Shaw, Camps, and Eccleston, 1967; Bourne,

Bunney, Colburn, Davis, Davis, Shaw, and Coppen, 1968; Pare, Young, Price, and Stacey, 1969). Their results are not quite identical, but they point in the same direction. Unfortunately, it was impossible to obtain sufficient data on the histories of these suicide victims and on the nature of their depressive illness. Psychopathological heterogeneity of the different groups perhaps explains why the results were not entirely in agreement.

B. CSF Studies

A second source of information is the concentration of MA metabolites in the cerebrospinal fluid (CSF).

Studies of this type are based on the following line of argument. The concentration of a MA metabolite in the CNS is related to the amount of the corresponding MA which is locally metabolized. Moreover, there is a relationship between the concentration of a MA metabolite in brain and spinal cord and that in CSF. In any case, this applies to CSF and adjacent parts of brain and spinal cord. According to this line of reasoning, the CSF concentration of a MA metabolite reflects the amount of the parent amine which is degraded in the CNS.

The concentrations of 5-HIAA and of the DA metabolite, homovanillic acid (HVA), in the CSF have been found to be decreased in depressive patients by several (but not by all) investigators. Reviews have been published by Mendels, Frazer, Fitzgerald, Ramsey, and Stokes (1972) and by Papeschi and McClure (1971). A survey of the different results shows fair agreement between values found in depressive patients, but substantial differences within control groups. This may be due to the fact that none of the investigators has so far been able to use normal controls. Controls in these studies are invariable nondepressive hospital patients, often originating from heterogeneous populations.

Our next consideration is NE metabolism in so far as it is reflected in the CSF. Whereas peripheral NE is largely degraded to vanilmandelic acid (VMA), the principal NE metabolite in the brain is 3-methoxy-4-hydroxyphenylglycol (MHPG). It was initially assumed that all MHPG excreted in the urine is of cerebral origin, and that consequently renal MPHG is a fairly exact index of central NE turnover. This has proved to be a fallacy. Recent studies in nonhuman primates suggest that about 50% of urinary MHPG originates from the periphery (Maas, Fawcett, and Dekirmenjian, 1972*b*). The peripheral NE pool, however, is much larger than the cerebral. In other words, while MHPG is the principal central NE metabolite, it is of only secondary importance as a peripheral NE metabolite. The conclusion that urinary MHPG gives information, albeit inexact, about central NE metabolism thus remains valid.

Such little research as has so far been carried out has not yielded clear-cut results. Normal as well as decreased MHPG concentration have been found in the CSF in depressive illness (e.g., Wilk, Shopsin, Gershon, and Suhl, 1972). In a well-designed study, Maas, Dekirmenjian, Garven, and Redmond (1972a) found renal MHPG excretion to be decreased in depression. It should be noted in this context that MHPG is a compound which is still difficult to determine reliably, so that relevant data should be regarded with circumspection.

C. Probenecid Technique

The introduction of the probenecid technique added a new dimension to the study of central MA metabolism in living human individuals. Probenecid is a substance which inhibits the transport of acid MA metabolites from the CNS to the bloodstream. Consequently these metabolites accumulate, and their rate of accumulation is a measure of the turnover of the corresponding parent amines.* Slow accumulation suggests low production of these metabolites, a low degree of degradation of parent amine, and consequently a restricted turnover of parent amine. Marked accumulation suggests the reverse: high production of metabolites, a high degree of degradation of parent amine, and therefore a high turnover of parent amine (see Bowers, 1972; Goodwin, Post, Dunner, and Gordon, 1973; van Praag, Korf, and Schut, 1973).

The response of acid MA metabolite concentrations in the CSF to probenecid is not only dependent on the rate at which they are produced but also on other factors, such as amount of probenecid which reaches the CNS (Korf and van Praag, 1971). The more probenecid is the CSF, the more marked the accumulation. This applies to 5-HIAA as well as to HVA. It is therefore necessary in the probenecid test to administer the maximum amount which can be tolerated. This amount was found to be 5 g in 5 hr. We have reason to

---

*Provided, of course, that there is a single end metabolite of a pathway; if there is a branched-chain, such as in the further oxidation or reduction of the intermediate aldehyde derived from a monoamine, things become more complicated. So far, however, there is no convincing evidence that metabolites other than 5-HIAA and HVA are of major importance in the breakdown of 5-HT and DA in the brain.

believe that at this dosage the transport of acid MA metabolites from the CNS is blocked completely (van Praag et al., 1973). I mention the results obtained in this way in 38 depressive patients and 12 nondepressive controls (van Praag et al., 1973). In the depressive group, 5-HIAA accumulation was significantly less than in the control group. A similar tendency was observed for HVA accumulation, which was decreased in the depression group compared with the controls (Table 1). Probenecid concentrations varied within the limits found in the control group. We interpret these findings as suggesting that the turnover of 5-HT and DA in the brain is lower in depressive patients than in controls. These results therefore point in the same direction as the postmortem findings.

TABLE 1. *Probenecid test results in depressed patients and controls*

Patients	No. of test subjects	CSF 5-HIAA (ng/ml)			CSF HVA (ng/ml)		
		Baseline	After Probenecid	Increase	Baseline	After Probenecid	Increase
Controls	12	29 ± 10.3	110 ± 46.7	81 ± 43.3	35 ± 13.4	170 ± 47.6	135 ± 46.5
Depressed	38	25 ± 8.5	76 ± 33.4	51 ± 32.1	37 ± 16.2	123 ± 51.7	86 ± 43.8

III. EXPLANATION OF THE CENTRAL MA DEFICIENCY

How do we explain the apparently decreased central MA turnover in depressive patients? So far as 5-HT metabolism is concerned, two observations may be relevant.

There are strong indications that the rate of cerebral 5-HT synthesis is largely determined by the amount of tryptophan which is locally available. Cerebral tryptophan concentration, in its turn, is closely related to the plasma concentration of free tryptophan, that is, the tryptophan fraction not bound to serum albumin (Knott and Curzon, 1972).

It was recently reported by Coppen, Eccleston, and Peet (1972a) and by Coppen, Brooksbank, and Peet (1972b) that the serum concentration of free tryptophan and the CSF concentration of tryptophan were lower in depressive patients than in a matched control group. Should the CSF concentration of tryptophan be representative of the situation in the brain, then Coppen's finding could explain the decreased 5-HT turnover suspected on the basis of the probenecid test.

There are also pointers to a different explanation, a reduced capacity to convert tryptophan to 5-HT (van Praag, Flentge, Korf, Dols, and Schut, 1973*a*) A series of depressive patients were submitted to an oral load of 5 g of L-tryptophan, whereupon CSF concentrations of tryptophan and 5-HIAA were determined. In different patients, a lumbar puncture was performed at different intervals after tryptophan loading. Eccleston, Ashcroft, Crawford, Stanton, Wood, and McTurk (1970) did the same in a group of neurological patients without psychiatric symptoms. In the depressive patients the tryptophan concentration had tripled 1 hr after loading; at that time there was little change in the nondepressive patients. Eight hours after loading, moreover, the 5-HIAA concentration in the CSF showed a less marked increase in the depressive than in the nondepressive group. More tryptophan in the CSF and less 5-HIAA seem to suggest that the tryptophan administered accumulates in the CNS in depressive patients, that it is less readily converted to 5-HT. This phenomenon, too, might explain a diminished 5-HT turnover.

IV. DEPRESSIVE ILLNESS WITH AND WITHOUT CENTRAL MA DEFICIENCY

We must now consider the second basic question. If disorders of central MA metabolism occur in depressive illness, are they encountered in all patients, in certain categories only, or distributed at random throughout the different categories?

Let us once more consider the results of the study with probenecid in 38 depressive patients mentioned above (van Praag, Korf, and Schut, 1973). Of these patients, 28 showed symptoms of endogenous depression and 10 those of neurotic depression. The entire depression group included 12 patients whose 5-HIAA response to probenecid was subnormal, that is, outside the range of the control group. All these patients were in the endogenous group (Fig. 1). From this fact, I am inclined to conclude that the group of endogenous depressions, although fairly homogeneous in psychopathological terms, was heterogeneous in biochemical terms.

The depression group included 14 patients with a subnormal HVA response to probenecid (Fig. 2); depression had been diagnosed as endogenous in 11 and neurotic in 3 of these patients. Nevertheless, all patients showed one common characteristic: motor retardation. A subnormal HVA response to probenecid has also been observed in Parkinson's disease (Olsson and Roos, 1968). This syndrome also

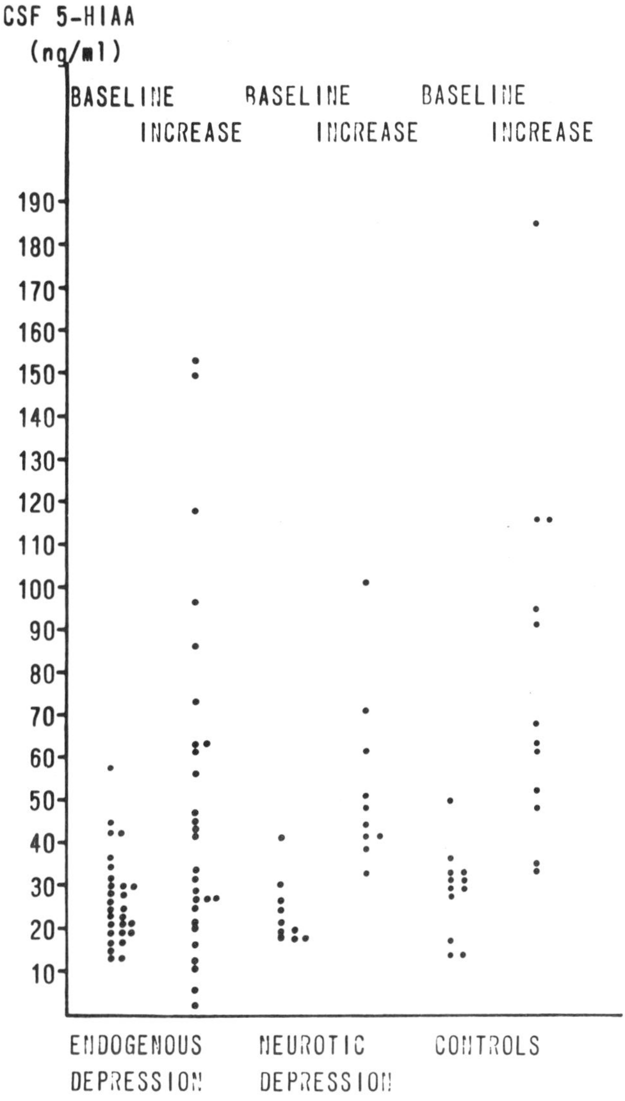

FIG. 1. Baseline 5-HIAA concentration in the CSF with increase after probenecid, in endogenous depressive, neurotic depressive, and control groups.

encompasses hypokinesia. It is therefore quite possible that diminished cerebral DA turnover (which the subnormal HVA suggests) is not specific of given disease entity (in this case Parkinson's disease), but rather characterizes a certain functional condition, in this case hypokinesia.

To summarize, these findings indicate that disorders in central MA

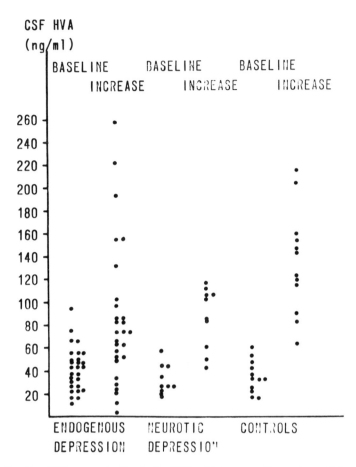

FIG. 2. Baseline HVA concentration in the CSF, with increase after probenecid, in endogenous depressive, neurotic depressive, and control groups.

metabolism are not a universal phenomenon in depressive illness, nor one randomly distributed through the group; rather, it seems to be confined to certain categories of depression.

V. CENTRAL MA METABOLISM AND ANTIDEPRESSANT MEDICATION

Is the responsiveness or resistance of a patient to antidepressants related to the presence or absence of disorders of cerebral MA metabolism? Do antidepressants (all potentiators of central monoaminergic activity, as already pointed out) abolish the supposed MA deficiency in the brain? So far, three studies have been devoted to this question and their results, although of a preliminary nature, certainly warrant further investigation of the hypothesis.

The first study was carried out in Groningen (van Praag, Korf, Dols and Schut, 1972). Of ten patients with severe endogenous depression, five were treated with a placebo for 3 weeks, while five were given DL-5-hydroxytryptophan (5-HTP), the 5-HT precursor which is transformed to 5-HT in the brain. The study was of double-blind design. The probenecid test was carried out in advance in all cases. The purpose was to test the hypothesis that, should 5-HTP have antidepressant properties, these would most likely become manifest in patients with a subnormal 5-HIAA response to probenecid. The five placebo patients showed no improvement. Of the five treated with 5-HTP, three showed improvement as judged by the physician in charge, the attending nurse, and themselves. In each, 5-HIAA accumulation after probenecid had been subnormal. Of the two patients who did not improve, one had shown normal and the other low-normal 5-HIAA accumulation (Table 2). The groups were too small to warrant definite conclusions being drawn but, stated prudently, these results are not inconsistent with the hypothesis that 5-HTP is an antidepressant and that 5-HIAA accumulation in the CSF after probenecid can be predictive of the chance of success of this medication.

TABLE 2. *Increase in CSF 5-HIAA in response to probenecid and effectiveness of 5-HTP*

Patients	Increase CSF 5-HIAA level* (ng/ml)	CSF probenecid level* (μg/ml)
5-HTP (improved)	16 (11 – 26)	7.2 (5.0 – 10.7)
5-HTP (not improved)	50 (39 – 68)	6.5 (5.5 – 7.5)
Nondepressive controls	80 (26 – 153)	9.1 (3.0 – 16.5)

*Group mean and range

Åsberg, Bertilsson, Tuck, Cronholm, and Sjövist (1973) studied the question of whether a relationship exists between 5-HIAA concentration in the CSF and the therapeutic response to nortriptyline. They found that the latter was considerably less in the group of patients with a 5-HIAA concentration less than 15 ng/ml than in those with a higher level. This was in spite of the fact that plasma nortriptyline level was well within the therapeutic range.

Nortriptyline is an antidepressant which differs in mechanism of action from the classical tricyclic compounds, imipramine (Tofranil®) and amitriptyline (Tryptizol®). Rather than exerting a similar inhibitory effect on the reuptake of catecholamines (NE and DA) and 5-HT, it shows a marked preference for the catecholamine "pump": 5-HT reuptake is much less markedly inhibited. To put it differently, nortriptyline potentiates the central activity of catecholamines much more intensively than of 5-HT. It is conceivable that such a compound loses some of its efficacy in patients where, to judge from CSF studies, 5-HT turnover seems to be diminished.

The third study to be mentioned in this context is that of Maas et al. (1972a), who demonstrated that patients excreting relatively low levels of MHPG prior to treatment with imipramine or desmethylimipramine respond better than patients with a relatively higher output.

VI. TOWARD A BIOCHEMICAL CLASSIFICATION OF DEPRESSION?

A tentative answer can now be given to the three basic questions I posed in the introduction. In depressive patients, cerebral MA turnover may be diminished. Such a disturbance does not occur in all patients, but seems to be confined to certain categories of depression. Whether central MA turnover is decreased or normal seems to be *a* factor which determines whether antidepressant medication will succeed or fail. The question why some patients respond while others do not respond to antidepressants is a very complex one. Numerous factors are probably involved: the symptomatology of the depression, its etiology, and pharmacokinetic factors such as the patient's ability to turn a given dose of an antidepressant into an adequate blood concentration. I have only been able to shed some light on one aspect of this complex problem, the metabolic.

Can factors related to central MA metabolism exert an influence on the therapeutic efficacy of antidepressants? The fact that I am inclined to answer this question in the affirmative has both theoretical and practical implications—theoretical because it has been established that MA-deficient and nondeficient patients are not always distinguishable on the basis of psychopathological symptoms. So far, we have been accustomed to using an etiological and a symptomatological classification of depressive illness. The findings I have discussed would seem to suggest that a biochemical classification is another possibility.

The practical implications are twofold. First, biochemical factors are likely to help us more clearly to define the indications for antidepressant medication in the not too distant future. Second, a biochemical classification of depression would provide a stimulus and give impetus to the search for antidepressants with a more specific action than those currently available. The future seems to promise the right drug for the right patient.

ACKNOWLEDGMENT

This study was supported by grants from the Netherlands Organization for Pure Research (Z.W.O.), the "Preventie Fonds," and the Merck, Sharp & Dohme Research Laboratories (U.S.A.).

REFERENCES

Åsberg, M., Bertilsson, L., Tuck, D., Cronholm, B., and Sjövist, F. (1973): Indoleamine metabolites in the cerebrospinal fluid of depressed patients before and during treatment with nortriptyline. *Clinical Pharmacology and Therapeutics,* 14:277–286.

Bourne, H. R., Bunney, W. E., Colburn, R. W., Davis, J. M., Davis, J. N., Shaw D. M., and Coppen, A. (1968): Noradrenaline, 5-hydroxytryptamine and 5-hydroxyindoleacetic acid in hind brains of suicidal patients. *Lancet,* II:805–808.

Bowers, M. B., Jr. (1972): Clinical measurements of central dopamine and 5-hydroxytryptamine metabolism: Reliability and interpretation of cerebrospinal fluid acid monoamine metabolite measures. *Neuropharmacology,* 11:101–111.

Coppen, A., Eccleston, E. J., and Peet, M. (1972a): Total and free tryptophan concentration in the plasma of depressive patients. *Lancet,* II:1415–1416.

Coppen, A., Brooksbank, B. W. L., and Peet, M. (1972b): Tryptophan concentration in the cerebrospinal fluid of depressive patients. *Lancet,* II:1393.

Eccleston, D., Ashcroft, G. W., Crawford, T. B. B., Stanton, J. B., Wood, D., and McTurk, P. H. (1970): Effect of tryptophan administration on 5-HIAA in cerebrospinal fluid in man. *Journal of Neurology, Neurosurgery and Psychiatry,* 33:269–272.

Goodwin, F. K., Post, R. M., Dunner, D. L., and Gordon, E. K. (1973): Cerebrospinal fluid amine metabolites in affective illness: The probenecid technique. *American Journal of Psychiatry,* 130:73–79.

Hordern, A., Burt, C. G., and Holt, N. F. (1965): *Depressive States. A Pharmacotherapeutic Study.* Charles C. Thomas, Springfield, Ill.

Klerman, J. L., and Cole, J. O. (1965): Clinical pharmacology of imipramine and related antidepressant compounds. *Pharmacological Reviews,* 17:101–141.

Knott, P. J., and Curzon, G. (1972): Free tryptophan in plasma and brain tryptophan metabolism. *Nature,* 239:452–453.

Korf, J., and van Praag, H. M. (1971): Amine metabolism in human brain: Further evaluation of the probenecid test. *Brain Research,* 35:221–230.

Maas, J. W., Dekirmenjian, H., Garven, D., and Redmond, D. E., Jr. (1972a): Excretion of MHPG after intraventricular 6-OH-DA. Presented at annual meeting, American Psychiatric Association, Dallas, Texas.

Maas, J. W., Fawcett, J. A., and Dekirmenjian, H. (1972b): Catecholamine metabolism, depressive illness and drug response. *Archives of General Psychiatry,* 26:252–262.

Mendels, J., Frazer, A., Fitzgerald, R. G., Ramsey, T. A., and Stokes, J. W. (1972): Biogenic amine metabolites in cerebrospinal fluid of depressed and manic patients. *Science,* 175:1380–1382.

Olsson, R., and Roos, B.-E. (1968): Concentrations of 5-hydroxyindoleacetic acid and homovanillic acid in the cerebrospinal fluid after treatment with probenecid in patients with Parkinson's disease. *Nature* 219:502–503.

Pare, C. M. B., Yeung, D. P. H., Price, K., and Stacey, R. S. (1969): 5-Hydroxytryptamine, noradrenaline and dopamine in brainstem, hypothalamus, and caudate nucleus of controls and of patients committing suicide by coal-gas poisoning. *Lancet*, II:133–135.

Papeschi, R., and McClure, D. J. (1971): Homovanillic and 5-hydroxyindoleacetic acid in cerebrospinal fluid of depressed patients. *Archives of General Psychiatry*, 25:354–358.

Shaw, D. M., Camps, F. E., and Eccleston, E. C. (1967): 5-Hydroxytryptamine in the hind brain of depressive suicides. *British Journal of Psychiatry*, 113:1407–1411.

van Praag, H. M., Korf, J., Dols, L. C. W., and Schut, T. (1972): A pilot study of the predictive value of the probenecid test in application of 5-hydroxytryptophan as antidepressant. *Psychopharmacologia*, 25:14–21.

van Praag, H. M., Korf, J., and Schut, T. (1973): Cerebral monoamines and depression. An investigation into their correlation with the aid of the probenecid technique. *Archives of General Psychiatry*, 28:829.

van Praag, H. M., Flentge, F., Korf, J., Dols, L. C. W., and Schut, T. (1973a): The influence of probenecid on the metabolism of serotonin, dopamine and their precursors in man. *Psychopharmacologia*, 33:141–151.

Wilk, S., Shopsin, B., Gershon, S., and Suhl, M. (1972): Cerebrospinal fluid levels of MHPG in affective disorders. *Nature*, 235:440–441.

Advances in Biochemical Psychopharmacology, Vol. 11
Raven Press, New York © 1974

Diagnosis of Manic-Depressive Psychosis from Cerebrospinal Fluid Concentration of 5-Hydroxyindoleacetic Acid

Rolf Sjöström

Psychiatric Research Center, University of Uppsala, Ulleråker Hospital, Uppsala, Sweden

I. INTRODUCTION

The purpose of this chapter is to discuss the possibility of using the measurement of acidic monoamine metabolites in cerebrospinal fluid (CSF) (especially after loading with probenecid) in the diagnosis of manic-depressive psychosis.

Probenecid induces an increase in concentration of 5-hydroxyindoleacetic (5-HIAA) and homovanillic (HVA) acids by inhibiting their active transport out of the CSF (Guldberg, Ashcroft, and Crawford, 1966). The increase in 5-HIAA or HVA level after probenecid has been used in animals to calculate the turnover of parent amine (Neff and Tozer, 1968). We have shown (Sjöström and Roos, 1972) that an increase of both 5-HIAA and HVA is small in both phases of manic-depressive psychosis in comparison with other psychiatric patients and healthy volunteers (Table 1). Other investigators have

TABLE 1. *Increase in 5-HIAA-HVA in CSF after probenecid (percent)*

	5-HIAA	HVA
Depression	27 ± 9 (24)[a]	83 ± 16 (10)[b]
Mania	20 ± 6 (21)[a]	73 ± 33 (13)[b]
Control patients	61 ± 13 (16)	240 ± 78 (8)
Healthy volunteers	66 ± 12 (13)	207 ± 40 (14)

Means ± SEM number of patients.
[a] Different from the control groups ($p < 0.01$).
[b] Different from the control groups ($p < 0.02$).

reported similar results (van Praag, Korf, and Puite, 1970; Goodwin, Post, Dunner, and Gordon, 1973). These findings were interpreted as signs of slow turnover of 5-hydroxytryptamine (5-HT) and dopamine in manic-depressive psychosis.

This interpretation has been questioned (Sjöström, 1972) because of great interindividual differences in probenecid concentration in plasma (Fig. 1) after the standard procedure we used (5 oral doses of 1 g over 2 to 5 days with lumbar puncture 2 hr after the final dose) and the high correlations both between probenecid concentrations in plasma and CSF, and between probenecid in CSF and 5-HIAA concentration (Fig. 2).

It may be that differences in rise of 5-HIAA or HVA level were due to pharmacokinetic variations in the disposition of probenecid between different groups of patients and not to altered monoamine turnover rates. However, probenecid pharmacokinetics seem to be similar in manic-depressive patients and controls (Sjöström, 1973): Similar concentrations of probenecid were present in plasma (total and free), and CSF of manic-depressive patients and controls after a standard oral load and biological half-life and volume of distribution of probenecid did not differ after an intravenous injection.

It has been shown (Sjöström, 1973) that the regression line for the

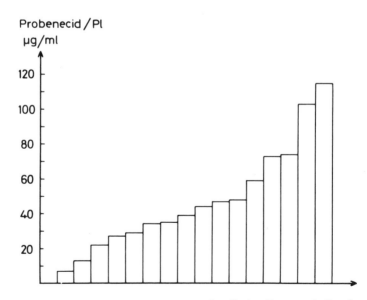

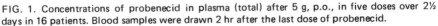

FIG. 1. Concentrations of probenecid in plasma (total) after 5 g, p.o., in five doses over 2½ days in 16 patients. Blood samples were drawn 2 hr after the last dose of probenecid.

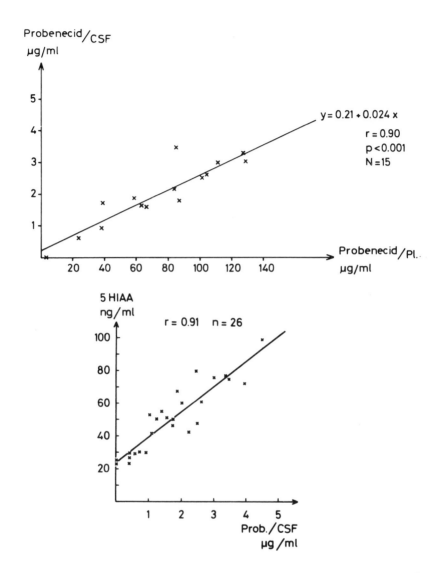

FIG. 2. The regression lines for concentrations of probenecid in plasma, on probenecid in CSF (top) and probenecid in CSF on 5-HIAA (bottom) after 5 g, p.o., over 2½ days.

relationship between probenecid concentration in the CSF (probenecid$_{CSF}$) and 5-HIAA follows a much steeper course in controls than in manic-depressive patients (Fig. 3). This finding implies that at comparable levels of probenecid$_{CSF}$ (equivalent degree of blockade of acid transport), manic-depressive patients have a low concentration of 5-HIAA$_{CSF}$ in comparison with controls. This finding supports the

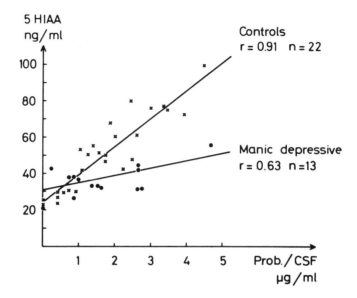

FIG. 3. The regression lines for probenecid in CSF on 5-HIAA in CSF in manic-depressive patients and controls. The slope of the regression line for controls is significantly steeper than for manic-depressive patients ($p < 0.001$). Crosses = controls. Dots = manic-depressive patients.

view that there is a low production of 5-HIAA in manic-depressive psychosis, perhaps as a result of slow turnover of 5-HT in the central nervous system.

II. DIAGNOSTIC IMPLICATIONS

A. Baseline Concentrations of 5-HIAA or HVA in CSF

Ashcroft, Eccleston, Crawford, Sharman, MacDougall, Stanton, and Binns (1966) were the first to report low concentrations of 5-HIAA in depression. This finding is controversial. Up to the beginning of 1973 there were nine reports in the literature concerning 5-HIAA levels in CSF in depression. Five could not confirm the results of Ashcroft's group (for review, see Sjöström, 1974). It seems obvious that single determinations of 5-HIAA or HVA are of no diagnostic help in the individual case.

B. 5-HIAA or HVA Concentrations in CSF after Probenecid

There is good agreement in the literature on the finding of a small increase only in 5-HIAA or HVA level in depression compared with controls after probenecid. Although there is a highly significant difference between the groups (Student's $t = 5,60\ n = 69$ for the difference in increase in both 5-HIAA and HVA between manic-depressive patients and controls), the discriminatory power is low in the individual case. A discriminant function analysis indicates that if there were a false positive diagnosis of manic-depressive psychosis in 10% of the controls, it would be possible to allocate only 50% of patients with a clinical diagnosis of manic-depressive psychosis to their correct category from the biochemical findings.

C. 5-HIAA or HVA Concentrations When Probenecid$_{CSF}$ Is Greater than 1.0 µg/ml

Figure 3 shows that differences in 5-HIAA concentration between manic-depressive patients and controls become evident when probenecid$_{CSF}$ is higher than 1.0 µg/ml. Table 2a shows the distribution of patients with probenecid$_{CSF} > 1.0$ µg/ml according to whether the concentration of 5-HIAA was greater or less than 45 ng/ml. Three of 29 patients (10%) would have been classified incorrectly (Sjöström, 1973, 1974). The test now approaches clinical usefulness. If concentrations of HVA are also taken into account and one considers combinations of concentrations of 5-HIAA and HVA with the best discriminatory power (Table 2b), the results are about equally good in a comparison between depressed patients and controls.

TABLE 2. *Distributions of 5-HIAA and HVA concentrations in patients with concentrations of probenecid $_{CSF} > 1.0/µg/ml$. Number of patients*

	a. 5-HIAA	
	< 45 ng/ml	> 45 ng/ml
Manic-depressive	9	1
Controls	1	13
Chi-square = 13.24; $p < 0.001$.		
	b. 5-HIAA < 50 ng/ml and HVA < 80 ng/ml	Other combinations
Depression	6	1
Controls	0	13

III. DIAGNOSIS AFTER REFINEMENT OF
THE PROBENECID TECHNIQUE

The results in Fig. 3 and Table 2 show it to be desirable that probenecid$_{CSF}$ be greater 1.0 μg/ml to achieve sufficient diagnostic accuracy. Figure 3 shows that several patients do not reach this probenecid$_{CSF}$ with our standard procedure. Such difficulties can be overcome with the help of modern pharmacokinetic techniques (Dost, 1968). The formula:

$$C = \frac{Dt_{1/2}}{V_d \quad \triangle t} \times 1.44$$

(where C = concentration in plasma under conditions of apparent steady state, $t_{1/2}$ = biological half-life for the drug, V_d = volume of distribution, $\triangle t$ = time interval between doses, D = administered dose) can be used to calculate the dose which it is necessary to administer to any patient to obtain any desired concentration of drug in plasma. The calculations require knowledge of $t_{1/2}$ and V_d, which can be obtained after an intravenous injection of probenecid.

We have tested the usefulness of this formula in the following way: a concentration of probenecid in plasma of 100 μg/ml (corresponding to about 2.5 μg/ml in CSF) was chosen as suitable. First, the standard dose (5 g over 2.5 days) was administered to seven patients and the plasma probenecid concentration was determined on the third day. Three days later, probenecid (1 g) was injected intravenously to the same group of seven patients and $t_{1/2}$ and V_d calculated. The individual dose necessary to obtain a plasma concentration of 100 μg/ml in each was determined and administered for 3 days. After the standard dose, concentrations of probenecid in plasma ranged from 34 to 156 μg/ml. Following the individual dose, the range was 103 to 152 μg/ml, with a median deviation from the "goal" (= 100 μg/ml) of 14%.

From the diagnostic point of view, a more effective probenecid test might be as follows:

Day 1:   Injection of 1 g of probenecid i.v. for calculation of $t_{1/2}$ and V_d and determination of the individual dose needed to obtain a suitable concentration of probenecid in plasma (80 to 100 μg/ml).

Day 2–4: Administration of the individual dose of probenecid.

Day 4:   Lumbar puncture for determination of 5-HIAA and HVA.

If our preliminary findings can be confirmed in a larger number of patients, it should be possible to predict whether or not a patient belongs to the manic-depressive group using the technique described above. However, there are still important unsolved problems, such as the fact that our data were obtained on highly selected material, clear-cut cases of manic-depressive psychosis, where doubtful cases had been excluded. It is not yet clear whether these conclusions would still be valid for mild or uncertain cases.

REFERENCES

Ashcroft, G. W., Eccleston, D., Crawford, T. B. B., Sharman, D. F., MacDougall, E. J., Stanton, J. B., and Binns, J. K. (1966): 5-Hydroxyindole compounds in the cerebrospinal fluid of patients with psychiatric or neurological diseases. *Lancet*, I:1049–1052.

Dost, F. H. (1968): *Grundlagen der Pharmakokinetik* 2nd ed. Thierna, Stuttgart.

Goodwin, F. K., Post, R. M., Dunner, D. L., and Gordon, E. K. (1973): Cerebrospinal fluid amine metabolites in affective illness: The probenecid technique. *American Journal of Psychiatry*, 130:73–99.

Guldberg, H. C., Ashcroft, G. W., and Crawford, T. B. B. (1966): Concentrations of 5-hydroxylindole-ylacetic acid and homovanillic acid in the cerebrospinal fluid of the dog before and during treatment with probenecid. *Life Sciences* 5: 1571–1575.

Neff, N. H., and Tozer, T. N. (1968): *In vivo* measurement of brain serotonin turnover. In: *Advances in Pharmacology. Vol. 6a*, Edited by Garattini and Shore, pp. 97–109. Academic Press, New York.

Sjöström, R., and Roos, B.-E. (1972): 5-Hydroxyindoleacetic acid and homovanillic acid in cerebrospinal fluid in manic-depressive psychosis. *European Journal of Clinical Pharmacology* 4:170–176.

Sjöström, R. (1972): Steady-state levels of probenecid and their relation to acid monoamine metabolites in human cerebrospinal fluid. *Psychopharmacologia*, 25:96–100.

Sjöström, R. (1973a): Cerebrospinal fluid content of 5-hydroxyindole acetic acid and homovanillic acid in manic-depressive psychosis. *Acta Universitatis Upsaliensis*, No. 154.

Sjöström, R. (1973b): 5-HIAA and HVA in cerebrospinal fluid in manic-depressive psychosis and the effect of probenecid treatment. *European Journal of Pharmacology* (in press).

van Praag, H. M., Korf, J., and Puite, J. (1970): 5-Hydroxyindoleacetic acid levels and the cerebrospinal fluid of depressive patients treated with probenecid. *Nature*, 225:1259–1260.

Advances in Biochemical Pschopharmacology, Vol. 11
Raven Press, New York © 1974

Serotonergic-Dopaminergic Interactions and Extrapyramidal Function

Thomas N. Chase

National Institute of Mental Health, Bethesda, Maryland 20014

Although the role of the nigrostriatal dopaminergic pathway has been accorded major emphasis in recent studies of extrapyramidal function, considerable evidence suggests that the interaction of several neuronal systems determines the regulatory influence of the basal ganglia on mammalian motor behavior. The concept of balanced neural pathways, each containing a characteristic monoamine, as determinants of specific brain functions is hardly new. More than 15 years ago, Brodie and Shore (1957) proposed that serotonin (5-HT) and norepinephrine (NE) modulated opposing systems in brain. Considering only extrapyramidal function, Barbeau (1962) later suggested that both the acetylcholine-dopamine (DA) and the histamine-5-HT systems functioned in equilibrium. Abundant clinical as well as preclinical data have since accumulated pointing to the interdependence of central dopaminergic and cholinergic systems (Andén and Bédard, 1971; Bartholini, Stadler, and Lloyd, 1973; Klawans and Rubovits, 1972), and more recently to the possible role of γ-aminobutyric acid (GABA) containing pathways in the modulation of cholinergic function (Kim, Bak, Hassler, and Okada, 1971). Pharmacologic evidence relevant to the possible interaction between 5-HT and DA-containing neuronal systems, especially in relation to the control of motor activity, will be considered here.

I. PRECLINICAL STUDIES

Since monoamine turnover may reflect nerve impulse activity within neural pathways containing these presumptive neurotransmitters (Alousi and Weiner, 1966; Bunney, Walters, Roth, and Aghajanian, 1973), an examination of the ability of drugs which act primarily on one monoaminergic system to influence amine metabolism within the other system might afford insight into their functional interactions.

D-Amphetamine, which enhances DA-mediated synaptic function, reportedly increases 5-HT turnover (Reid, 1970). On the other hand, neuroleptics which markedly accelerate DA metabolism, possibly as a result of DA-receptor blockade, exert little or no effect on 5-HT turnover (Pletscher and Da Prada, 1967). Parachlorophenylalanine, and inhibitor of tryptophan hydroxylase, also reduces DA synthesis (Tagliamonte, Tagliamonte, Corsini, Mereu, and Gessa, 1973), but α-methylparatyrosine, which interferes with tyrosine hydroxylase, accelerates 5-HT turnover (Jouvet and Pujol, *This Volume*). The interpretation of such results is complicated, however, by the possibility that a direct drug effect rather than an indirect response of one neurohumoral system to functional changes in another might account for the observed metabolic changes. Indeed, the former possibility is supported by the observation that drugs believed to act as direct DA-receptor agonists, such as ET 495 (Corrodi, Farnebo, Fuxe, Hamberger, and Ungerstedt, 1972), or 5-HT agonists, such as lysergic acid diethylamide (LSD) (Leonard and Tonge, 1969), have no significant effect on the central turnover of the other amine.

An examination of the motor response to amine-active drugs provides a second approach to the study of serotonergic-dopaminergic interactions. Experiments involving the direct injection of drugs into DA- or 5-HT-sensitive areas of cat caudate nucleus suggest that the motor effects induced by changes in dopaminergic mechanisms may be modulated by the activity of 5-HT neurons (Cools, 1973). Surgically induced tremor in monkeys can be ameliorated by either L-DOPA or 5-hydroxytryptophan (5-HTP), and the simultaneous administration of both monoamine precursors exerts an additive effect (Goldstein, Battista, Nakatani, and Anagnoste, 1969). Similarly, either L-DOPA or 5-HTP can potentiate the gnawing compulsion induced by apomorphine in monoamine oxidase (MAO) inhibitor pretreated rats (Fekete and Kurti, 1970). On the other hand, neither L-tryptophan, parachlorophenylalanine, nor methysergide influences stereotyped motor behavior occurring in rats treated with apomorphine, a DA-receptor agonist (Rotrosen, Angrist, Wallach, and Gershon, 1972). Moreover, parachlorophenylalanine pretreatment does not appear to affect the ability of L-DOPA to reverse neuroleptic-induced catalepsy (Maj, Kapturkiewicz, and Sarnek, 1972).

Animals with surgically or chemically induced brain lesions provide another approach to the study of functional relationships between 5-HT- and DA-containing neural systems. For example, stereotaxic lesions placed in the substantia nigra of rats or dogs can induce a marked reduction in the DA content of the ipselateral striatum without altering striatal 5-HT levels, while lesions in the midbrain raphe of rats produce a substantial loss of striatal 5-HT but no change in DA concentration (Goldstein, Himwich, Leiner, and Stout, 1971; Gumulka, del Angel, Samanin, and Valzelli, 1970). In such experiments the stability of amine levels within the anatomically intact pathway does not necessarily exclude functional alterations in response to the interruption of another monoaminergic system. Indeed, 6-hydroxy-dopamine-induced degeneration of central catecholamine neurons is attended by relatively little alteration in brain 5-HT levels (Uretsky and Iversen, 1970), despite a significant rise in 5-HT turnover (Lloyd and Bartholini, 1974).

II. CLINICAL STUDIES

Difficulties associated with the interpretation of data deriving from studies of functional interactions between monoaminergic systems in the experimental animal are compounded by attempts to examine these systems in the central nervous system of man. Limiting factors include the ethical considerations attending the conduct of human experimentation as well as the uncertain validity of the indirect methodology currently available. Two approaches to the study of dopaminergic-serotonergic relationships have been employed. First, the ability of pharmacologic agents, generally believed to act primarily on one neurohumoral system, to influence monoamine metabolism in the other has been studied. Second, 5-HT metabolism has been examined in the central nervous systems of patients with naturally occurring disorders known to be associated with alterations in dopaminergic function. Studies involving either approach are based on the assumption that changes in homovanillic acid (HVA) and 5-hydroxyindoleacetic acid (5-HIAA) levels in lumbar cerebrospinal fluid (CSF) during probenecid loading provide an index of central turnover of their parent amines. Evidence in support of this supposition has been extensively discussed in this symposium (Goodwin, *This Volume*; Sjöström, *This Volume*; van Praag, *This Volume*). As in the preclinical studies, it is also assumed that 5-HT and DA turnover reflect overall nerve impulse activity in neural pathways containing these amines.

With the exception of L-DOPA, pharmacological agents which influence DA-mediated function have little apparent effect on central 5-HT metabolism in man. Acute intravenous administration of L-DOPA produces a transient rise in lumbar CSF levels of 5-HIAA as well as HVA (Chase, 1974), while chronic treatment with this amino acid diminishes central 5-HT turnover as estimated by the probenecid technique (Goodwin, Post, Dunner, and Gordon, 1973). These findings are consistent with the results of preclinical studies suggesting that DA formation during L-DOPA loading may displace 5-HT from its central binding sites (Ng, Chase, Colburn, and Kopin, 1972a), thus causing an initial release of the indoleamine, while competition between L-DOPA and precursors of 5-HT for uptake and metabolism in central neurons (Bartholini, Da Prada, and Pletscher, 1968; Karobath, Diaz, and Huttunen, 1972) undoubtedly contributes to the longer term inhibitory effects of L-DOPA on 5-HT metabolism.

The possibility that L-DOPA treatment alters 5-HT metabolism directly rather than as an indirect, interneuronal response is supported by observations in patients receiving other DA-active drugs. Neither α-methylparatyrosine, haloperidol, nor ET 495 produces any consistent alteration in central 5-HT metabolism at doses which markedly influence both DA turnover and extrapyramidal function (Goodwin et al., 1973; Chase, 1973a). Similarly, treatment with L-tryptophan, 5-HTP, or p-chlorophenylalanine does not appear markedly to alter central DA turnover despite substantial changes in 5-HT metabolism and in some cases in motor function (Chase, Ng, and Watanabe, 1972; Goodwin et al., 1973). Since probenecid-induced changes in lumbar CSF levels of 5-HIAA were used as an index of 5-HT turnover, alterations in 5-HT function confined to pathways terminating in the basal ganglia cannot, however, be excluded.

Studies of central 5-HT metabolism in patients with neurologic disorders in which DA-mediated function appears abnormal have also yielded generally inconclusive results. Although this approach to the clinical study of dopaminergic-serotonergic relationships eliminates the problem of distinguishing direct from indirect drug effects, it imposes the analogous difficulty of ascertaining whether alterations in 5-HT metabolism occurring in patients with diseases involving dopaminergic systems constitute a direct or an indirect disease effect. In some instances pathologic or pharmacologic observations provide assistance in dealing with this issue.

Parkinson's disease now stands as the most extensively studied human disorder of dopaminergic function. Direct biochemical assay of postmortem tissues has demonstrated a characteristic depletion of both DA and HVA (Hornykiewicz, 1972), while probenecid loading studies suggest that the degree of reduction in central DA turnover (Table 1) correlates closely with the overall severity of parkinsonian signs (Chase and Ng, 1972). Somewhat smaller, yet no less consistent, reductions in the cerebral content of 5-HT (Hornykiewicz, 1972) and in 5-HT turnover (Table 1), as estimated by the probenecid technique (Chase and Ng, 1972), also occur in parkinsonian patients. As in the case of DA metabolism, there is a highly significant inverse correlation between overall parkinsonian severity and central 5-HT turnover (Chase and Ng, 1972).

Although it is generally assumed that the changes in DA metabolism in Parkinson's disease arise as a consequence of the degeneration of

TABLE 1. *Probenecid-induced accumulation of monoamine metabolites in untreated patients with extrapyramidal disease and in control subjects*

	Number of patients	HVA (ng/ml)	5-HIAA (ng/ml)
Controls	8	166 ± 24	84 ± 8.2
Parkinson's disease	20	62 ± 11[a]	43 ± 5.4[a]
Huntington's chorea	12	100 ± 14[b]	86 ± 9.3
Dystonia musculorum deformans	8	154 ± 19	86 ± 15

The technique of probenecid loading and CSF assay were as previously described (Chase and Ng, 1972).

[a] $p < 0.001$.

[b] $p < 0.05$ for difference from control levels by t test for unpaired data.

DA-containing neurons comprising the nigrostriatal pathway, alterations in 5-HT metabolism might reflect either primary structural damage to serotonergic neurons or a functional response of serotonergic systems to the reduction in DA-mediated activity. If the former were true, then replenishment of 5-HT stores by precursor loading might be expected to ameliorate parkinsonism in the same way that L-DOPA is believed to alleviate the symptoms of this disorder. Clinical experience, however, has shown the opposite to be the case. Relatively high doses of L-5-HTP given in combination with a peripheral decarboxylase inhibitor (MK 486), or L-tryptophan given with pyridoxine, tend to exacerbate rather than diminish parkinsonian signs (Chase et al., 1972; Hall, Weiss, Morris, and Prange, 1972). Conceivably these results may be attributable to a further reduction in dopaminergic function due to competition between precursors of 5-HT and DA for uptake, metabolism, and storage within central neurons (Fuxe, Butcher, and Engel, 1971; Ng, Chase, Colburn, and Kopin, 1972*b*).

The foregoing results might, however, be taken to favor the possibility that 5-HT-mediated neuronal function is diminished as a compensatory response to a primary reduction in dopaminergic activity. Indirect support for this contention derives from the apparent lack of characteristic pathological changes in brainstem raphe areas thought to be rich in 5-HT-containing nerve cell bodies. Moreover, the administration of MAO inhibitors to parkinsonian patients reportedly fails to alter brain DA concentrations, but increases the cerebral content of 5-HT to normal or above normal levels (Hornykiewicz, 1972). Although this result might be attributable to the differential effects of MAO inhibitors on the catabolism of DA and 5-HT or to changes in serotonergic systems whose function is not altered in parkinsonian patients, it is also consistent with the possibility that the changes in 5-HT metabolism in Parkinson's disease reflect secondary functional rather than primary structural alterations in serotonergic systems.

Abnormalities in central DA metabolism have also been reported in patients with Huntington's chorea. Both striatal DA levels (Bernheimer and Hornykiewicz, 1973) and probenecid-induced accumulations of HVA in lumbar CSF (Chase, 1973b) appear reduced. Unlike Parkinson's disease, however, pathological changes in patients with Huntington's chorea characteristically involve nerve cells in the caudate nucleus and putamen; cell bodies of neurons comprising the nigrostriatal dopaminergic pathway are largely spared. These findings suggest that the decrease in DA turnover may constitute a secondary functional alteration of neurons containing this amine. Conceivably, the striatal nerve cells which degenerate in Huntington's chorea patients receive inhibitory inputs from neurons comprising the nigrostriatal system (Chase, 1973b). Reduced nerve impulse activity in this dopaminergic system might thus occur as a compensatory response to the primary degeneration of the striatal neurons. In further contrast to Parkinson's disease, 5-HT metabolism in Huntington's chorea appears normal (Table 1). Moreover, neither the administration of L-tryptophan nor of p-chlorophenylalanine influences choreatic severity (Chase, 1973b).

The foregoing clinical observations suggest that alterations in central 5-HT metabolism, and thus presumably in the activity of neural pathways containing this monoamine, are not obligatory concomitants of disorders associated with abnormalities of dopaminergic function and that pharmacologically induced alterations in motor behavior associated with changes in either the DA or the 5-HT system are not necessarily attended by functional alterations in both systems. In patients who are free of central nervous system disease, DA and 5-HT turnover rates, as judged by the probenecid technique, appear to be highly correlated (Table 2). This relationship is maintained in individ-

TABLE 2. *Correlation (r) between the probenecid-induced accumulation of homovanillic acid and 5-hydroxyindoleacetic acid in lumbar spinal fluid*

	Number of patients	Slope	r	p
Controls	8	0.31	0.93	<0.001
Parkinson's disease	20	0.34	0.73	<0.001
Huntington's chorea	12	0.38	0.55	NS
Dystonia musculorum deformans	8	0.47	0.60	NS

uals with Parkinson's disease, where both DA and 5-HT turnover are reduced in proportion to clinical severity, as well as in patients with dystonia musculorum deformans, in whom no characteristic alteration in central DA or 5-HT metabolism has been detected (Table 1). The correlation between DA and 5-HT is lost, however, in Huntington's chorea, where 5-HT metabolism remains unchanged despite alteration in DA turnover (Table 2).

III. CONCLUSIONS

The limited and often inconsistent data now available relevant to possible functional relationships between 5-HT- and DA-containing pathways within the central nervous system discourage even the most noncommittal and speculative comment. These data, however, appear to offer little support to the widely held concept that a simple, balanced arrangement between central 5-HT and DA systems serves as a critical determinant of extrapyramidal function. It is likely that the various neural pathways comprising the mammalian brain function in a highly interactive manner. Continued efforts to elucidate the nature of these interactions would thus seem crucial to any comprehensive view of central nervous system function.

REFERENCES

Alousi, A., and Weiner, N. (1966): The regulation of norepinephrine synthesis in sympathetic nerves: Effect of nerve stimulation, cocaine, and catecholamine-releasing agents. *Proceedings of the National Academy of Sciences* (U.S.), 56:1491–1496.

Andén, N.-E., and Bédard, P. (1971): Influences of cholinergic mechanisms on the function and turnover of brain dopamine. *Journal of Pharmacy and Pharmacology*, 23:460–462.

Barbeau, A. (1962): The pathogenesis of Parkinson's disease: A new hypothesis. *Canadian Medical Association Journal*, 87:802–807.

Bartholini, G., Da Prada, M., and Pletscher, A. (1968): Decrease of cerebral 5-hydroxytryptamine by 3,4-dihydroxyphenylalanine after inhibition of extracerebral decarboxylase. *Journal of Pharmacy and Pharmacology*, 20:228–229.

Bartholini, G., Stadler, H., and Lloyd, K. G. (1973): Cholinergic-dopaminergic interactions in the extrapyramidal system. In: *Advances in Neurology*, Vol. 3, edited by D. B. Calne. Raven Press, New York.

Bernheimer, H., and Hornykiewicz, O. (1973): Brain amines in Huntington's chorea. In: *Advances in Neurology*, Vol. 1, edited by A. Barbeau, T. N. Chase, and G. W. Paulson. Raven Press, New York.

Brodie, B. B., and Shore, P. A. (1957): A concept for a role of serotonin and norepinephrine as chemical mediators in the brain. *Annals of the New York Academy of Sciences*, 66:631–642.

Bunney, B. S., Walters, J. R., Roth, R. H., and Aghajanian, G. K. (1973): Dopaminergic neurons: Effect of antipsychotic drugs and amphetamine on single cell activity. *Journal of Pharmacology and Experimental Therapeutics*, 185:560–571.

Chase, T. N. (1973a): Central monoamine metabolism in man: Effect of putative dopamine receptor agonists and antagonists. *Archives of Neurology*, 29:349–351.

Chase, T. N. (1973b): Biochemical and pharmacologic studies of monoamines in Huntington's chorea. In: *Advances in Neurology*, Vol. 1, edited by A. Barbeau, T. N. Chase, and G. W. Paulson. Raven Press, New York.

Chase, T. N. (1974): Serotonergic mechanisms and extrapyramidal function in man. *Advances in Neurology*, Vol. 5, edited by F. McDowell and A. Barbeau. Raven Press, New York.

Chase, T. N., and Ng, L. K. Y. (1972): Central monoamine metabolism in Parkinson's disease. *Archives of Neurology*, 27:486—491.

Chase, T. N., Ng, L. K. Y., and Watanabe, A. M. (1972): Parkinson's disease: Modification by 5-hydroxytryptophan. *Neurology*, 22:479—484.

Cools, A. R. (1973): The caudate nucleus and neurochemical control of behavior. The function of dopamine and serotonin in the caput nuclei caudati of cats. Drukkerij Brakkenstein, Nijmegen.

Corrodi, H., Farnebo, L.-O., Fuxe, K., Hamberger, B., and Ungerstedt, U. (1972): ET 495 and brain catecholamine mechanisms: Evidence for stimulation of dopamine receptors. *European Journal of Pharmacology*, 20:195—204.

Fekete, M., and Kurti, A. M. (1970): On the dopaminergic nature of the gnawing compulsion induced by apomorphine in mice. *Journal of Pharmacy and Pharmacology*, 22:377—379.

Fuxe, K., Butcher, L. L., and Engel, J. (1971): DL-5-Hydroxytryptophan-induced changes in central monoamine neurons after peripheral decarboxylase inhibition. *Journal of Pharmacy and Pharmacology*, 23:420—424.

Goldstein, M., Battista, A. F., Nakatani, S., and Anagnoste, B. (1969): Drug-induced relief of the tremor in monkeys with mesencephalic lesions. *Nature*, 224:382—384.

Goldstein, S., Himwich, W. A., Leiner, K., and Stout, M. (1972): Psychoactive agents in dogs with bilateral lesions in subcortical structures. *Neurology*, 21:847—852.

Goodwin, F. K., Post, R. M., Dunner, D. L., and Gordon, E. K. (1973): Cerebrospinal fluid amine metabolites in affective illness: The probenecid technique. *American Journal of Psychiatry*, 130:73—79.

Gumulka, W., del Angel, A. R., Samanin, R., and Valzelli, L. (1970): Lesions of substantia nigra: Biochemical and behavioral effects in rats. *European Journal of Pharmacology*, 10:79—82.

Hall, C. D., Weiss, E. A., Morris, C. E., and Prange, A. J. (1972): Rapid deterioration in patients with parkinsonism following tryptophan-pyridoxine administration. *Neurology*, 22:231—237.

Hornykiewicz, O. (1972): Neurochemistry of parkinsonism. In: *Handbook of Neurochemistry*, Vol. VII, edited by A. Lajtha. Plenum Press, New York.

Karobath, M., Diaz, J.-L., and Huttunen, M. (1972): Serotonin synthesis with rat brain synaptosomes. Effects of L-DOPA, L-3-methoxytyrosine and catecholamines. *Biochemical Pharmacology*, 21:1245—1251.

Kim, J. S., Bak, I. J., Hassler, R., and Okada, Y. (1971): Role of γ-aminobutyric acid (GABA) in the extrapyramidal motor system. 2. Some evidence for the existence of a type of GABA-rich strio-nigral neurons. *Experimental Brain Research*, 14:95—103.

Klawans, H. L. and Rubovits, R. (1972): Central cholinergic-anticholinergic antagonism in Huntington's chorea. *Neurology*, 22:107—116.

Leonard, B. E., and Tonge, S. R. (1969): The effect of some hallucinogenic drugs upon the metabolism of noradrenaline. *Life Sciences* (Part I), 8:815—825.

Maj, J., Kapturkiewicz, K., and Sarnek, J. (1972): The effect of L-DOPA on neuroleptic-induced catalepsy. *Journal of Pharmacy and Pharmacology*, 24:735—737.

Ng, L. K. Y., Chase, T. N., Colburn, R. W., and Kopin, I. J. (1972a): L-DOPA in parkinsonism. *Neurology*, 22:688—696.

Ng, L. K. Y., Chase, T. N., Colburn, R. W., and Kopin, I. J. (1972b): Release of [^{3}H]dopamine by L-5-hydroxytryptophan. *Brain Research*, 45:499—505.

Pletscher, A., and Da Prada, M. (1967): Mechanism of action of neuroleptics. In: *Neuropsychopharmacology*, edited by H. Brill. Excerpta Medica Foundation, Amsterdam.

Reid, W. D. (1970): Turnover rate of brain 5-hydroxytryptamine increased by D-amphetamine. *British Journal of Pharmacology*, 40:483–491.

Rotrosen, J., Angrist, B. M., Wallach, M. B., and Gershon, S. (1972): Absence of serotonergic influence on apomorphine-induced stereotypy. *European Journal of Pharmacology*, 20:133–135.

Tagliamonte, A., Tagliamonte, P., Corsini, G. U., Mereu, G. P., and Gessa, G. L. (1973): Decreased conversion of tyrosine to catecholamines in the brain of rats treated with p-chlorophenylalanine. *Journal of Pharmacy and Pharmacology*, 25:101–103.

Uretsky, N. J., and Iversen, L. L. (1970): Effects of 6-hydroxydopamine on catecholamine containing neurons in the rat brain. *Journal of Neurochemistry*, 17:269–278.

Advances in Biochemical Psychopharmacology, Vol. 11
Raven Press, New York © 1974

Serotonin and 5-Hydroxyindoleacetic Acid in Discrete Areas of the Brainstem of Suicide Victims and Control Patients

Kenneth G. Lloyd, Irene J. Farley, John H. N. Deck, and
Oleh Hornykiewicz

*Department of Psychopharmacology, Clarke Institute of Psychiatry and the
Departments of Pathology and Pharmacology, University of Toronto,
Toronto, Ontario, Canada*

I. INTRODUCTION

The hypothesis that alterations in serotonin (5-HT)-containing neuronal systems of the brain are part of the endogenous component of depression has been under consideration for many years (cf. Coppen, 1972). Previous biochemical analyses of 5-HT in brain tissue of depressed patients have been limited to the examination of whole lower brainstems (hindbrains) from suicide victims. In these studies, Shaw, Camps, and Eccleston (1967) and Pare, Yeung, Price, and Stacey (1969) reported a lowering of 5-HT levels in comparison to controls. In contrast, Bourne, Bunney, Colburn, Davis, Davis, Shaw, and Coppen (1968) failed to find any difference in lower brainstem 5-HT levels between controls and suicides, but observed a significant decrease of 5-hydroxyindoleacetic acid (5-HIAA) in suicides as compared to controls. Thus, in spite of some contradictions, these previous investigations pointed to an alteration in 5-hydroxyindoles in the lower brainstem of suicide victims. However, it was not convincingly demonstrated that these alterations were not bound up with differences in the mode of death of the suicides compared with controls, or to the large age differences between the two groups. The significance of the results of all three studies was also weakened by the fact that no attempt was made to study discrete areas of the lower brainstem, so that small but possibly important disturbances might not have been detected. The present study was planned to investigate these possibilities further.

II. MATERIALS AND METHODS

A. *Human Material*

A substantial number of neurologically normal patients (controls) have previously been examined (Farley, Lloyd, and Hornykiewicz, *in preparation*), from whom five patients with cardiovascular disease have been selected (on the basis of suitable age, interval between death and autopsy, and storage time) for comparison with five suicide victims (see Table 2). Three of these five suicide victims died from drug overdose, had a previous history of suicidal attempts, and had received psychiatric care; drugs were not involved in the other two cases. Another two nondrug suicides, both 24 years of age, were excluded from the statistical comparison in Table 1 because controls of comparable age were not available.

B. *Dissection of Material*

All brains were received within 30 min of autopsy and were immediately processed. After the removal and freezing of superficially lying areas other than cortex (e.g., hypothalamus), the cerebellum was removed and the whole lower brainstem separated from the rest of the brain at the precollicular level. The forebrain (higher brainstem and telencephalon) and the lower brainstem were then divided into halves by a midsagittal section and frozen on Dry Ice. The cerebellum was frozen as a whole. These major subdivisions were stored on Dry Ice until their dissection into 3-mm slices, from which the discrete brain regions were obtained. Isolation of the forebrain areas was performed from slices cut at different planes (cf. Lloyd and Hornykiewicz, 1972). Reliability of the dissection technique is extremely important in the case of the unfixed lower brainstem, for which no procedure seems to have been described in the neurochemical literature. In our hands, a highly reproducible dissection procedure proved to be the sectioning of each half of the lower brainstem into four sagittal slices of approximately 3 mm thickness in which nuclear areas could be well distinguished (Farley, Lloyd, and Hornykiewicz, *in preparation*). This dissection procedure has been found particularly suited to the isolation of the various nuclei of the raphe system (contained in the most median slice), the detailed biochemical analysis of which is reported in this study for the first time (for details of the anatomy of the lower brainstem in man, see Olszewski and Baxter, 1954).

C. *Determination of 5-HT and 5-HIAA*

The concentrations of 5-HT and 5-HIAA were estimated in the same tissue sample by a method to be published in detail (Farley, Lloyd and Hornykiewicz, *in preparation*). In brief, the samples were weighed and then homogenized in ice-cold 0.1 N HCl. Following protein precipitation, 5-HIAA was extracted into ether and returned to a buffer phase at pH 7.3. 5-HT was extracted from the salt-saturated aqueous solution into butanol at pH 10 and subsequently back-extracted into acid by addition of excess heptane. Concentrations of 5-HT and 5-HIAA were estimated by reading the native fluorescence in the presence of strong acid at 295/540 nm. All samples were corrected for recovery (80%) and expressed in terms of micrograms per gram of tissue.

III. RESULTS AND DISCUSSION

A. *Factors Possibly Influencing 5-HT and 5-HIAA Concentrations*

1. Interval between death and autopsy. Most of the brains were received between 10 and 20 hr after death; the shortest autopsy interval was 6 hr, the longest 13 hr. We could not detect any differences in brain 5-HT and 5-HIAA levels for autopsy intervals between 6 and 33 hr. These results confirm those of a previous study (MacLean, Nicholson, Pare, and Stacey, 1965), which noted that total lower brainstem 5-HT levels did not vary significantly from 12 to 92 hr postmortem.

2. Patient age. As suicide victims are often quite young, age-matching with controls can be difficult. In an earlier study (Robinson, Nies, Davis, Bunney, Davis, Colburn, Bourne, Shaw, and Coppen, 1972) which utilized 55 whole lower brainstems, the authors failed to see a significant correlation between age and 5-HT level. In the present examination of isolated brainstem nuclei (e.g., red nucleus, Fig. 1) from 9 suicides (aged 24 to 76 years) and 14 controls (aged 28 to 93 years), there was no obvious correlation of either 5-HT or 5-HIAA level with age. Similarly, for the forebrain structures examined (e.g., caudate nucleus) 5-HT and 5-HIAA concentrations did not change with age after young adulthood (Farley, Lloyd, and Hornykiewicz, *unpublished observations*).

3. Storage of material. The effect of storage is an important factor to consider. It is known that brain 5-HT is stable for short periods *in situ* and that losses are minimal if the material is frozen to at least $-20°C$ immediately after death (Joyce, 1962). Figure 2 shows that storage of material for up to 200 days had no effect on the levels of 5-HT and 5-HIAA in 24 hypothalami. These results do not support the finding of Dowson (1969), who reported an exponential fall of both compounds within the first 25 days of storage to about two-thirds of the initial concentrations. Our results indicate that under the present storage conditions 5-HT and 5-HIAA are stable for at least 200 days at $-40°C$. A

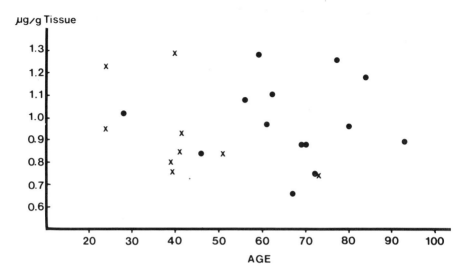

FIG. 1. Concentrations of 5-HT in the red nucleus of control patients (●) and suicide victims (x) with ages ranging from 24 to 93 years.

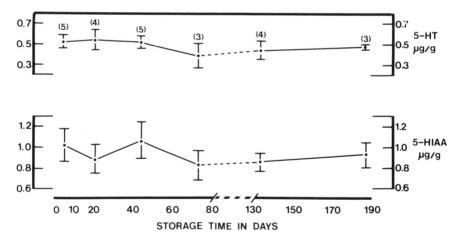

FIG. 2. Mean concentrations (± SEM) of 5-HT and 5-HIAA in 24 hypothalami following different storage times (Dry Ice). Figures in parentheses indicate the number of samples analyzed to obtain each value.

possible explanation for the discrepancy is a stricter screening of control patients in the present study, with a resulting greater homogeneity of material. A similar stability of 5-HT and 5-HIAA levels with storage has been demonstrated with better-defined regions (e.g., red nucleus).

B. 5-HT and 5-HIAA in the Brains of Control Patients and Suicide Victims

In the series of controls the regions of the brain which were conspicuous for their high 5-HT and 5-HIAA content were the tegmentum of the midbrain and the pons-medulla. Within this part of the lower brainstem, those areas situated closest to the midline (i.e., the areas composed of the raphe nuclei) had by far the greatest concentrations of 5-hydroxyindoles. Thus, the raphe areas containing the raphe nuclei dorsalis and centralis superior had 5-HT concentrations (2.22 and 2.25 μg/g, respectively) which were four times higher than those in the hypothalamus (0.50 μg/g); this difference was even more pronounced for 5-HIAA, whose levels in the above raphe areas were 9 and 15 times (7.04 and 12.37 μg/g, respectively) those measured in the hypothalamus (0.84 μg/g). The concentrations of 5-HT and 5-HIAA fell sharply in areas of the lower brainstem tegmentum situated lateral to the corresponding raphe regions. In general, the rostral mesencephalic portion of the lower brainstem exhibited a distinctly steeper mediolateral gradient in 5-HT than did the caudal pons-medulla part. As a rule, the concentrations of 5-HT and 5-HIAA in regions of the higher brainstem and telencephalon decreased in proportion to the distance of a given area from the mesencephalic raphe (Farley, Lloyd, and Hornykiewicz, *in preparation*).

1. Lower brainstem. Table 1 summarizes the results obtained for the raphe nuclei in five controls (mean age 66 years, average time between death and autopsy, 17 hr) and five suicide victims (mean age 42 years, average time between death and autopsy, 20 hr). Storage time for the ten brains ranged from 6 to 150 days; average storage time was 70 days. Individuals with very short and longer storage times were evenly distributed in both groups.

The 5-HIAA concentrations in the raphe nuclei of the suicide victims were, on the whole, within the control range. A possible exception may be the raphe nucleus obscurus, the 5-HIAA level of which in suicides

TABLE 1. Concentrations of 5-HIAA and 5-HT in the raphe nuclei of controls and suicides

Raphe nuclei	5-HIAA (µg/g)				5-HT (µg/g)			
	Controls Mean ± SEM	n	Suicides Mean ± SEM	n	Controls Mean ± SEM	n	Suicides Mean ± SEM	n
Centralis superior	12.37 ± 0.65 (11.25–14.53)	5	12.39 ± 0.91 (9.19–14.61)	5	2.25 ± 0.19 (1.88–2.97)	5	1.86 ± 0.16 (1.41–2.26)	5
Dorsalis	7.04 ± 0.59 (5.85–8.63)	5	6.14 ± 0.52 (5.00–7.79)	5	2.22 ± 0.13 (1.92–2.56)	5	1.55 ± 0.12[a] (1.23–1.90)	5
Pontis	7.66 ± 0.57 (6.21–8.92)	5	7.12 ± 0.68 (5.42–9.55)	5	1.34 ± 0.21 (0.80–1.89)	5	1.04 ± 0.13 (0.65–1.38)	5
Centralis inferior	4.49 ± 0.43 (3.63–5.77)	5	4.01 ± 0.32 (3.14–4.67)	5	1.32 ± 0.12 (0.94–1.47)	5	0.95 ± 0.07[b] (0.69–1.09)	5
Obscurus	2.86 ± 0.24 (2.00–3.46)	5	3.36 ± 0.31 (2.46–4.24)	5	1.07 ± 0.14 (0.62–1.39)	5	1.02 ± 0.14 (0.75–1.56)	5
Pallidus	2.28 ± 0.16 (1.85–2.67)	5	2.25 ± 0.20 (1.90–2.74)	4	0.61 ± 0.09 (0.44–0.91)	5	0.42 ± 0.12 (0.28–0.76)	4

n is the number of brains examined. Ranges are given in parentheses. Significantly different from controls: [a] $p < 0.01$; [b] $p < 0.05$.

was approximately 18% above control level; however, this difference was not significant on a statistical basis.

In contrast to 5-HIAA, the concentrations of 5-HT were decreased throughout the raphe nuclei of the suicides, as compared to controls. In the raphe nuclei dorsalis and centralis inferior, these differences in 5-HT levels reached statistical significance ($p < 0.01$ and $p < 0.05$, respectively). The decrease of 5-HT in the area containing the raphe nucleus dorsalis is of particular interest, since this region has been shown to be part of the midbrain extension of the limbic system (Nauta, 1958). Areas situated laterally to the raphe system (including the substantia nigra and the red nucleus; cf. Table 3) did not show any similar changes.

One possible explanation for lowered 5-HT levels in suicides might be the use of drugs at death. In fact, for two previous studies (Shaw et al., 1967; Bourne et al., 1968), the noted differences in 5-HT between suicides and controls could have been related to the barbiturates used by the suicides. To obviate this possibility, we have separated seven suicides into two subgroups—those whose death was not related to a drug overdose and those who committed suicide by the use of drugs. Table 2 shows the data for 5-HT, in these two subgroups of suicides, for two raphe nuclei areas: the area of the raphe nucleus dorsalis, where a significantly subnormal 5-HT concentration was observed in our suicide material, and the area of the raphe nucleus centralis superior, where no significant change in 5-HT level was apparent. With regard to the raphe nucleus dorsalis, the mean concentration of 5-HT was 1.55 μg/g for the nondrug group and 1.48 μg/g for the drug-overdose group. With respect to the raphe nucleus centralis superior, the mean levels were 1.66 μg/g and 1.87 μg/g, respectively. Thus for the cases where no drugs were involved, the 5-HT levels were the same as for the drug-overdose group. In both subgroups the 5-HT concentrations in the nucleus dorsalis were significantly lower than the controls ($p < 0.01$), and were no different from the controls in the nucleus centralis superior. Differences between suicide victims and controls with respect to the suddenness of death is another factor to consider when comparing the two groups. However, as can be seen in Table 2 ("cause of death" column), both our groups comprised mixed cases: Some can be assumed to have died suddenly (ruptured aortic aneurysm; exsanguination) and some slowly (chronic heart disease; drug overdose). Thus, it is unlikely that this factor was reponsible for the differences observed between our two groups.

2. Higher brainstem and telencephalon. An investigation of 5-HT and 5-HIAA in higher brainstem and telencephalic regions of the suicides failed to disclose any trend toward an alteration of 5-hydroxyindole levels. In most of the regions examined (Table 3), the levels were well within the range of the control group. However, certain regions

TABLE 2. 5-HT concentrations in the nuclei raphe dorsalis and centralis superior of individual controls and suicides

| Patient Data | | | | Raphe nucleus 5-HT (μg/g) | |
Age	Sex	Autopsy interval (hr)	Cause of death	Dorsalis	Centralis superior
A. Controls					
56	M	33	Myocardial infarct	2.50	2.06
62	M	15	Chronic congestive heart failure	2.56	2.97
77	F	10	Chronic arteriosclerotic heart disease	1.92	1.88
69	M	19	Ruptured aortic aneurysm	2.10	2.32
67	M	6	Ruptured aortic aneurysm	2.04	2.00
B. Suicides					
40	M	19	Carbon monoxide poisoning	1.44	1.41
51	M	24	Exsanguination—cut wrists	1.90	2.26
24	M	22	Asphyxia due to suspension	1.25	1.33
24	F	16	[a]Jumped under train	1.61	1.62
39	F	24	[a]Diazepam (+ other drugs?)	1.51	1.73
39	M	13	[a]Diazepam + propoxyphene	1.23	2.18
41	F	20	[a]Diazepam + oxazepam + alcohol	1.71	1.71

[a]Known history of depression.

TABLE 3. Concentrations of 5-HIAA and 5-HT in selected areas of the mes-, di-, and telencephalon of controls and suicides

Brain region	5-HIAA (µg/g)				5-HT (µg/g)			
	Controls Mean ± SEM	n	Suicides Mean ± SEM	n	Controls Mean ± SEM	n	Suicides Mean ± SEM	n
Substantia nigra	3.11 ± 0.16 (2.44 − 3.55)	10	3.52 ± 0.43 (2.76 − 5.59)	6	1.02 ± 0.09 (0.64 − 1.68)	10	0.91 ± 0.09 (0.70 − 1.20)	6
Red nucleus	4.56 ± 0.27 (2.81 − 6.23)	13	5.33 ± 0.42 (3.84 − 6.36)	6	0.98 ± 0.05 (0.66 − 1.28)	13	0.99 ± 0.09 (0.80 − 1.29)	6
Subthalamic nucleus	2.04 ± 0.22 (1.20 − 3.36)	8	2.16 ± 0.26 (1.21 − 3.09)	7	0.60 ± 0.11 (0.30 − 1.09)	8	0.49 ± 0.03 (0.34 − 0.57)	7
Hypothalamus	0.84 ± 0.10 (0.56 − 1.33)	8	0.91 ± 0.11 (0.61 − 1.45)	7	0.50 ± 0.05 (0.35 − 0.70)	8	0.47 ± 0.06 (0.18 − 0.60)	7
Amygdala	0.85 ± 0.03 (0.76 − 0.97)	8	0.75 ± 0.10 (0.33 − 1.06)	6	0.37 ± 0.04 (0.27 − 0.46)	8	0.31 ± 0.04 (0.19 − 0.47)	6
Putamen	1.03 ± 0.08 (0.60 − 1.47)	12	1.05 ± 0.05 (0.88 − 1.25)	6	0.37 ± 0.04 (0.21 − 0.67)	12	0.35 ± 0.03 (0.29 − 0.48)	6
Caudate nucleus	0.60 ± 0.08 (0.37 − 1.24)	12	0.58 ± 0.09 (0.40 − 0.83)	5	0.32 ± 0.03 (0.16 − 0.47)	12	0.29 ± 0.02 (0.23 − 0.32)	5
Hippocampus	0.29 ± 0.01 (0.25 − 0.32)	7	0.34 ± 0.08 (0.21 − 0.65)	5	0.11 ± 0.01 (0.07 − 0.17)	7	0.12 ± 0.02 (0.06 − 0.17)	5

n is the number of brains examined. Ranges are given in parentheses.

forming part of the forebrain limbic system (e.g., the mammillary bodies) have shown a significant increase in 5-HIAA concentration. This observation is presently being investigated in greater detail.

The significance of lowered 5-HT levels in certain raphe nuclei, and the possibility of increased 5-HIAA levels in parts of the forebrain limbic system of the suicide victims, is at present unclear. Since it is known that the 5-HT containing raphe nuclei receive a dense noradrenergic innervation which might influence the functional state of the 5-HT neurons in question, a study of the behavior of norepinephrine in discrete areas of the lower brainstem in controls and suicide victims has been initiated. At the present, sufficient data are not available to permit us to classify all our suicide victims with regard to mental state at the time of suicide; however, it is known that four of our cases had previously been receiving psychiatric treatment for depression. If—for the sake of argument—the assumption is made that in our material an underlying cause of suicide was depression, our results would suggest that alterations of brain 5-HT metabolism in depression might be of a very discrete nature, affecting only certain functional systems rather than the whole brain or such gross anatomical subdivisions as the whole of the lower brainstem ("hindbrain").

IV. CONCLUSIONS

5-HT and 5-HIAA concentrations have been examined in discrete regions from the brains of neurologically normal patients and from suicide victims. In the lower brainstem, 5-HT levels were significantly lower in the raphe nuclei dorsalis and centralis inferior; 5-HIAA levels were essentially unaltered as compared to controls. In the higher brainstem and telencephalon, 5-HT levels were apparently normal in the suicides whereas 5-HIAA levels may have been increased in certain limbic structures (mammillary bodies). The present study has demonstrated the necessity of performing biochemical estimations in anatomically discrete regions which attempt to reflect, as correctly as possible, the intricate morphology of the human brain; it is hoped that such an approach may result in a better understanding of the neurochemistry of psychic disorders.

ACKNOWLEDGMENTS

This study was supported by the Clarke Institute of Psychiatry and the Eaton Laboratories, Norwich, N.Y.

REFERENCES

Bourne, H. R., Bunney, W. E., Jr., Colburn, R. W., Davis, J. M., Davis, J. N., Shaw, D. M., and Coppen, A. J. (1968): Noradrenaline, 5-hydroxytryptamine, and 5-hydroxyindoleacetic acid in hindbrains of suicidal patients. *Lancet*, II:805–808.

Coppen, A. J. (1972): Indoleamines and affective disorders. *Journal of Psychiatric Research*, 9:163–171.

Dowson, J. H. (1969): The significance of brain-amine concentrations. *Lancet*, II:596–597.

Joyce, D. (1962): Changes in the 5-hydroxytryptamine content of rat, rabbit and human brain after death. *British Journal of Pharmacology*, 18:370–380.

Lloyd, K. G., and Hornykiewicz, O. (1972): Occurrence and distribution of aromatic L-amino acid (L-DOPA) decarboxylase in the human brain. *Journal of Neurochemistry*, 19:1549–1559.

Maclean, R., Nicholson, W. J., Pare, C. M. B., and Stacey, R. S. (1965): Effect of monoamineoxidase inhibitors on the concentrations of 5-hydroxytryptamine in the human brain. *Lancet*, II:205–208.

Nauta, W. J. H. (1958): Hippocampal projections and related neural pathways to the mid-brain in the cat. *Brain*, 81:319–340.

Olszewski, J., and Baxter, D. (1954): *Cytoarchitecture of the Human Brain Stem.* Lippincott, Philadelphia and Montreal.

Pare, C. M. B., Yeung, D. P. H., Price, K., and Stacey, R. S. (1969): 5-Hydroxytryptamine, noradrenaline, and dopamine in brainstem, hypothalamus and caudate nucleus of controls and of patients committing suicide by coal-gas poisoning. *Lancet*, II:133–135.

Robinson, D. S., Nies, A., Davis, J. N., Bunney, W. E., Davis, J. M., Colburn, R. W., Bourne, H. R., Shaw, D. M., and Coppen, A. J. (1972): Ageing, monoamines and monoamine oxidase levels. *Lancet*, I:290–291.

Shaw, D. M., Camps, F. E., and Eccleston, E. G. (1967): 5-Hydroxytryptamine in the hindbrain of depressive suicides. *British Journal of Psychiatry*, 113:1407–1411.

Advances in Biochemical Psychopharmacology, Vol. 11
Raven Press, New York © 1974

The Effect of L-5-Hydroxytryptophan Alone and in Combination with a Decarboxylase Inhibitor (Ro 4-4602) in Depressive Patients

N. Matussek, J. Angst, O. Benkert, M. Gmür, M. Papousek,
E. Rüther, B. Woggon

*Psychiatrische Klinik der Universität München, Munich, Germany, and
Psychiatrische Universitätsklinik Zürich, Zurich, Switzerland*

I. INTRODUCTION

Currently, there is a large body of experimental evidence to indicate that disturbances in brain 5-hydroxytryptamine (5-HT) metabolism may be an important factor in affective disorders. There may well be a relative or absolute reduction in the effective availability of 5-HT at receptor sites (Coppen, 1967; Coppen, Prange, Whybrow, and Noguera, 1972). As a consequence, it seemed logical to compensate for this postulated deficit by administration of the serotonin precursors, L-tryptophan or 5-HTP. Initial experience with L-tryptophan alone was reported to be useful by some investigators, but this has not been borne out by subsequent work (Carroll, 1971). Early studies using DL-5-HTP were negative, but shortly after L-5-HTP became available there followed an enthusiastic report of its efficacy in depression (Sano, 1972a, b), and it was reported to be of particular benefit in some patients in whom a low cerebrospinal fluid (CSF) level of 5-hydroxy-indoleacetic acid (5-HIAA) was found (van Praag, Korf, Dols, and Schut, 1972).

To try to clarify the situation, we have investigated the administration of L-5-HTP alone and in combination with a decarboxylase inhibitor, Ro 4-4602 [N^1-(DL-seryl)-N^2-(2,3,4-trihydroxybenzyl)-hydrazine], in an open trial in depressed patients.

II. PATIENTS AND METHODS

Patients with various depressive syndromes were included in this study, the age range being 20 to 60 yrs. Patients with known cardiac, hepatic, or renal disease were excluded. As will be seen in Table 2, patients in both groups were severely depressed as measured by the Hamilton Rating Scale.

The groups included the following types of depression:

Group I:
Number of patients	4	10	1
ICD no.	296.0	296.2	295.7

Group II:
Number of patients	4	3	1
ICD no.	296.0	296.2	296.3

296.0 = involutional melancholia
296.2 = manic-depressive illness, depressed type (endogenous depression)
296.3 = manic-depressive illness, circular type
295.7 = schizo-affective psychosis

No other psychotropic drugs were administered with the exception of chloral hydrate, and its use was restricted to those patients with severe sleep disturbances. To provide a standard of comparison for the level of depression throughout the study, we used the Hamilton Rating Scale, the AMP System (Angst, Battegay, Bente, Berner, Broeren, Cornu, Dick, Engelmeier, Heimann, Heinrich, Helmchen, Hippius, Pöldinger, Schmidlin, Schmitt, and Weis, 1969), and the self-rating scale of von Zerssen, Koeller, and Rey (1970). The results of the latter two rating scales will be published later. Thirteen patients were rated by two independent observers at the same interview and ten patients by one observer only, with the Hamilton scores doubled. In group I, Ro 4-4602 was always given 20 min prior to L-5-HTP. Table 1 lists further details of the patient sample and the dosages given.

TABLE 1. *Data concerning treatment and type of patient*

Group	Treatment	Dosage (mg/day)	Treatment days	Number of patients	Females	Males	Age
I	Ro 4-4602 + L-5-HTP	3 ×125 100 – 300	5 – 20	15	10	5	49.6 ± 11.9
II	L-5-HTP	150 – 300	4 – 17	8	5	3	53.8 ± 6.7

III. RESULTS

In group I (combination of decarboxylase inhibitor and 5-HTP) and in group II (5-HTP only) there is a decrease in Hamilton score on the fourth to fifth day as compared with day 0 (Table 2). This decrease is

TABLE 2. *Hamilton scores before treatment and on selected days thereafter*

Group	Day 0	Days 4 to 5	Days 9 to 11	Days 13 to 15	Day 20
I	61.8 ± 17.5 (15)	49.5 ± 27.1 (14)	46.7 ± 31.9 (14)	50.1 ± 32.7 (13)	50.4±35.7 (10)
II	54.8 ± 12.3 (8)	45.4 ± 15.1 (8)	45.2 ± 17.2 (6)[a]	26 ± 21.2 (2)	

[a] Rating, for this group only, on seventh day of treatment.

primarily due to the fact that patients appeared to be without symptoms after 4 to 5 days of treatment (two in group I; one in group II) (Table 3). These two patients in group I showed no evidence of depression after 5 days of treatment. In one of these patients 14 days of treatment was followed by placebo for 3 days. After placebo the treatment was discontinued but the patient did not relapse, so that we are in some doubt as to whether improvement was due to 5-HTP treatment. The one patient in group II was free of symptoms after 3 days as determined by his self-rating score, but this impression was not corroborated by the clinical observers until the 12th to 14th day of treatment. On the 17th day he was started on placebo for 6 days, during which time he exhibited symptoms of depression, particularly in the morning. On a subsequent treatment course with dibenzepine

TABLE 3. *Results after treatment*

Group	Symptom free	Good improvement	Slight improvement	No improvement	Worse
I	2	4	2	5	2
II	1	0	2	4	1

Without symptoms = 0 to 9 Hamilton score at the end of the treatment.
Good improvement = more than 25 points decrease in the Hamilton score.
Slight improvement = between 25 and 5 points decrease in the Hamilton score.
No improvement = ±5 points change in the Hamilton score.
Worse = more than 5 points increase in the Hamilton score.

(Noveril®), improvement was not as great as during treatment with 5-HTP. We therefore think that this patient's improvement was due to 5-HTP treatment. Substantial improvement was seen in four patients in group I, but in none in group II (Table 3). Two patients in each group were slightly improved. In group I five patients showed no improvement and two got worse, whereas in group II four patients showed no improvement and one got worse.

In general, the only side effects encountered were gastrointestinal and those were minimal—mild bloating and/or epigastric distress and occasional short-lived diarrhea. The only exception was the appearance of severe vomiting in one patient on combined treatment (5-HTP + decarboxylase inhibitor) after 2 days, which made it necessary to discontinue this therapy.

IV. DISCUSSION

In this small open trial of 5-HTP, we have been unable to obtain therapeutic results as good as those reported by Sano (1972a, b). In his group of 107 patients, 40 showed "dramatic" responses. This means, in his terms, that therapeutic effects were obtained in a few days, mostly by the third day of treatment. Our three patients who became symptom-free also reacted within the first five days of treatment (Table 3), one of them only in the self-rating scale. The percentage of the highest category of responders quoted in Sano's paper is much higher (38%) as compared to both our groups (13%). In addition to the rapid responders in Sano's study, there was a sizable group of patients who showed significant improvement during the 14 days of treatment, but after the first few days. When these patients are combined with the rapid responders, a total of 74 patients (69%) who were either cured or showed marked improvement is obtained. In our study we did not obtain as large a percentage of patients who improved after the first days of treatment. These differences cannot be explained on the basis of significant differences in the types of depression or different dosages of 5-hydroxytryptophan. Sano has treated both monopolar and bipolar depressive patients, while most patients in our trial were monopolar depressives (ICD no. 296.2). Like Sano, we found few side effects during 5-HTP treatment.

We were interested in observing the effect of 5-HTP on sleep disturbances in our patients. In normal sleeping subjects, 5-HTP and tryptophan have been shown to reduce sleep latency and to increase slow-wave sleep (Williams, Lester, and Coulter, 1969; Williams and Salamy, 1971) as well as REM sleep and REM activity (Oswald, Ashcroft, Berger, Eccleston, Evans, and Thacore, 1966; Hartmann, Chung, and Chien, 1971; Wyatt, Zarkone, Englemann, Dement, Snyder, and Sjoerdsman, 1971). Initial data suggest that, within our dose range, 5-HTP brought about no change in the subjective quality of sleep as measured by a sleep questionnaire. There was some indication, however, of sleep improvement related to relief of depression. All night polygraphic records point to similar findings. These data will be reported in detail elsewhere (Rüther and Papousek, *unpublished*).

On the basis of our results, we are somehat sceptical about the effectiveness of L-5-HTP in the treatment of depressed patients in general. However, there may be depressed patients belonging to a particular subgroup who react well to L-5-HTP administration. We did not measure 5-HIAA concentration in CSF as van Praag (1972) did, but according to the literature, a high percentage of depressed patients show low 5-HIAA levels in CSF (Coppen 1967; Coppen et al., 1972). It therefore seems reasonable to assume that most of the 23 patients we treated not only the three who showed good improvements, had low CSF 5-HIAA levels.

Despite negative results with L-DOPA administration in depressed patients, it is not possible to conclude that there is no disturbance of brain norepinephrine metabolism in depression. Similarly, negative results obtained using 5-HTP do not allow us to conclude that no disturbance in serotonin metabolism exists in depressed patients.

ACKNOWLEDGMENTS

We thank Hoffmann-La Roche Company for supplies of Ro 4-4602 and L-5-HTP, and Dr. R. Ilaria for translation of the manuscript.

REFERENCES

Angst, J., Battegay, R., Bente, D., Berner, P., Broeren, W., Cornu, F., Dick, P., Engelmeier, M. P., Heimann, H., Heinrich, K., Helmchen, H., Hippius, H., Pöldinger, W., Schmidlin, P., Schmitt, W., and Weis, P. (1969): Das Dokumentationszentrum der Arbeitsgemeinschaft für Methodik und Dokumentation in der Psychiatrie. *Arzneimittel-Forschung*, 19:399–405.

Carroll, J. (1971): Monoamine precursors in the treatment of depression. *Clinical Pharmacology and Therapeutics*, 12:743–761.

Coppen, A. (1967): The biochemistry of affective disorders. *British Journal of Psychiatry*, 113:1237–1267.

Coppen, A., Prange, A. J., Whybrow, P. C., and Noguera, R. (1972): Abnormalities of indolamines in affective disorders. *Archives of General Psychiatry*, 26:474–478.

Hartmann, E. R., Chung, R., and Chien, C. P. (1971): L-Tryptophan and sleep. *Psychopharmacologia*, 19:114–127.

Oswald, I., Ashcroft, G. W., Berger, R. J., Eccleston, D., Evans, J. I., and Thacore, V. R. (1966): Some experiments in the chemistry of normal sleep. *British Journal of Psychiatry*, 112:391–399.

Sano, I. (1972a): L-5-Hydroxytryptophan (L-5-HTP)-Therapie bei endogener Depression. *Münchener Medizinische Wochenschrift*, 114:1713–1716.

Sano, I. (1972b): L-5-Hydroxytryptophan-Therapie. *Folia Psychiatrica et Neurologica Japonica*, 26:7–17.

van Praag, H. M., Korf, J., Dols, L. C. W., and Schut, T. (1972): A pilot study of the predictive value of the probenecid test in application of 5-hydroxytryptophan as antidepressant. *Psychopharmacologia,* 25:14—21.

von Zerssen, D., Koeller, D. M., and Rey, E. R. (1970): Die Befindlichkeitsskala (B.-S.)—ein einfaches Instrument zur Objektivierung im Rahmen von Längsschnittuntersuchungen. *Arzneimittel-Forschung,* 20:915—918.

Williams, H. L., Lester, B. K., and Coulter, L. D. (1969): Monoamines and the EEG stages of sleep. *Activitas Nervosa Superior,* 11:188—192.

Williams, H. L., and Salamy, A. (1972): Alcohol and sleep. In: *The Biology of Alcoholism, Vol. II, Physiology and Behavior,* edited by B. Kissin and H. Begleiter. Plenum Press, New York.

Wyatt, R. J., Zarkone, V., Engelmann, K., Dement, W. C., Snyder, F., and Sjoerdsma, A. (1971): Effects of 5-hydroxytryptophan on the sleep of normal human subjects. *Electroencephalography and Clinical Neurophysiology,* 30:505—509.

Advances in Biochemical Psychopharmacology, Vol. 11
Raven Press, New York © 1974

Tryptophan-Induced Suppression of Conditioned Avoidance Behavior in Rats

Jörgen Engel and Kjell Modigh

*Department of Pharmacology, University of Göteborg,
Göteborg, Sweden*

I. INTRODUCTION

L-Tryptophan has been shown by many investigators to induce sedation and changes in sleep patterns in both humans and experimental animals (for reviews see Hartmann, Chung, and Chien, 1971; Wyatt, 1972). Experiments in humans have revealed that the tryptophan-induced sedation is not prevented if the patients are pretreated with the tryptophan hydroxylase inhibitor parachlorophenylalanine (PCPA) (Wyatt, Engelman, Kupfer, Fram, Sjoerdsma, and Snyder, 1970). Furthermore, the tryptophan-induced effects on sleep patterns in man are the opposite of those caused by the immediate precursor of 5-hydroxytryptamine (5-HT), that is, 5-hydroxytryptophan (5-HTP), which are enhanced rather than curtailed by pretreatment with PCPA (Wyatt, 1972). These findings may suggest that in man the actions of tryptophan on sleep are not mediated via its metabolite 5-HT. Tryptophan is also directly decarboxylated to tryptamine, and a quantitatively important metabolic route starts with the oxidation via liver tryptophan pyrrolase to formylkynurenine. It has recently been shown that tryptophan reduces locomotor activity in mice, an effect which persists even after pretreatment with drugs that block tryptophan hydroxylase (PCPA), aromatic L-amino acid decarboxylase (NSD 1015; 3-hydroxybenzylhydrazine HCl), or tryptophan pyrrolase (allopurinol) (Modigh, 1973). It thus appears that tryptophan has behavioral effects which are mediated neither via 5-HT nor by tryptamine nor through kynurenine metabolites.

The aim of the present study was to investigate qualitatively the behavioral suppression induced by tryptophan. For this purpose we have used conditioned avoidance behavior, which is affected differently by neuroleptic and hypnotic drugs.

II. EFFECT OF VARIOUS DOSES OF
L-TRYPTOPHAN ON CONDITIONED
AVOIDANCE BEHAVIOR

Rats of the Sprague-Dawley strain were trained to avoid an electric shock in a two-way shuttle-box with the sound of a house buzzer as warning stimulus. The shuttle-box was divided into two compartments by a wooden partition with an opening. The box was housed in a sound-attenuated chamber and the rats were observed through a one-way mirror. The behavioral experiments were performed between 9 a.m. and 3 p.m. The rats were kept under regulated light/dark conditions (light period was from 6 a.m. to 6 p.m.) The conditioned stimulus (CS) was the sound of the house buzzer. The unconditioned stimulus (UCS) consisted of an intermittent shock, delivered through the grid floor of either compartment. In each trial the CS was presented for 10 sec, followed by the CS plus the UCS for another 10 sec. The following variables were recorded. *Conditioned avoidance response (CAR):* The rat crossed through the opening within 10 sec after the CS had been presented. *Escape response (ER):* A cross within 10 sec after the shock (UCS) had been delivered. *No response (NR):* The rat remained in the same compartment for the entire trial (> 20 sec). Prior to the first training session the rats were allowed to adapt to the shuttle-box for 30 min. Training sessions lasted for 20 min and consisted of 20 trials. The rats were trained until at least three consecutive sessions with 90% CAR had been achieved.

The administration of L-tryptophan in a dose of 400 mg/kg, i.p., had no significant effect on the CAR when the rats were tested 1 hr, 2 hr, 3 hr, and 6 hr after injection of the amino acid (Fig. 1). L-Tryptophan, when given in doses of 600 or 800 mg/kg, caused a statistically significant ($p < 0.05$, Wilcoxon's t test) suppression of the CAR at the 1-hr, 2-hr, and 3-hr intervals, and a corresponding increase in the ER was observed; thus no NR was registered. At the 6-hr interval the avoidance performance of the rats had returned to their predrug levels. As assessed by gross observation, the animals appeared to be somewhat sedated, but reacted to sound and painful stimuli such as pinching the tail in an adequate way. As assessed by gross observation, no signs of ataxia were observed.

It is possible to find doses of neuroleptic drugs (e.g., chlorpromazine and reserpine) which cause a marked depression of the CAR, but no or a small degree of NR, whereas hypnotic drugs (e.g., barbiturates) generally produce a large amount of NR when given in doses which produce a marked suppression of the avoidance behavior. In the present experiment we have found that L-tryptophan caused a dose-dependent, marked suppression of the CAR without producing any NR, thus

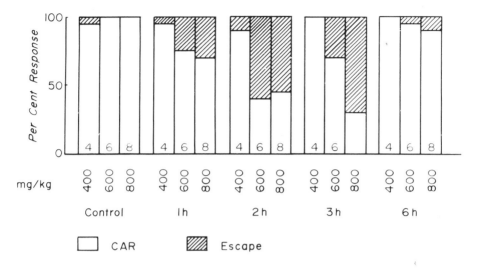

FIG. 1. The effect of various doses (400, 600, and 800 mg/kg, i.p.) of L-tryptophan on conditioned avoidance responses (CAR, open bars) and escape responses (hatched bars) in rats, when tested 1 hr, 2 hr, 3 hr, and 6 hr after the injection. The control represents the test session immediately before the administration of L-tryptophan. Percent responses are represented by the median value of the number of rats as indicated on the bottom of each bar.

indicating that L-tryptophan has a neuroleptic-like effect on the CAR in rats.

In rats maximal accumulation of brain 5-HT after inhibition of monoamine oxidase is elicited by the intraperitoneal administration of 120 mg/kg of L-tryptophan (Grahame-Smith, 1971), thus indicating a saturation of brain tryptophan hydroxylase after this dose of tryptophan. Consequently, our finding that a dose of L-tryptophan higher than 400 mg/kg, i.p., was required to suppress the CAR suggests that mechanisms other than the increased concentrations of brain 5-HT are involved in this behavioral effect of L-tryptophan.

III. ANTAGONISM OF L-DOPA ON L-TRYPTOPHAN-INDUCED CAR-SUPPRESSION

The suppressant effect of neuroleptic drugs on conditioned behavior is in all probability due to their effects on central catecholamine neurons (see Engel, 1972). Furthermore, high doses of L-tryptophan have been shown to inhibit the synthesis of brain catecholamines when measured as the accumulation of L-DOPA after inhibition of aromatic L-amino acid decarboxylase by means of NSD 1015 (cf. Carlsson, Davis, Kehr, Lindqvist, and Atack, 1972); for example, 800 mg/kg, i.p., of L-tryptophan given to mice reduces the catecholamine synthesis by approximately 50% (Modigh, *unpublished data*). Taken together, these findings prompted us to investigate the effect of L-3,4-dihydroxy-phenylalanine (L-DOPA), the immediate precursor of the catechol-amines, on the L-tryptophan-induced suppression of the CAR (see Fig. 2). Since peripheral factors may contribute significantly to the overall behavioral effects of L-DOPA (Butcher and Engel, 1969*a, b*) an inhibitor of the peripheral DOPA-decarboxylase, N^1-(DL-seryl)-N^2-(2,3,4-trihydroxybenzyl)hydrazine (Ro 4-4602; Bartholini, Burkard, Pletscher, and Bates, 1967; Bartholini and Pletscher, 1968) was used.

The administration of L-tryptophan, 600 mg/kg, i.p., followed 30 min later by Ro 4-4602, 50 mg/kg, i.p., caused a significant suppression of the CAR 1 hr, 2 hr, and 3 hr after injection of L-tryptophan ($p < 0.05$ as compared to predrug performance, Wilcoxon's t test). The same drug treatment followed by L-DOPA, 100 mg/kg, i.p., 30 min after Ro 4-4602, had no significant effect on the CAR. Thus L-DOPA completely antagonized the L-tryptophan-induced suppression of the

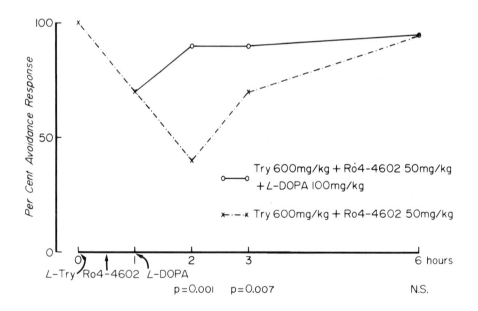

FIG. 2. The effect of L-DOPA (100 mg/kg, i.p.) and L-tryptophan (600 mg/kg, i.p.) after peripheral DOPA-decarboxylase inhibition by Ro 4-4602 (50 mg/kg, i.p.) on the conditioned avoidance response (CAR) in rats. L-Tryptophan was given immediately after the control (0-hr) test session and L-DOPA was injected immediately after the 1-hr test session. Percent conditioned avoidance responses are represented by the median value of six rats (Try + Ro 4-4602 group) and seven rats (Try + Ro 4-4602 + L-DOPA group). The statistical analysis for differences in the CAR between the two groups was performed by the Mann-Whitney U test.

CAR. One explanation for this antagonism is that the impaired catecholamine synthesis (see above) is overcome by L-DOPA.

In conclusion, L-tryptophan has a neuroleptic-like effect on conditioned avoidance behavior in rats, an effect which seems to involve mechanisms other than the tryptophan-induced increase in the concentrations of brain 5-HT. Furthermore, this effect of L-tryptophan is completely antagonized by the administration of the immediate precursor of the catecholamines, that is, L-DOPA. The mechanism for the L-tryptophan-induced suppression of CAR remains to be elucidated. However, the present findings suggest that central catecholamine neurons may be involved in this effect.

ACKNOWLEDGMENTS

The work has been supported by grants from the Swedish Board for Technical Development, Medical Faculty, University of Göteborg, and Magnus Bergwalls Stiftelse. The technical assistance of Mr. Kenn Johannessen is gratefully acknowledged.

REFERENCES

Bartholini, G., Burkard, W. P., Pletscher, A., and Bates, H. M. (1967): Increase of cerebral catecholamines caused by 3,4-dihydroxyphenylalanine after inhibition of peripheral decarboxylase. *Nature* (London), 215:852–853.
Bartholini, G., and Pletscher, A. (1968): Cerebral accumulation and metabolism of C^{14}-DOPA after selective inhibition of peripheral decarboxylase. *Journal of Pharmacology and Experimental Therapeutics*, 161:14–20.
Butcher, L. L., and Engel, J. (1969a): Peripheral factors in the mediation of the effects of L-DOPA on locomotor activity. *Journal of Pharmacy and Pharmacology*, 21:614–616.
Butcher, L. L., and Engel, J. (1969b): Behavioural and biochemical effects of L-DOPA after peripheral decarboxylase inhibition. *Brain Research*, 15:233–242.
Carlsson, A., Davis, J. N., Kehr, W., Lindqvist, M., and Atack, C. V. (1972): Simultaneous measurement of tyrosine and tryptophan hydroxylase activities in brain *in vivo* using an inhibitor of the aromatic amino acid decarboxylase. *Naunyn-Schmiedeberg's Archives of Pharmacology*, 275:153–168.
Engel, J. (1972): *Neurochemistry and Behaviour. A Correlative Study with Special Reference to Central Catecholamines.* Elanders Boktryckeri AB, Göteborg.
Grahame-Smith, D. G. (1971): Studies *in vivo* on the relationship between brain tryptophan, brain 5-HT synthesis and hyperactivity in rats treated with a monoamine oxidase inhibitor and L-tryptophan. *Journal of Neurochemistry*, 18:1053–1066.
Hartmann, E., Chung, R., and Chien, C. (1971): L-Tryptophan and sleep. *Psychopharmacologia* (Berlin), 19:114–127.
Modigh, K. (1973): Effects of L-tryptophan on motor activity in mice. *Psychopharmacologia* (Berlin), 30:123–134.
Wyatt, R. J. (1972): The serotonin-catecholamine-dream bicycle: A clinical study. *Biological Psychiatry*, 5:33–64.
Wyatt, R. J., Engelman, K., Kupfer, D. J., Fram, D. H., Sjoerdsma, A., and Snyder, F. (1970): Effects of L-tryptophan (a natural sedative) on human sleep. *Lancet*, II:842–846.

Advances in Biochemical Psychopharmacology, Vol. 11
Raven Press, New York © 1974

A Genetic Analysis of Behavior: A Neurochemical Approach

A. Oliverio, C. Castellano, A. Ebel, and P. Mandel

Laboratorio di Psicobiologia e Psicofarmacologia (C.N.R.), Rome, Italy, and Centre de Neurochimie du CNRS, Strasbourg, France

The findings presented in this paper concern a genetic analysis of different learning abilities and their neurochemical correlates in the mouse.

In the beginning an approach based on the use of inbred strains was adopted, while more recently more refined genetic analyses based on the use of recombinant inbred strains were used. By this means evidence was recently obtained that the levels of plasma serotonin are controlled in the mouse mainly by a single major gene and that this means also controls active avoidance learning (Eleftheriou and Bailey, 1972; Oliverio, Eleftheriou, and Bailey, 1973*b*). The existence of a line differing for just a single gene for higher levels of serotonin from its control inbred strain will allow different research approaches in the future.

This conclusion has been reached by following two different approaches. A first study was based on the use of three inbred strains of mice. A number of experiments indicate that these three inbred strains are different for their avoidance and discrimination learning abilities (Bovet, Bovet-Nitti, and Oliverio, 1969; Elias, 1970; Oliverio, Castellano, and Messeri, 1972) or activity patterns (Abeelen, 1966; McClearn, Wilson, and Meredith, 1970). In particular, C57 mice are characterized by high activity and low avoidance levels, while DBA and SEC mice are more sluggish but attain high levels of avoidance learning. A biometrical analysis of avoidance, maze learning, and wheel activity has been conducted in mice belonging to these three strains and to their F_1 and F_2 hybrids (Fig. 1).

In general, the two C57XSEC and C57XDBA hybrid mice are very similar for avoidance, maze, and activity patterns to their SEC or C57 parents, respectively (Fig. 2). Also, if these similarities do not determine that the genetic differences affect learning ability in the broader sense, an analogy between the brain mechanisms characterizing each of these hybrids and its dominant parent is conceivable. The results

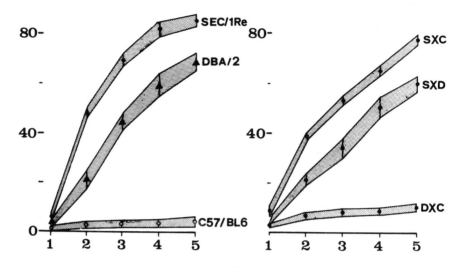

FIG. 1. Mean percentage avoidances (± 95% confidence limits of the mean) during sessions 1 to 5 in three inbred strains of mice and their F_1 offspring. C = C57BL/6J; D = DBA/2J; S = SEC/1ReJ (From Oliverio et al., 1972.)

of different biochemical estimates, indicating clear differences among the brain chemistry of these strains and of their hybrids, suggest that these lines and their crosses are an interesting model for correlations between individual differences in learning mechanisms and brain biochemistry.

A number of findings indicate in fact that these strains differ not only in behavior but also in brain chemistry. Pryor, Schlesinger, and Calhoun (1966) have measured brain acetylcholinesterase (AChE) activity in five strains of mice. They found large differences among the

FIG. 2. Genetic triangles representing the observed and expected population means in three strains of inbred mice (C, D, and S) and their F_1 and F_2 progeny. The mean values for the nonsegregating populations form the corner points of a triangle which is inscribed in a square. The mean measurement for F_1 lies on the vertical line which bisects the square, at a distance d above the midparent (M). The expected means for F_2 (open circles) are at a distance $d/2$ above the midparent. The observed F_2 values are shown in filled-in circles and an arrow points from observed to the expected means. The S strain is dominant over the C strain, while D mice are recessive in relation to the C line. (From Oliverio et al., 1972.)

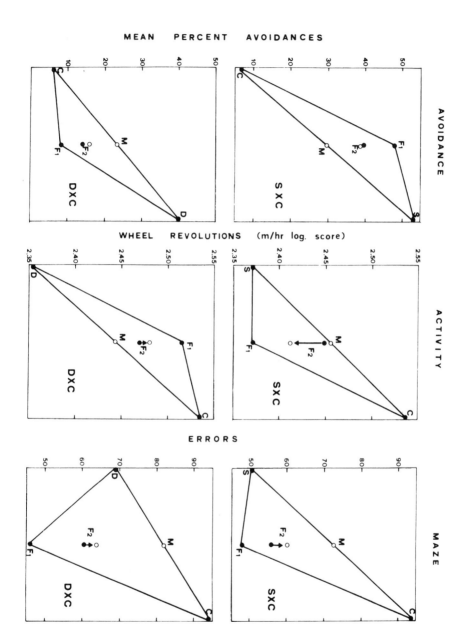

TABLE 1. *AChE activity in different regions in two strains of inbred mice and their F_1 hybrids (mmol/g/hr)*

	DBA/2J	C57BL/6J	F_1
Frontal	2.15 ± 0.37	1.94 ± 0.39	2.20 ± 0.30
Temporal	3.98 ± 0.62[a]	2.26 ± 0.49	3.62 ± 0.66
Limbic	3.76 ± 0.40	2.93 ± 0.51	3.21 ± 1.01
Occipital	2.23 ± 0.31	1.87 ± 0.24	1.55 ± 0.15
Parietal	2.05 ± 0.55	1.88 ± 0.37	1.75 ± 0.12
Overall	2.83 ± 0.45	2.17 ± 0.40	2.46 ± 0.44

[a]Significantly different from the C57BL/6J value ($p < 0.001$).
From Ebel et al., 1973.

strains, and these differences become even more pronounced if a regional analysis is performed. It is particularly interesting to note that the two strains characterized by high avoidance levels (A and DBA/2) also show a higher AChE activity than the two low avoiding strains (C3H and C57BL). A more detailed analysis carried out by Ebel, Hermetet, and Mandell (1973) in DBA and C57BL and their F_1 revealed that DBA mice have higher AChE (Table 1) and choline-acetyl-transferase (ChA) activities in the temporal lobe than C57 mice (Table 2).

All these findings indicate that behavioral differences must reflect some biochemical differences at the brain level. The problem is to know whether a given biochemical trait is related to learning ability or simply exhibits covariation with it.

TABLE 2. *Cholineacetyltransferase activity in different brain regions in two strains of inbred mice and their F_1 hybrids (mmol/g/hr)*

	DBA/2J	C57BL/6J	F_1
Frontal	16.62 ± 3.01[a]	11.83 ± 1.93	12.87 ± 1.61
Temporal	32.09 ± 2.71[b]	17.72 ± 3.52	21.43 ± 1.95
Limbic	15.85 ± 1.45	12.80 ± 3.00	11.17 ± 1.04
Occipital	13.67 ± 3.06	11.33 ± 1.42	11.09 ± 0.89
Parietal	13.37 ± 2.55	12.83 ± 4.11	13.35 ± 2.28
Overall	18.32 ± 2.55	13.30 ± 2.79	13.98 ± 1.55

[a]Significantly different from the C57BL/6J value ($p < 0.01$).
[b]Significantly different from the C57BL/6J value ($p < 0.001$).
From Ebel et al., 1973.

A recently developed powerful method of genetical analysis, the recombinant inbred (RI) strains (Bailey, 1971), seems to be extremely useful and promising for tracing gene-behavior pathways and for a biochemical approach to mechanism-specific behaviors.

RI strains are derived from the cross of two unrelated but highly inbred progenitor strains, for example, C57BL/6 and BALB/c mice, and then maintained independently from the F_2 generation under a regimen of strict inbreeding. This procedure genetically fixes the chance recombinations of genes, since full homozygosity is approached in the generations following the F_1 generation.

In addition, a battery of congenic lines developed from a cross of C57BL/6 and BALB/c mice and backcrossed to C57 for many generations was used. This procedure results in different congenic lines, each of which differs from C57 itself by only an introduced chromosomal segment, including a BALB allele at a distinctive locus. Each of these congenic lines was tested against the RI strain by means of skin grafts to find which of the RI strains carries the BALB allele and which the C57 allele at a given histocompatibility locus; this procedure allows establishment of *strain distribution patterns (SDP)*. Once it has been found that a given trait (for example, the level of brain chemical) has the same distribution pattern of that already determined for a given histocompatibility locus, this indicates that we are dealing with two closely linked genes. If one takes the existing congenic line that presents the same SDP evident for the new trait considered and if this congenic line differs from the control C57 strain, then it is possible to link the gene responsible for the trait under study. In addition, an extremely useful line is available since it differs from the control line for just *the* single gene which controls the trait. Of course, this is the case only if the trait under study is controlled by just a single or a major gene and not by many.

It has been possible, by using this procedure, in a first study, to show that a major gene controls the exploratory behavior of mice (Oliverio, Eleftheriou, and Bailey, 1973a). The analysis of the data obtained in the testing of all strains indicated that these data were distinctly grouped into two major categories. This finding was confirmed by data from backcrosses showing two distinct groupings and a clear bimodal distribution typical for single-gene segregation (Fig. 3). It was possible to link this locus to H(w26), which is located on chromosome 4, and to suggest the symbol *Exa* to designate the two alleles determining high or low exploratory activity. In a similar way, it has been possible to identify and locate the gene responsible for the central effect of scopolamine on activity, exploratory behavior, and similar behavioral patterns.

To reach finally the point related to serotonin and behavior, our

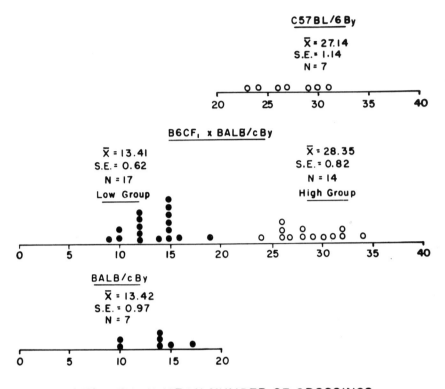

ACTIVITY IN MEAN NUMBER OF CROSSINGS

FIG. 3. Distribution of mean activity (mean number of crossings in toggle-floor box) in the backcross B6CF₁ X BALB/cBy and in the parent strains BALB/cBy and C57BL/6By. Each point represents the performance of a different mouse. (From Oliverio et al., 1973a.)

findings have shown that by using a similar procedure it has been possible to ascribe to a single gene a major effect on a complex behavior trait such as active avoidance learning. In fact, RI strains exhibited either high or low levels of active avoidance learning, the two groups being similar to each of the progenitor strains, BALB/c and C57BL/6, respectively (Fig. 4). The SDP of the strains permitted to link the gene responsible for active avoidance learning *(Aal)* to H(w56) on chromosome 9, with *Aal^h* to designate the allele determining the high level, and *Aal* the allele determining low levels of avoidance (Oliverio, Eleftheriou, and Bailey, 1973*b*).

It is particularly interesting to note that the same gene, or a gene that segregates in association to it, seems also responsible for the levels of

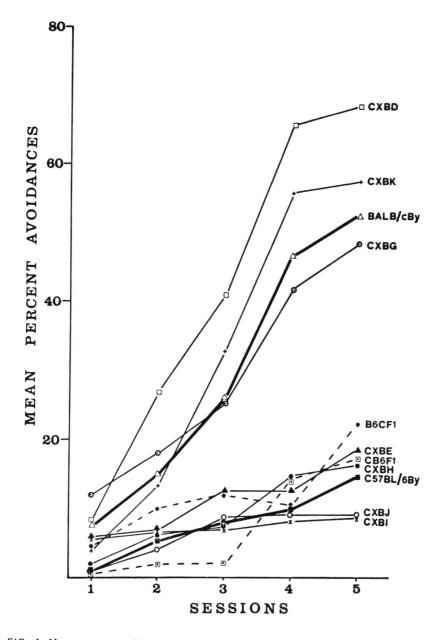

FIG. 4. Mean percent avoidances per session (days 1 to 5) for seven RI strains, their two progenitor strains, and two reciprocal F_1 hybrids. The statistical analyses indicate the existence of two main groupings (e.g., high avoiding or low avoiding strains) which are BALB/c or C57BL/6 type, respectively. This distribution is typical of a major gene effect. (From Oliverio et al., 1973*b*.)

plasma serotonin (Eleftheriou and Bailey, 1972). The levels of serotonin in the line differing by a single gene from control C57BL/6 mice are in fact clearly lower than those of normal mice (C57BL/6By = 2.07 ± 0.2 μg/ml; H(w56) = 0.72 ± 0.09 μg(ml).

Preliminary findings obtained by Drs. Gessa and Tagliamonte seem to indicate that these differences are also evident at the brain level.

In general, these findings indicate that plasma brain serotonin and a complex behavioral trait such as avoidance learning are controlled by a single major gene effect. These two traits do not correlate at random, and we believe that these data indicate the existence of a real relationship between serotonin and the modulation of some learning processes. Of course, more data are necessary in order to indicate which is the role of serotonin in relation to other amines and which is its turnover in mice characterized by enhanced learning processes. However, a reliable model is available for a series of biochemical, histological, and pharmacological studies.

REFERENCES

Abeelen, J. H. F., van (1966): Effects of genotype on mouse behaviour. *Animal Behaviour*, 14:218–225.

Bailey, D. W. (1971): Recombinant inbred strains. *Transplantation*, 11:325–327.

Bovet, D., Bovet-Nitti, F., and Oliverio, A. (1969): Genetic aspects of learning and memory in mice. *Science*, 163:139–149.

Ebel, A., Hermetet, J. C., and Mandel, P. (1974): Étude comparee de l'acetyl-cholinesterase et de la cholineacetyltransferase des souris des deux souches DBA et C57. *Science* (Paris) (*in press*).

Eleftheriou, B. E., and Bailey, D. W. (1972): A gene controlling plasma serotonin levels in mice. *Journal of Endocrinology*, 55:225–226.

Elias, M. F. (1970): Differences in reversal learning between two inbred mouse strains. *Psychiatric Science*, 20:179–180.

McClearn, G. E., Wilson, J. R., and Meredith, N. (1970): The use of isogenic and heterogenic mouse stocks in behavioral research. In: *Contributions to Behavior—Genetic Analysis. The Mouse as a Prototype*, edited by G. Lindzey and D. D. Thiessen. Appleton–Century–Crofts, New York.

Oliverio, A., Castellano, C., and Messeri, P. (1972): A genetic analysis of avoidance, maze, and wheel running behaviors in the mouse. *Journal of Comparative and Physiological Psychology*, 79:459–473.

Oliverio, A., Eleftheriou, B. E., and Bailey, D. W. (1973a): Exploratory activity: Genetic analysis of its modification by scopolamine and amphetamine. *Physiology and Behavior*, 10:893–899.

Oliverio, A., Eleftheriou, B. E., and Bailey, D. W. (1973b): A gene influencing active avoidance performance in mice. *Physiology and Behavior*, 2:497–501.

Pryor, G. T., Schlesinger, K., and Calhoun, W. H. (1966): Differences in brain enzymes among five inbred strains of mice. *Life Sciences*, 5:2105–2111.

SUBJECT INDEX

A

B